AF680559

CHAOS IN ELECTRIC DRIVE SYSTEMS

CHAOS IN ELECTRIC DRIVE SYSTEMS

ANALYSIS, CONTROL AND APPLICATION

K.T. CHAU
The University of Hong Kong, Hong Kong, China

ZHENG WANG
Southeast University, China

IEEE

John Wiley & Sons (Asia) Pte Ltd

This edition first published 2011
© 2011 John Wiley & Sons (Asia) Pte Ltd

Registered office
John Wiley & Sons (Asia) Pte Ltd, 2 Clementi Loop, # 02-01, Singapore 129809

For details of our global editorial offices, for customer services and for information about how to apply for permission to reuse the copyright material in this book please see our website at www.wiley.com.

All Rights Reserved. No part of this publication may be reproduced, stored in a retrieval system or transmitted, in any form or by any means, electronic, mechanical, photocopying, recording, scanning, or otherwise, except as expressly permitted by law, without either the prior written permission of the Publisher, or authorization through payment of the appropriate photocopy fee to the Copyright Clearance Center. Requests for permission should be addressed to the Publisher, John Wiley & Sons (Asia) Pte Ltd, 2 Clementi Loop, #02-01, Singapore 129809, tel: 65-64632400, fax: 65-64646912, email: enquiry@wiley.com.

Wiley also publishes its books in a variety of electronic formats. Some content that appears in print may not be available in electronic books.

Designations used by companies to distinguish their products are often claimed as trademarks. All brand names and product names used in this book are trade names, service marks, trademarks or registered trademarks of their respective owners. The Publisher is not associated with any product or vendor mentioned in this book. This publication is designed to provide accurate and authoritative information in regard to the subject matter covered. It is sold on the understanding that the Publisher is not engaged in rendering professional services. If professional advice or other expert assistance is required, the services of a competent professional should be sought.

Library of Congress Cataloging-in-Publication Data

Chau, K. T.
Chaos in electric drive systems : analysis, control and application / K.T. Chau, Zheng Wang.
p. cm.
Includes bibliographical references and index.
ISBN 978-0-470-82633-1 (cloth)
1. Electric driving–Automatic control. 2. Chaotic behavior in systems. I. Wang, Zheng, 1979- II. Title.
TK4058.C37 2011
621.46–dc22 2010051061

Print ISBN: 978-0-470-82633-1
E-PDF ISBN: 978-0-470-82634-8
O-book ISBN: 978-0-470-82635-5
E-Pub ISBN: 978-0-470-82836-6

Set in 9/11pt Times by Thomson Digital, Noida, India.

To our parents, families, colleagues and friends worldwide

Contents

PART II ANALYSIS OF CHAOS IN ELECTRIC DRIVE SYSTEMS

PART III CONTROL OF CHAOS IN ELECTRIC DRIVE SYSTEMS

PART IV APPLICATION OF CHAOS IN ELECTRIC DRIVE SYSTEMS

Preface

Chaos is a phenomenon that occurs in nature, from as large as the universe to as tiny as a particle. The concepts of chaology have penetrated into virtually all branches of science and engineering. In the field of electrical and electronic engineering, recent research has covered a wide spectrum, including the analysis of chaos, the stabilization of chaos, the stimulation of chaos, and the application of chaos.

There are many books that deal with the field of chaology, focusing on the theoretical analysis of chaotic systems and the mathematical formulation of chaotic behaviors. In recent years, some books have begun to deal with chaos in electronic engineering, especially in the areas of electronic circuits and telecommunications. Although chaos and its practical application in electric drive systems have been widely published as papers in learned journals, a book that comprehensively discusses chaos in electric drive systems is highly desirable.

The purpose of this book is to provide a comprehensive discussion on chaos in electric drive systems, including the analysis of their chaotic behaviors, the control of their chaotic characteristics, and the application of their chaotic features. Contrary to other books which usually involve intensive mathematics or idealized experiments, this book aims to use a minimum of mathematical treatments, extensive computer simulations, and realistic experimentations to discuss chaos in various electric drive systems, including DC drive systems, AC drive systems, and switched reluctance drive systems. Also, while other books consider that the relevant application of chaos or chaos theory is to model or simply explain some strange behaviors, this book aims to discuss and explore the realistic application of chaos in electric drive systems, especially the utilization of chaotic motion for compactors, mixers, washers, HVAC devices, and grinders.

While an electric drive can be chaoized by a simple parameter and stabilized by a feedback controller, my life is also chaoized by a naughty boy and stabilized by a wonder woman. I would therefore like to take this opportunity to express my heartfelt gratitude to my son, Aten Man-ho, and my wife, Joan Wai-yi, for their existence in my life.

K.T. Chau
The University of Hong Kong, Hong Kong, China

Since its first introduction by Poincaré in the 1890s, chaos has been discovered in many disciplines. Although it was previously identified as a scientific problem, chaos is being paid more and more attention by engineers today. As a person engaged in electrical engineering, I got to know chaos when I read a book about nonlinearity in power systems. My first impression was that chaos was interesting, but so complicated, and with increasing knowledge I began to realize that chaos was actually a "conservative" guy with a messy outlook but a beautiful intrinsic property.

I became familiar with chaos when I started my PhD study at The University of Hong Kong (HKU) in 2004. Encouraged by my supervisor, Professor K.T. Chau, I was connected closely with the area of chaos in electric drives. By that time, Professor Chau had already developed a lot of pioneering work on the identification, analysis, and stabilization of chaos in electric drives. As in other disciplines, the innovative idea of the positive utilization of chaos in electric drives had just started in Professor Chau's group. My work then focused on chaotic drives, in particular on their industrial applications. When Professor Chau told me about his book proposal, I felt very happy as some of the creative work by our group on this topic could be introduced systematically worldwide. Most importantly, we hope that more people might pay attention to this multidisciplinary area, not only scientifically, but also in an engineering capacity. We also hope that more colleagues join us in this area!

Finally, I would like to take this opportunity to express my appreciation to Professor Chau for his guidance and for allowing me to participate in the writing of this book; to Professors Jie Wu and Ming Cheng for their support during my master's degree and work at the Southeast University of China (SEU); and to my group fellows in HKU and SEU, whose work has greatly excited me. I also wish to acknowledge the genuine support and unselfish care I have received from my parents at all times.

Zheng Wang
Southeast University, China

Organization of this Book

This book is a happy marriage of two fields of research – chaology and electrical engineering. Chaology has always been tagged as an abstract field that involves intensive mathematics but lacks practical application. On the other hand, electrical engineering has been well recognized as a practical field that usually transforms innovative technology into commercial products, thus improving our living standards. Chaos in electric drive systems is a representative of this marriage, enabling chaos to exhibit realistic behavior and provide a practical application. It also fuels electrical engineering with a new breed of technology.

The book covers the multidisciplinary aspects of chaos and electric drive systems, and is written for a wide range of readers, including students, researchers, and engineers. It is organized into four parts:

- Part I presents an introduction to the book. It contains Chapters 1 and 2, which will provide an overview of chaos with an emphasis on electric drive systems. These chapters will also introduce the basic theory of chaos and a fundamental knowledge of electric drive systems.
- Part II is a core section of the book – namely, how to analyze chaos in various electric drive systems. It consists of Chapters 3, 4, and 5, which will discuss the analysis of chaos in DC drive systems, AC drive systems, and switched reluctance drive systems.
- Part III is another core section which explains how chaos in various electric drive systems can be controlled. It comprises two chapters, Chapters 6 and 7, which will discuss various methods of controlling chaos, including the stabilization and stimulation of chaos. It should be noted that this book adopts the general perception of the meaning of control, rather than use the jargon of chaos theory where 'control' and 'anticontrol' represent 'stabilize' and 'destabilize', respectively.
- Part IV – which is probably the most influential part of the book – unveils and proposes some promising applications of chaos for electric drive systems. It contains three chapters (Chapters 8–10) that will be devoted to describing various applications of chaos, including the application of chaos stabilization, the application of chaotic modulation, and the application of chaotic motion.

Since these four parts have their individual themes, readers have the flexibility to select and read those parts that they find most interesting. The suggestions for reading are as follows:

- Undergraduate students taking a course dedicated to electric drive systems may be particularly interested in Parts I, II, and IV.
- Postgraduate students taking a course dedicated to advanced electric drive systems may find all parts interesting.
- Researchers in the areas of chaos and/or electric drive systems may also be interested in all parts. In particular, they may have special interest in Parts III and IV, which involve newly explored research topics.

- Practicing engineers for product design and development may be more interested in Parts I and IV, in which new ideas can be triggered by the overview, and commercial products can be derived from the proposed applications.
- General readers may be interested in all parts. They are advised to read the book from beginning to end, page by page, and will find the book to be most enjoyable.

The book contains 10 chapters, each of which has various sections and subsections. In order to facilitate a reading selection, an outline of each chapter is given below:

- Chapter 1 gives an overview of chaos, including the definition of chaos, the development of chaology, and the research of chaos in the field of electrical engineering, with an emphasis on electric drive systems.
- Chapter 2 introduces the necessary background knowledge for this book – namely, a description of the basic theory of chaos and the fundamentals of electric drive systems.
- Chapter 3 is devoted to analyzing chaos in DC drive systems, including both of the voltage-controlled mode and the current-controlled mode. The corresponding modeling, analysis, simulation, and experimentation will be discussed.
- Chapter 4 is devoted to analyzing chaos in AC drive systems, including the induction drive system, the permanent magnet synchronous drive system, and the synchronous reluctance drive system. The corresponding modeling, analysis, simulation, and experimentation will be discussed.
- Chapter 5 is devoted to analyzing chaos in switched reluctance drive systems, including the voltage-controlled mode and the current-controlled mode. Relevant discussion with verification will be given.
- Chapter 6 describes various control approaches to stabilize the chaos that occurs in both DC and AC drive systems. A relevant discussion with verification will be given.
- Chapter 7 describes various control approaches to stimulate chaos operating at various electric drive systems. Both of the control-oriented chaoization and the design-oriented chaoization will be discussed.
- Chapter 8 presents the stabilization of chaos in various applications, including automotive wiper systems, centrifugal governor systems, and rate gyro systems. The corresponding modeling, analysis, and stabilization will be elaborated.
- Chapter 9 presents how to apply chaotic modulation to PWM inverter drive systems, hence reducing the corresponding audible noise and mechanical vibration. Open-loop and closed-loop control will both be discussed.
- Chapter 10 presents a new breed of chaos application, namely the electrically-chaoized motion – simply known as chaotic motion. Various promising applications of chaotic motion, including compaction, mixing, washing, HVAC, and grinding, will be unveiled and elaborated.

Acknowledgments

The material presented in this book is a collection of many years of research and development by the authors in the Department of Electrical and Electronic Engineering, The University of Hong Kong.

We are grateful to all members of our research group, especially Mr Wenlong Li, Miss Jiangui Li, Miss Shuang Gao, and Miss Diyun Wu, for their help in the preparation of this work. We must express our sincere gratitude to our PhD graduates, namely Dr Jihe Chen, Dr Yuan Gao, and Dr Shuang Ye, who have made enormous contributions to the area of chaos in electric drive systems.

We are deeply indebted to our colleagues and friends worldwide for their continuous support and encouragement over the years. We appreciate the reviewers of this book for their thoughtful and constructive comments, and thank the editors at John Wiley & Sons for their patience and effective support.

Last but not least, we thank our families for their unconditional support and absolute understanding during the writing of this book.

About the Authors

K. T. Chau received his BSc (Eng) degree with First Class Honors, MPhil degree, and PhD degree all in Electrical and Electronic Engineering from The University of Hong Kong. He joined his alma mater in 1995, and currently serves as Professor in the Department of Electrical and Electronic Engineering. He is a Chartered Engineer and Fellow of the Institution of Engineering and Technology. He has served as editor and editorial board member of various international journals as well as chair and organizing committee member of many international conferences. His teaching and research interests are electric drives, electric vehicles, and renewable energy. In these areas, he has published over 300 refereed technical papers. Professor Chau has received many awards, including the Chang Jiang Chair Professorship; the Environmental Excellence in Transportation Award for Education, Training and Public Awareness; the Award for Innovative Excellence in Teaching, Learning, and Technology; and the University Teaching Fellow Award.

Zheng Wang received his BSc (Eng) degree and Master (Eng) degree in Electrical Engineering from the Southeast University of China, and his PhD degree in Electrical and Electronic Engineering from The University of Hong Kong. After working as a postdoctoral fellow in Ryerson University of Canada, he joined the School of Electrical Engineering of the Southeast University in 2009, and currently serves as an associate professor. He is a member of the Institute of Electrical and Electronics Engineers. His teaching and research interests are power electronics, electric drives, and renewable energy techniques. He has published several technical papers and industrial reports in these areas. Dr Wang has received some awards including the postdoctoral fellowship financially supported by the Canadian NSERC/Rockwell, the Hong Kong Electric Co. Ltd Electrical Energy Postgraduate Scholarship, and the Outstanding Young Teacher Award of Southeast University.

Part One

Introduction

1

Overview of Chaos

This chapter gives an overview of chaos, including the definition of chaos, the development of chaology, and the research of chaos. In particular, the research and development of chaos in the field of electrical engineering – with an emphasis on electric drive systems – are discussed in detail.

1.1 What is Chaos?

The etymology of the word "chaos" is a Greek word "$\chi\alpha'\varsigma$" (Nagashima and Baba, 1999) which means "the nether abyss, or infinite darkness," and was personified as "the most ancient of the gods." Namely, the god Chaos was the foundation of all creation. From this god arose Gaea (god of the earth), Tartarus (god of the underworld) and Eros (the god of love). Eros drew Chaos and Gaea together so that they could produce descendants, the first born of whom was Uranus (the god of the sky). This also resulted in the creation of the elder gods known as Titans. The interaction of these gods resulted in the creation of other gods, including such well-known figures as Aphrodite, Hades, Poseidon, and Zeus.

There is a Chinese myth of chaos (Liu, 1998), taken from one of the ancient Chinese classics *Chuang-Tzu*: "The god of the Southern Sea was called *Shu* (Change), the god of the Northern Sea was called *Hu* (Suddenness), and the god of the Central was called *Hun-tun* (Chaos). *Shu* and *Hu* often came together for a meeting in the Central, and *Hun-tun* treated them generously. *Shu* and *Hu* determined to repay his kindness, and said, 'Mankind has seven holes for seeing, hearing, eating and breathing; but *Hun-tun* has none of them; let us bore the holes for him!' So, every day they bored one hole in his head. On the seventh day, *Hun-tun* died." This myth not only indicates the disorder-like or random-like behavior of chaos, but also implies that chaos is the natural state of the world and should not be disrupted by a sudden change.

There are many myths relating to the god of chaos in different cradles of civilization, such as Greece, China, Egypt, and India, but in the modern world chaos is no longer a god. In 1997, its meaning in the *Oxford English Dictionary Online* was updated as "Behavior of a system which is governed by deterministic laws but is so unpredictable as to appear random, owing to its extreme sensitivity to changes in parameters or its dependence on a large number of independent variables; a state characterized by such behavior" (Simpson, 2004).

The general perception on chaos is equivalent to disorder or even random. It should be noted that chaos is not exactly disordered, and its random-like behavior is governed by a rule – mathematically,

Chaos in Electric Drive Systems: Analysis, Control and Application, First Edition. K.T. Chau and Zheng Wang.
© 2011 John Wiley & Sons (Asia) Pte Ltd. Published 2011 by John Wiley & Sons (Asia) Pte Ltd.

a deterministic model or equation that contains no element of chance. Actually, the disorder-like or random-like behavior of chaos is due to its high sensitivity on initial conditions.

Similar to many terms in science, there is no standard definition of chaos. Nevertheless chaos has some typical features:

- **Nonlinearity:** Chaos cannot occur in a linear system. Nonlinearity is a necessary, but not sufficient condition for the occurrence of chaos. Essentially, all realistic systems exhibit certain degree of nonlinearity.
- **Determinism:** Chaos must follow one or more deterministic equations that do not contain any random factors. The system states of past, present and future are controlled by deterministic, rather than probabilistic, underlying rules. Practically, the boundary between deterministic and probabilistic systems may not be so clear since a seemingly random process might involve deterministic underlying rules yet to be found.
- **Sensitive dependence on initial conditions:** A small change in the initial state of the system can lead to extremely different behavior in its final state. Thus, the long-term prediction of system behavior is impossible, even though it is governed by deterministic underlying rules.
- **Aperiodicity:** Chaotic orbits are aperiodic, but not all aperiodic orbits are chaotic. Almost-periodic and quasi-periodic orbits are aperiodic, but not chaotic.

1.2 Development of Chaology

Chaology means the study of chaos theory and chaotic systems. The development of chaology started in mathematics and physics, and expanded into chemistry, biology, engineering, and social sciences. In particular, there have been growing interests in practical applications based on various aspects of chaotic systems (Ditto and Munakata, 1995).

Jules Henri Poincaré first introduced the idea of chaos in 1890 when he participated in a contest sponsored by King Oscar II of Sweden. Although the King Oscar prize was granted to the first person who could solve the n-body problem to prove the stability of the solar system, Poincaré won the prize by being closest to a solution of the problem and discovered that the orbit of a three-body celestial system can exhibit unpredictable behavior. To show how visionary Poincaré was, the description of chaos – sensitive dependence on initial conditions – in his 1903 book *Science and Method* (Poincaré, 1996) is quoted below:

> A very small cause which escapes our notice determines a considerable effect that we cannot fail to see, and then we say that that effect is due to chance. If we knew exactly the laws of nature and the situation of the universe at the initial moment, we could predict exactly the situation of that same universe at a succeeding moment. But, even if it were the case that the natural laws had no longer any secret for us, we could still only know the initial situation **approximately**. If that enabled us to predict the succeeding situation **with the same approximation**, that is all we require, and we should say that the phenomenon had been predicted, that it is governed by laws. But it is not always so; it may happen that small differences in the initial conditions produce very great ones in the final phenomena. A small error in the former will produce an enormous error in the latter. Prediction becomes impossible, and we have the fortuitous phenomenon.

Until Edward Norton Lorenz rediscovered a chaotic deterministic system in 1963, Poincaré's finding did not receive the attention it deserved. Lorenz studied air convection in the atmosphere, for which he built a simple mathematical model. He discovered that the weather did not always change in accordance with prediction, and observed that small changes in the initial conditions of variables in his primitive computer weather model could result in very different weather patterns. This model was described by a simple system of equations (Lorenz, 1963):

$$\begin{cases} \dfrac{dx}{dt} = -\sigma x + \sigma y \\ \dfrac{dy}{dt} = -xz + rx - y \\ \dfrac{dz}{dt} = xy - bz \end{cases} \tag{1.1}$$

where σ is the Prandtl number and r is the Rayleigh number. The system exhibits the well-known Lorenz attractor as shown in Figure 1.1, when $\sigma = 10$, $b = 8/3$, and $r = 28$.

This sensitive dependence on initial conditions is the essence of chaos, and is known as the "butterfly effect", which is often ascribed to Lorenz. In a 1963 paper for the New York Academy of Sciences, Lorenz said:

> One meteorologist remarked that if the theory were correct, one flap of a seagull's wings would be enough to alter the course of the weather forever.

By the time of his 1972 speech at the meeting of the American Association for the Advancement of Science in Washington, DC, Lorenz used a more poetic butterfly statement (Hilborn, 1994):

> Predictability: Does the Flap of a Butterfly's Wings in Brazil set off a Tornado in Texas?

Yoshisuke Ueda first encountered chaos in analog simulations of a nonlinear oscillator at the end of 1961 when he was a doctoral student. With his subsequent studies of the Duffing equation, he published pioneering articles in 1970 and 1973 (Ueda, 1992), illustrating the tangled invariant manifold structure in both its transient and steady-state manifestations. Ueda utilized the following Duffing equation to describe a series-resonant circuit containing a saturable inductor under the supply of DC and sinusoidal voltage:

$$\begin{cases} \dfrac{dx}{dt} = y \\ \dfrac{dy}{dt} = -ky - x^3 + B\cos t + B_0 \end{cases} \tag{1.2}$$

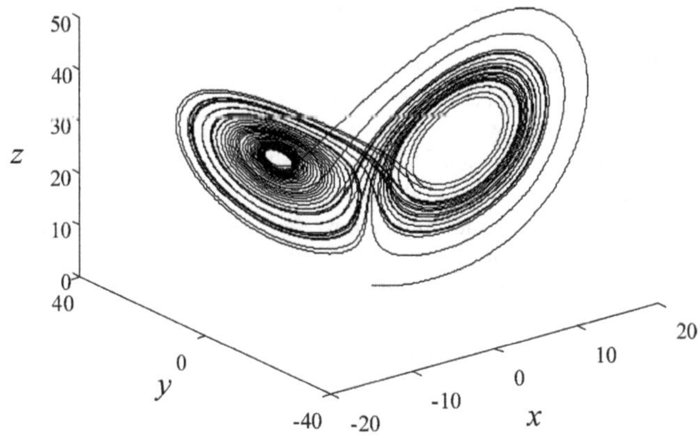

Figure 1.1 The Lorenz attractor

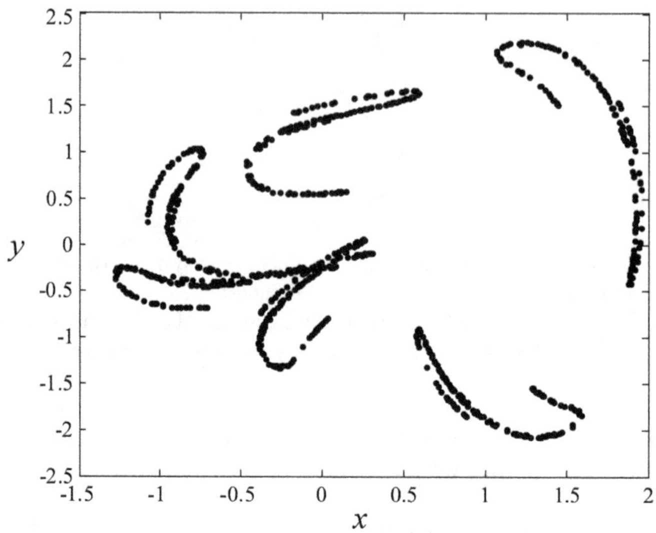

Figure 1.2 The Ueda attractor

where $k = 0.2$, $B = 1.2$, and $B_0 = 0.85$. Figure 1.2 shows the corresponding strange attractor. The fundamental nature of steady-state chaos was described in terms of unstable periodic motions (Abraham and Ueda, 2000).

Robert M. May introduced the idea of chaos to population biology in 1974, which was actually the first, modern, technical use of the term "chaos." He published a paper "Biological populations with non-overlapping generations: stable point, stable cycles and chaos" in *Science* (May, 1974) in which he outlined that the "evolution of populations (in this case) can be apparently random, even when the underlying equations have nothing random about them." In his population study, the outside influences (such as climate, food and health) could all factor into the randomness of the population changes. By applying the logistic equation to model the relationship population, May discovered that the population growth exhibited chaotic behavior – the "sensitive dependence on initial conditions." Namely, if the initial population level was changed, all other successive populations would change; and even where two initial population levels were extremely close, their later population levels could become severely different (May, 1976).

In 1975, Tien-Yien Li and James A. Yorke first coined the mathematical definition of chaos in a ground-breaking paper "Period three implies chaos" in the *American Mathematical Monthly* (Li and Yorke, 1975). This Li–Yorke Theorem states that any continuous one-dimensional system which exhibits a period-three orbit must also display orbits of every other length, as well as completely chaotic orbits. Although A.N. Sharkovskii had published the equivalent of the first part of the Li–Yorke Theorem a decade earlier, the latter part of the Li–Yorke Theorem thoroughly unveiled the nature and characteristics of chaos: the sensitive dependence on initial conditions and the resulting unpredictable nature.

In 1975, Mitchell Jay Feigenbaum discovered that the ratio of the difference between the values at which successive period-doubling bifurcations occur tends to be a constant of around 4.6692, based on his primitive HP-65 computer. He then mathematically proved that the same constant would occur in a wide class of nonlinear mappings when approaching a chaotic level. The logistic map is a well-known example of the mappings that Feigenbaum studied in his famous 1978 paper "Quantitative universality for a class of nonlinear transformations" (Feigenbaum, 1978). Named after Feigenbaum, there are two constants which are depicted in the bifurcation diagram of the logistic map, as shown in Figure 1.3.

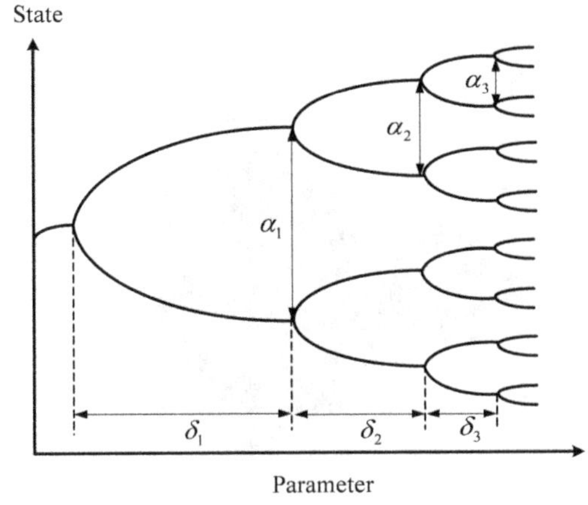

Figure 1.3 The definition of Feigenbaum constants

The first Feigenbaum constant is given by:

$$\delta = \lim_{k \to \infty} \frac{\delta_k}{\delta_{k+1}} = 4.66920160910299067185320382\cdots \tag{1.3}$$

which is the ratio between successive bifurcation intervals.

The second Feigenbaum constant is given by:

$$\alpha = \lim_{k \to \infty} \frac{\alpha_k}{\alpha_{k+1}} = 2.502907875095892822283902873218\cdots \tag{1.4}$$

which is the ratio between successive tine widths.

Benoît B. Mandelbrot almost single-handedly created the fractal geometry that is critical to an understanding of the laws of chaos. He also showed that fractals have fractional dimensions, rather than whole number dimensions. With the aid of computer graphics, he plotted the ground-breaking image of a simple mathematical formula: $z_{n+1} = z_n^2 + c$, which is now called the Mandelbrot set. His work was first fully elaborated in his book *The Fractal Geometry of Nature* in 1982. The Mandelbrot set is an iterative calculation of complex numbers with zero as the starting point. The order behind the chaotic generation of numbers can only be seen by a graphical portrayal of these numbers; otherwise, the generated numbers seem to be random. As shown in Figure 1.4, in which the stable points are represented by black points, the order is strange and beautiful, exhibiting self-similar recursiveness over an infinite scale (Mandelbrot, 1982).

Edward Ott, Celso Grebogi, and James A. Yorke kicked off the control of chaos (Ott, Grebogi, and Yorke, 1990). They developed a method known (after their initials) as the OGY technique, which functions to apply tiny disturbances to chaotic attractors so that the selected unstable orbits can be stabilized – namely, from chaos to order. This OGY technique has been widely accepted since it requires no analytical model and all necessary dynamics are estimated from past observations of the system.

In 1997, Guanrong Chen coined the term "anticontrol" of chaos, and discussed its potential in nontraditional applications (Chen, 1997). Instead of stabilizing a chaotic system, the anticontrol of chaos (namely, from order to chaos) has received great interest. The idea of the anticontrol of chaos was first proposed experimentally to prevent the periodic behavior of the neuronal population bursting of an *in vitro* rat brain (Schiff *et al.*, 1994). It is interesting to note that the new words "chaotify" and "chaotification" were created to represent "to generate chaos" and "the generation of chaos," respectively.

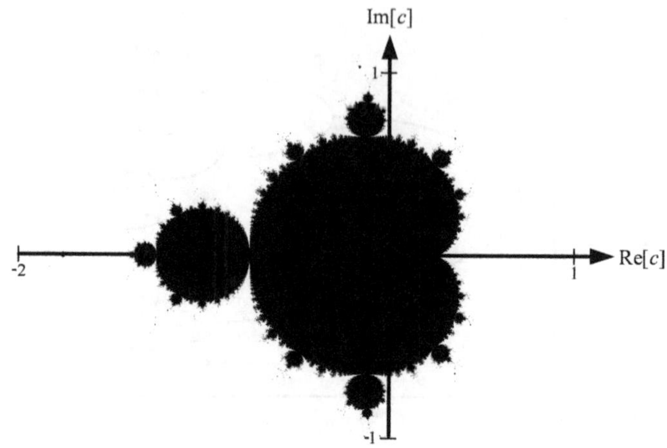

Figure 1.4 An image of the Mandelbrot set

Table 1.1 Milestones in chaology

1890	In a contest sponsored by the King Oscar II of Sweden, Poincaré was awarded the prize for his discovery that the orbit of a three-body celestial system is deterministic but can exhibit unpredictable behavior – the first realization of chaos.
1963	Lorenz developed a simple system of equations to model weather, hence displaying the first chaotic attractor, which he then poetically termed the well-known butterfly effect.
1970	Ueda published a pioneering paper to illustrate the tangled invariant manifold structure of the Duffing equation.
1974	May identified chaotic behavior in population biology, and first introduced the term "chaos" as is technically used today.
1975	Li and Yorke published a ground-breaking paper "Period three implies chaos," and coined the mathematical definition of chaos.
1978	Feigenbaum unveiled the Feigenbaum constant of a wide class of nonlinear mappings prior to the onset of chaos.
1982	Mandelbrot coined the word "fractal" and laid the foundation of fractal geometry to elaborate chaos.
1990	Ott, Grebogi, and Yorke opened the research theme on the control of chaos.
1997	Chen coined the term "anticontrol" of chaos, and began his research in this direction.

Actually, there are formal words "chaoize" and "chaoization" that carry these meanings, although they were rarely used.

The aforementioned milestones in chaology are summarized in Table 1.1 in which they are focused on theoretical developments and systems, rather than practical developments and systems. As the practicability of chaology covers many disciplines and fields, our discussion will focus on the field of electrical engineering.

1.3 Chaos in Electrical Engineering

In electrical engineering, research on chaos has covered a very wide range of areas, including, but not limited to, electronic circuits, telecommunications, power electronics, power systems, and electric drive systems. Overviews will be given to these selected areas, with emphasis on the chaos in electric drive systems.

1.3.1 Chaos in Electronic Circuits

The electronic circuit was the natural extension from theoretical chaos to practical electrical engineering. Basically, the investigation of chaos in electronic circuits can be grouped as one-dimensional map circuits, higher-dimensional map circuits, continuous-time autonomous circuits, and continuous-time non-autonomous circuits (Van Wyk and Steeb, 1997).

There were some electronic circuits that can be described by one-dimensional maps (Mishina, Kohmoto, and Hashi, 1985). Among them, the switched-capacitor circuit was one of the simplest, which is composed of a voltage source, a linear capacitor, a nonlinear switched-capacitor component and three analog switches (Rodríguez-Vázquez, Huertas, and Chua, 1985). For a specific choice of circuit parameters, its dynamics is equivalent to the well-known logistic map which is actually the simplest chaotic polynomial discrete map.

There were many electronic circuits that can be described by higher-dimensional maps. Among them, the infinite impulse response (IIR) digital filter, which utilized a two's complement adder with overflow, exhibits nonlinear dynamics ranging from limit cycles to chaos (Chua and Lin, 1988; Kocarev and Chua, 1993). Also, the adaptation algorithm for the filter weights was demonstrated to be one of the elements which may cause the filter to behave chaotically (Macchi and Jaidane-Saidane, 1989).

The continuous-time autonomous circuits have no external inputs so that such chaotic circuits are a kind of oscillators. The famous Chua circuit is the representative of this kind of oscillators. It is considered to be the simplest electronic circuit that can exhibit chaos and bifurcation phenomena (Chua, 2007), which have been verified by theoretical analysis (Chua, Komuro, and Matsumoto, 1986), computer simulation (Matsumoto, 1984) and experimental measurement (Zhong and Ayrom, 1985). The Chua circuit was invented in 1983 (Chua, 1992), which is shown in Figure 1.5. It is composed of five circuit elements, namely four passive elements, including an inductor, a resistor and two capacitors, as well as an active nonlinear element termed the Chua diode. This Chua diode is characterized by a piecewise-linear odd-symmetric characteristic, which can be realized by various circuits (Cruz and Chua, 1993; Gandhi *et al.*, 2009). The Chua circuit exhibits a number of distinct routes to chaos and a chaotic attractor termed the double-scroll attractor which is multi-structural. Other well-known autonomous circuits that show chaotic behavior include the Shinriki circuit (Shinriki *et al.*, 1981), the Saito circuit (Saito, 1985), and the Matsumoto circuit (Matsumoto, Chua, and Kobayashi, 1986).

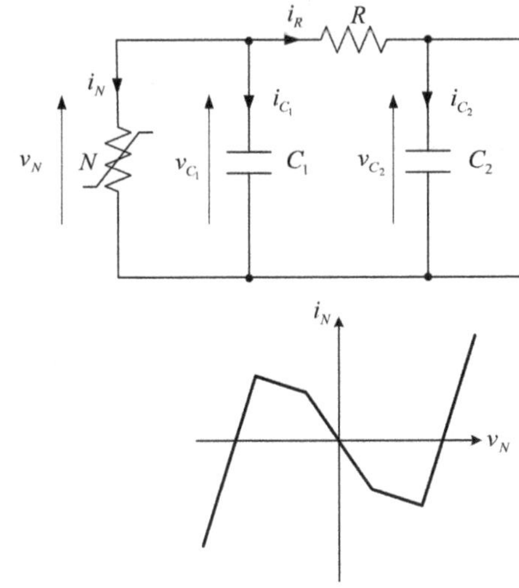

Figure 1.5 The Chua circuit

The continuous-time non-autonomous circuits refer to those circuits that are driven by an external source. The well-known phase-locked loop circuit is the representative of these types of driven circuit. It consists of three basic functional blocks, namely a voltage-controlled oscillator, a phase detector, and a loop filter, where it can perform various tasks such as frequency modulation and demodulation, frequency synthesis, and data synchronization. It has been identified that a second-order phase-locked loop circuit used as a frequency modulation demodulator exhibits chaotic behavior (Endo and Chua, 1988; Endo, Chua, and Narita, 1989). Other well-known non-autonomous circuits that show chaotic behavior include the periodically driven ferro-resonant circuit (Chua *et al.*, 1982), the triggered astable multivibrator (Tang, Mees, and Chua, 1983), the automatic gain control loop circuit (Chang, Twu, and Chang, 1993), and the cellular neural network circuit (Zou and Nossek, 1991).

1.3.2 Chaos in Telecommunications

As opposed to other areas in electrical engineering in which the investigation was started from the identification of chaos, the study of chaos in telecommunications was based on practical applications. In recent years, there has been tremendous interest in the use of chaos in telecommunications, as depicted in Figure 1.6. Chaotic signals possess inherent wideband characteristics, which can readily encode and spread narrowband information, resulting in spread-spectrum signals having the merits of difficulty in uninformed detection, mitigation of multipath fading, and an antijamming capability. Moreover, chaos provides the definite advantage of low-cost implementation. Basically, the chaos-based telecommunications can be grouped into three main types: chaotic masking, chaotic modulation, and chaotic switching.

In chaotic masking (Kocarev *et al.*, 1992; Cuomo and Oppenheim, 1993), the information signal is added to a chaotic signal at the transmitter. At the receiver, the original chaotic signal is reconstructed using chaos synchronization to allow the information signal to be extracted by subtracting the reconstructed chaotic signal from the incoming signal. Consequently, various techniques were proposed such as the feedback-based chaotic masking (Milanović and Zaghloul, 1996) and observer-based chaotic masking (Boutayeb, Darouach, and Rafaralahy, 2002; Liu and Tang, 2008).

In chaotic modulation (Kocarev and Parlitz, 1993; Itoh and Murakami, 1995; Tao and Chua, 1996), the communication is based on a modulated transmitter parameter – that is, the information signal is fed into a chaotic system to modulate its parameter, hence varying its dynamics. At the receiver, the change in dynamics of the chaotic signal is tracked in order to retrieve the information signal (Song and Yu, 2000; Feng and Tse, 2001).

In chaotic switching (Lau and Tse, 2003), the basic principle is to map bits or symbols to the basis functions of chaotic signals emanating from one or more chaotic attractors. One of the earliest chaotic-switching techniques is based on chaos shift keying (CSK) which maps different symbols to different basis functions of chaotic signals by varying a bifurcation parameter. If synchronized copies of the basis functions are available at the receiver, coherent detection can easily be achieved by evaluating the synchronization error (Parlitz *et al.*, 1992; Dedieu, Kennedy, and Hasler, 1993) or on the basis of

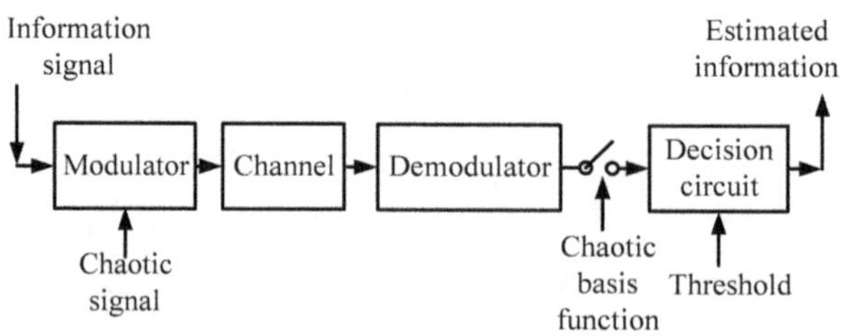

Figure 1.6 A chaos-based communication system

correlation (Kolumbán, Kennedy, and Chua, 1998). On the other hand, if there are no synchronized copies of the basis functions at the receiver, detection needs to be done by noncoherent means (Kolumbán and Kennedy, 2000). Another widely accepted chaotic-switching technique is based on differential chaos shift keying (DCSK) (Kolumbán *et al.*, 1996) which contains a basis function in half of its symbol period serving as the reference signal, thus avoiding synchronization between the transmitter and the receiver. Consequently, many other chaotic-switching techniques are derived from the CSK and DCSK, such as chaotic on-off-keying (Kolumbán, Kennedy, and Kis, 1997), correlation delay shift keying, symmetric CSK (Sushchik, Tsimring, and Volkovskii, 2000), quadrature CSK (Galias and Maggio, 2001), and frequency-modulated DCSK (Kolumbán, Jako, and Kennedy, 1999; Min *et al.*, 2010).

1.3.3 Chaos in Power Electronics

Power electronic systems are essentially piecewise-switched circuits. Their operation is characterized by the cyclic switching of circuit topologies, and as this results in a wide variety of nonlinear dynamics, they naturally prefer the use of nonlinear methods for design and analysis.

Starting from the late 1980s, chaos has been identified to be a real phenomenon in power electronics (Hamill and Jefferies, 1988; Wood, 1989). Furher discussions on instability and chaos in power electronic circuits and systems were reported in 1990 (Krein and Bass, 1990; Deane and Hamill, 1990), and since then much interest has been aroused in the investigation of the chaotic behavior of power electronic systems, especially DC–DC converters (Di Bernardo and Tse, 2002).

The milestone of investigations of chaos in power electronics was the paper by Hamill, Deane, and Jefferies (1992) which analyzed the occurrence of chaos in a simple buck converter, as shown in Figure 1.7. By using iterated nonlinear mappings, the occurrence of chaos was successfully modeled and simulated. This derivation was further extended to study the chaos in a current-mode controlled boost converter (Deane, 1992). The investigation was then extended to chaos in a DC–DC converter operating in discontinuous conduction mode (Tse, 1994) and chaos in a fourth-order Ćuk (Tse, Fung, and Kwan, 1996).

Further works on chaos in the buck converter were reported. For example, the bifurcation behavior under different variations of circuit parameters, including storage inductance, output capacitance, and load resistance, was specifically studied (Chakrabarty, Poddar, and Banerjee, 1996), while a detailed analytical description of converter dynamics – and, hence, its chaotic attractor – were discussed (Fossas and

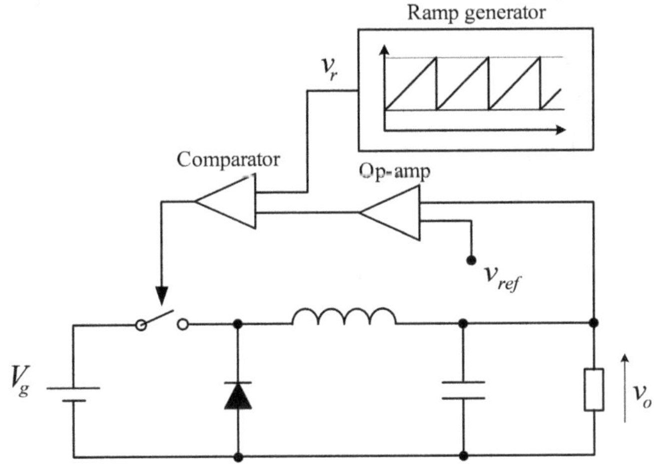

Figure 1.7 Chaos in a simple buck converter

Olivar, 1996). Moreover, some investigations on the sudden jump transition from periodic solution to chaos were conducted (Di Bernardo *et al.*, 1998; Yuan *et al.*, 1998), showing that it is due to a special class of bifurcation (the border-collision bifurcation) that is unique to switching systems.

Apart from the aforementioned analysis of chaos, the control of chaos in DC–DC converters was also investigated. By using parameter perturbation or by changing the switching instant, the chaos that occurred in the current-mode controlled buck converter was successfully stabilized (Poddar, Chakrabarty, and Banerjee, 1998). Also, a resonant parametric perturbation was applied to control chaos in the buck converter (Zhou *et al.*, 2003) and the buck-boost converter (Kavitha and Uma, 2008). Moreover, a recurrence plot analysis was employed to identify the set of unstable periodic orbits and, hence, use a resonant parameter perturbation to control chaos in the buck converter (Ivan and Serbanescu, 2009). On the other hand, the well-known time-delay feedback control was successfully applied to stabilize chaos in the buck converter (Batlle, Fossas, and Olivar, 1999). Also, by applying the control input to the system intermittently, the partial time-delay feedback control was proposed to save the control energy consumption (Bouzahir *et al.*, 2008).

In recent years, some applications of chaotic power electronic systems have been identified. For instance, chaos was positively employed to spread the noise spectrum and hence improve the electromagnetic interference in DC–DC converters (Deane and Hamill, 1996; Tse *et al.*, 2003). Another possible application is the use of chaotic transitions between different controlled states of the current-mode controlled boost converter, hence achieving very quick targeting (Aston, Deane, and Hamill, 1997).

1.3.4 Chaos in Power Systems

The investigation of chaos in power systems probably began in 1990, and spread over different areas, including power system stability and control, power flow optimization, unit commitment scheduling, load forecasting, and fault analysis.

Firstly, by decreasing the frequency of excitation in a single-machine quasi-infinite busbar system, it was shown that oscillatory solutions might lose their stability through period-doubling bifurcations, leading to chaos and unbounded motions (Hamdan, Nayfeh, and Nayfeh, 1990). Chaotic behavior was then observed in a simple power system over a range of reactive power loading conditions. As shown in Figure 1.8, this simple power system was composed of two generator buses and a load bus at which the load is modeled by an induction motor with a constant $P-Q$ load in parallel. By taking Q as the bifurcation parameter, four different kinds of bifurcation were identified, namely the subcritical Hopf

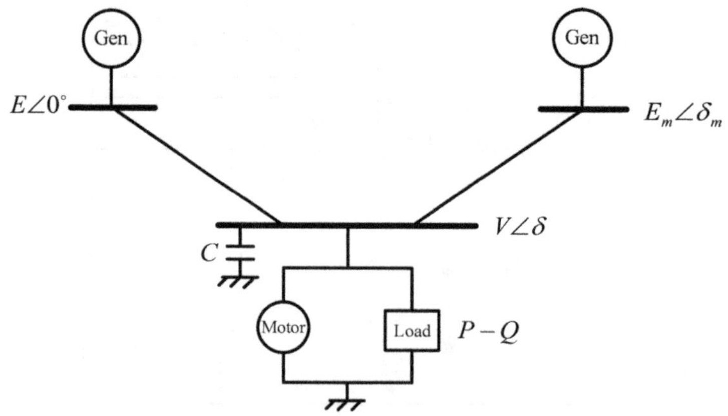

Figure 1.8 Chaos in a simple power system

bifurcation, period-doubling bifurcation, supercritical Hopf bifurcation, and saddle-node bifurcation. Hence, the period-doubling routes to chaos were confirmed by calculating the relevant Lyapunov exponents and broad-band spectrum (Chiang *et al.*, 1993). Similar period-doubling routes to chaos in a power system were also observed and discussed (Lee and Ajjarapu, 1993). Rather than varying the excitation or loading conditions, the Hopf bifurcation, the period-doubling bifurcation, and the chaos caused by the line resistance were investigated (Niu and Qiu, 2002). On the other hand, the banded chaos, namely the strange attractor with a band-like structure, was also reported in an actual power system with ferro-resonance (Ben-Tal, Kirk, and Wake, 2001).

Facing the occurrence of chaos in power systems, measures on how to avoid or control such chaotic behavior were brought up, aiming to reduce the risk of catastrophic failure (Wildberger, 1994). Then, the use of flexible AC transmission system (FACTS) devices was proposed to damp out the Hopf bifurcation and chaos in power systems (Srivastava and Srivastava, 1998). Moreover, an improved OGY method was also proposed to stabilize chaos in a power system (Okuno, Takeshita, and Kanari, 2002).

Although chaos has been identified to be undesirable for system stability, it can be positively utilized to benefit the power system. Since traditional optimization techniques generally suffer from the problem that optimal power flow is easily trapped by a local minimum solution, a chaos optimization algorithm was proposed to provide more robust convergence (Liu, Wang, and Hou, 2003). Similarly, a chaos search algorithm was proposed to provide the most economical solution for unit commitment scheduling (Liao and Tsao, 2004).

Furthermore, chaos time series analysis was employed to capture the characteristics of the complicated load behavior in power systems, hence performing short-term load forecasting (Mori and Urano, 1996). Similarly, fractal geometry was also employed to analyze the chaotic properties of high impedance faults during which phase currents and voltages exhibit a certain degree of chaotic behavior (Mamishev, Russell, and Benner, 1996).

1.3.5 Chaos in Electric Drive Systems

The investigation of chaos in electric drive systems can be categorized as three themes, namely the analysis of chaotic phenomena, the control of chaotic behaviors, and the application of chaotic characteristics. The milestones of chaos in electric drive systems are summarized in Table 1.2.

Chaos in electric drive systems was firstly identified in induction drive systems in 1989. That is, the bifurcation of induction motor drives was studied (Kuroe and Hayashi, 1989), which was actually an extension of the instability analysis of pulse-width-modulation (PWM) inverter systems. The bifurcation and chaos resulting from the tolerance-band PWM inverter-fed induction drive system were then investigated (Nagy, 1994; Sütő, Nagy, and Masada, 2000). It was also identified that saddle-node bifurcation, or even Hopf bifurcation, might occur in induction drive systems under indirect field-oriented control (Bazanella and Reginatto, 2000) and, consequently, the control of chaos in induction drive systems was investigated. An attempt was made to use a neural network stabilizing chaos during speed control of induction drive systems (Asakura *et al.*, 2000). On the other hand, an attempt was made to use periodic speed command to stimulate the chaotic motion of induction drive systems (Gao and Chau, 2003a).

Starting from 1997, the investigation of chaos has been accelerated. In 1997, the chaotic behavior in a simple DC drive system was unveiled (Chau *et al.*, 1997a); the dynamic bifurcation in DC drive systems was studied (Chau, Chen, and Chan, 1997b); and the subharmonics and chaos in DC drive systems were analytically modeled (Chau *et al.*, 1997c). Hence, the analysis and experimentation of their chaotic behaviors, including the voltage-mode controlled operation (Chen, Chau, and Chan, 1999) and the current-mode controlled operation (Chen, Chau, and Chan, 2000a), could be conducted. In 2000, the research was extended to the stabilization of chaos in DC drive systems by using a time-delay feedback control (Chen *et al.*, 2000b).

Table 1.2 Milestones of chaos in electric drive systems

1989	Kuroe and Hayashi identified bifurcation of a PWM inverter-fed induction drive system.
1994	Nagy identified bifurcation and chaos resulting from the tolerance-band PWM inverter-fed induction drive system.
1994	Hemati identified strange attractors in a PM brushless DC drive system.
1997	Chau *et al.* identified chaotic behavior and dynamic bifurcation in a simple DC drive system.
1998	Ito and Narikiyo applied chaotic motion for vertical spindle surface grinding.
1999	Chau *et al.* identified subharmonics and chaos in a SR drive system.
2000	Chen *et al.* utilized a time-delay feedback control to stabilize chaos in a DC drive system.
2000	Asakura *et al.* utilized a neural network to stabilize chaos in an induction drive system.
2001	Bellini *et al.* applied a chaotic map to generate chaotic PWM for an induction drive system.
2002	Gao and Chau utilized a time-delay feedback control to stimulate chaotic motion in a PM synchronous drive system.
2003	Gao and Chau utilized a periodic speed command to stimulate the chaotic motion of an induction drive system.
2004	Gao and Chau identified Hopf bifurcation and chaos in a synchronous reluctance drive system.
2004	Gao and Chau identified spontaneous chaotic behavior in a DSPM drive system.
2004	Chau *et al.* applied electrical chaoization to a DC drive system to generate chaotic motion for mixing.
2005	Ye and Chau utilized PM design parameters to stimulate chaotic motion in a PM synchronous drive system.
2005	Huang *et al.* utilized a control strategy to stimulate chaotic motion in a SR drive system.
2005	Chau and Wang applied electrical chaoization to a DC drive system to generate chaotic motion for compaction.
2005	Gao *et al.* applied electrical chaoization to a single-phase induction drive system to generate chaotic motion for cooling.
2006	Ren and Liu utilized nonlinear feedback control to stabilize chaos in a PM synchronous drive system.
2006	Chau and Wang utilized PM design parameters to stimulate chaotic motion in a DSPM drive system.
2006	Wang and Chau applied an extended time-delay auto-synchronization to stabilize chaos in a wiper system.
2006	Ye *et al.* applied electrical chaoization to a single-phase induction drive system for washing machines.

In 1999, the subharmonics and chaos in switched reluctance (SR) drive systems were first identified (Chau *et al.*, 1999) and the corresponding modeling was then developed (Chen *et al.*, 2000c; Chen *et al.*, 2002). Hence, the analysis of their chaotic behaviors under voltage PWM regulation (Chen, Chau, and Jiang, 2001) and current hysteresis regulation (Chau and Chen, 2002) was conducted. Furthermore, the experimentation of chaos in a practical SR drive system was first presented in the literature (Chau and Chen, 2003). In 2005, the research was extended to the stimulation of chaos in SR drive systems by using a control strategy combining piecewise proportional feedback and time-delay feedback (Huang, Chen, and Chau, 2005).

Without taking power electronic switching into consideration, it was identified that the permanent magnet (PM) brushless DC drive system could be transformed into a Lorenz system, which is well known to exhibit a Hopf bifurcation and chaotic behavior (Hemati, 1994). Also, chaotic behaviors have been identified in PM brushless AC drive systems or the so-called PM synchronous drive systems (Li *et al.*, 2002; Gao and Chau, 2003b). Consequently, the nonlinear feedback control method was developed to control the chaos in a PM synchronous drive system (Ren and Liu, 2006). Furthermore, both set-point and tracking output regulation of PM synchronous drive systems can be achieved by using a simple linear output feedback controller, provided that the operating point for the quadrature axis current is adequately chosen (Loría, 2009). On the other hand, the PM synchronous drive system was chaoized to produce chaotic motion by using a time-delay feedback control (Gao and Chau, 2002) or stator flux regulation (Wang, Chau, and Jian, 2008a). Instead of using control-oriented chaoization, the design-oriented chaoization was applied to a PM synchronous drive system which can spontaneously generate chaotic motion (Ye and Chau, 2005b).

By removing the PM materials and increasing the saliency of the PM synchronous motor, the resulting synchronous reluctance motor gives the definite advantages of high robustness and low cost. Similar to the PM brushless DC drive system, the synchronous reluctance drive system exhibits Hopf bifurcation and chaos (Gao and Chau, 2004a). On the other hand, by incorporating the concept of an SR motor into the PM brushless DC motor, the resulting doubly-salient PM (DSPM) motor offers the advantages of high robustness and immunity from PM thermal problems. It has been shown that the chaos in DSPM drive systems can occur spontaneously, depending on the initial design parameters of the PMs used (Gao and Chau, 2004b). Hence, based on the design of the PMs, the DSPM drive system can be purposely chaoized to produce chaotic motion (Chau and Wang, 2006).

The application of chaos in electric drive systems has focused on the practical use of the control of chaos, including the stabilization of chaos and the stimulation of chaos. For instance, chaotic vibration in an automotive wiper system not only decreases the wiping efficiency but also causes harmful distraction to the drivers (Suzuki and Yasuda, 1998). Thus, the corresponding chaos was directly stabilized by applying an extended time-delay auto-synchronization control to its DC drive (Wang and Chau, 2006; Wang and Chau, 2009a). This approach can be realized experimentally because the armature current of the DC motor can be easily measured by a Hall sensor and the perturbations on the feed-in motor voltage can be readily produced by a power converter.

Recently, there has been increasing attention to the emission of electromagnetic radiation from electric drive systems, which directly affects the electromagnetic interference (EMI) and electromagnetic compatibility, and indirectly creates acoustic noise and mechanical vibration. The EMI of an induction drive system was significantly suppressed by applying a chaotic map, namely the 4-way Bernoulli shift, to generate the so-called chaotic PWM (Bellini *et al.*, 2001). Then, a Chua circuit was employed to generate the desired chaotic sequence for chaotic PWM (Cui *et al.*, 2006). Consequently, a chaotically frequency-modulated signal was proposed to modulate the switching frequency of the PWM, which not only suppresses the peaky EMI, but also avoids the occurrence of low-order noises and mechanical resonance (Wang, Chau, and Liu, 2007). It was then extended to the space vector PWM which offers the additional advantages of less harmonic distortion, less switching loss, and better utilization of the DC supply voltage (Wang and Chau, 2007). Moreover, it was further extended to propose the chaotically amplitude-modulated signal to modulate the switching frequency of the space vector PWM for a closed-loop vector-controlled induction drive system (Wang, Chau, and Cheng, 2008b).

In recent years, chaotic mixing has been proposed to improve the energy efficiency and degree of homogeneity by using mechanical means (Alvarez-Hernández *et al.*, 2002) that are essentially based on the design of impeller vanes to produce chaotic motion. In order to offer the advantages of high flexibility and high controllability, the electrical chaoization was proposed to generate the desired chaotic motion for industrial mixing (Chau *et al.*, 2004; Ye and Chau, 2007). It applied a time-delay feedback control to the DC drive system which serves as the agitator. Similarly, destabilization control was also applied to the DC drive system which electrically generates chaotic motion for mixing (Ye and Chau, 2005a). Furthermore, the design-oriented chaoization was also applied to a PM synchronous drive system that can generate the desired chaotic motion for mixing (Ye and Chau, 2005b).

The application of chaos to compaction was initiated in a mechanical vibrator for compacting soft soil (Long, 2001). In order to offer high flexibility and high controllability, the electrical chaoization was proposed to generate the desired chaotic motion for compaction (Chau and Wang, 2005; Wang and Chau, 2008). Essentially, the PM DC drive system, which directly couples with an eccentric mass to translate the rotational motion to up–down motion, was chaoized by using a proportional time-delay feedback control. The use of a chaotic reference control was then introduced to further improve the ability to control the desired chaotic compaction (Wang and Chau, 2009b). On the other hand, based on the design-oriented chaoization, the DSPM drive system was chaoized to spontaneously produce chaotic motion for compaction (Chau and Wang, 2006).

Although there were many other applications of chaotic motion, most of them relied on a mechanical means to produce chaotic motion, while the electric drive was used as a driving force only. Nevertheless, some of them adopted electrical chaoization for the direct stimulation of chaotic motion, such as the chaoization of a single-phase induction drive system for washing machines (Ye, Chau, and Niu, 2006) and cooling fans (Gao, Chau, and Ye, 2005) as well as a DC drive system for vertical spindle surface grinders (Ito and Narikiyo, 1998).

References

Abraham, R.H. and Ueda, Y. (2000) *The Chaos Avant-Garde: Memoirs of the Early Days of Chaos Theory*, World Scientific, Singapore.

Alvarez-Hernández, M.M., Shinbrot, T., Zalc, J., and Muzzio, F.J. (2002) Practical chaotic mixing. *Chemical Engineering Science*, **57**, 3749–3753.

Asakura, T., Yoneda, K., Saito, Y., and Shioya, M. (2000) Chaos detection in velocity control of induction motor and its control by using neural network. Proceedings of IEEE International Conference on Signal Processing, pp. 1633–1638.

Aston, P.J., Deane, J.H.B., and Hamill, D.C. (1997) Targeting in systems with discontinuities, with applications to power electronics. *IEEE Transactions on Circuits and Systems – I: Fundamental Theory and Applications*, **44**, 1034–1039.

Batlle, C., Fossas, E., and Olivar, G. (1999) Stabilization of periodic orbits of the buck converter by time-delayed feedback. *International Journal of Circuit Theory and Applications*, **27**, 617–631.

Bazanella, A.S. and Reginatto, R. (2000) Robustness margins for indirect field-oriented control of induction motors. *IEEE Transactions on Automatic Control*, **45**, 1226–1231.

Bellini, A., Franceschini, G., Rovatti, R. *et al.* (2001) Generation of low-EMI PWM patterns for induction motor drives with chaotic maps. Proceedings of IEEE Industrial Electronics Society Annual Conference, pp. 1527–1532.

Ben-Tal, A., Kirk, V., and Wake, G. (2001) Banded chaos in power systems. *IEEE Transactions on Power Delivery*, **16**, 105–110.

Boutayeb, M., Darouach, M., and Rafaralahy, H. (2002) Generalized state space observers for chaotic synchronization and secure communication. *IEEE Transactions on Circuits and Systems – I: Fundamental Theory and Applications*, **49**, 345–349.

Bouzahir, H., El Guezar, F., El Aroudi, A., and Ueta, T. (2008) Partial time delayed feedback control of chaos in a hybrid model of a DC-DC converter. Proceedings of International Symposium on Communications, Control and Signal Processing, pp. 158–161.

Chakrabarty, K., Poddar, G., and Banerjee, S. (1996) Bifurcation behavior of the buck converter. *IEEE Transactions on Power Electronics*, **11**, 439–447.

Chang, F.-J., Twu, S.-H., and Chang, S. (1993) Global bifurcation and chaos from automatic gain control loops. *IEEE Transactions on Circuits and Systems I: Fundamental Theory and Applications*, **40**, 403–412.

Chau, K.T. and Chen, J.H. (2002) Analysis of chaotic behavior in switched reluctance motors using current hysteresis regulation. *Electric Power Components and Systems*, **30**, 607–624.

Chau, K.T. and Chen, J.H. (2003) Modeling, analysis and experimentation of chaos in a switched reluctance drive system. *IEEE Transactions on Circuits and Systems – I: Fundamental Theory and Applications*, **50**, 712–716.

Chau, K.T. and Wang, Z. (2005) Application of chaotic motion to industrial compactors. Proceedings of International Conference on Electrical Machines and Systems, pp. 1644–1649.

Chau, K.T. and Wang, Z. (2006) Design of permanent magnets to chaoize doubly salient PM motors for electric compaction. *Journal of Applied Physics*, **99**, 80R306:1–80R306:3.

Chau, K.T., Chen, J.H., and Chan, C.C. (1997b) Dynamic bifurcation in DC drives. Proceedings of IEEE Power Electronics Specialists Conference, pp. 1330–1336.

Chau, K.T., Chen, J.H., Chan, C.C., and Chan, D.T.W. (1997c) Modeling of subharmonics and chaos in DC motor drives. Proceedings of IEEE Industrial Electronics Conference, pp. 523–528.

Chau, K.T., Chen, J.H., Chan, C.C., and Jiang, Q. (1999) Subharmonics and chaos in switched reluctance motor drives. Proceedings of IEEE International Electric Machines and Drives Conference, pp. 661–663.

Chau, K.T., Chen, J.H., Chan, C.C. *et al.* (1997a) Chaotic behavior in a simple DC drive. Proceedings of IEEE Power Electronics and Drive Systems Conference, pp. 473–479.

Chau, K.T., Ye, S., Gao, Y., and Chen, J.H. (2004) Application of chaotic-motion motors to industrial mixing processes. Proceedings of IEEE Industry Applications Society Annual Meeting, pp. 1874–1880.

Chen, G. (1997) Control and anticontrol of chaos. Proceedings of International Conference on Control of Oscillations and Chaos, pp. 181–186.

Chen, J.H., Chau, K.T., and Chan, C.C. (1999) Chaos in voltage-mode controlled DC drive systems. *International Journal of Electronics*, **86**, 857–874.

Chen, J.H., Chau, K.T., and Chan, C.C. (2000a) Analysis of chaos in current-mode controlled DC drive systems. *IEEE Transactions on Industrial Electronics*, **47**, 67–76.

Chen, J.H., Chau, K.T., and Jiang, Q. (2001) Analysis of chaotic behavior in switched reluctance motors using voltage PWM regulation. *Electric Power Components and Systems*, **29**, 211–227.

Chen, J.H., Chau, K.T., Chan, C.C., and Jiang, Q. (2002) Subharmonics and chaos in switched reluctance motor drives. *IEEE Transactions on Energy Conversion*, **17**, 73–78.

Chen, J.H., Chau, K.T., Jiang, Q. *et al.* (2000c) Modeling and analysis of chaotic behavior in switched reluctance motor drives. Proceedings of IEEE Power Electronics Specialists Conference, pp. 1551–1556.

Chen, J.H., Chau, K.T., Siu, S.M., and Chan, C.C. (2000b) Experimental stabilization of chaos in a voltage-mode DC drive system. *IEEE Transactions on Circuits and Systems – I: Fundamental Theory and Applications*, **47**, 1093–1095.

Chiang, H.D., Liu, C.W., Varaiya, P.P. *et al.* (1993) Chaos in a simple power system. *IEEE Transactions on Power Systems*, **8**, 1407–1417.

Chua, L., Hasler, M., Neirynck, J., and Verburgh, P. (1982) Dynamics of a piecewise-linear resonant circuit. *IEEE Transactions on Circuits and Systems*, **29**, 535–547.

Chua, L.O. (1992) The Genesis of Chua's Circuit. *Archiv für Elektronik und Ubertragung-stechnik*, **46**, 250–257.

Chua, L.O. (2007) Chua circuit. *Scholarpedia*, **2**, 1488, http://www.scholarpedia.org/article/Chua_circuit.

Chua, L.O. and Lin, T. (1988) Chaos in digital filters. *IEEE Transactions on Circuits and Systems*, **35**, 648–658.

Chua, L.O., Komuro, M., and Matsumoto, T. (1986) The double scroll family. *IEEE Transactions on Circuits and Systems*, **33**, 1072–1118.

Cruz, J.M. and Chua, L.O. (1993) An IC Chip of Chua's circuit. *IEEE Transactions on Circuits and Systems-II*, **10**, 596–613.

Cui, W., Chau, K.T., Wang, Z., and Jiang, J.Z. (2006) Application of chaotic modulation to ac motors for harmonic suppression. Proceedings of IEEE International Conference on Industrial Technology, pp. 2343–2347.

Cuomo, K.M. and Oppenheim, A.V. (1993) Circuit implementation of synchronized chaos with applications to communications. *Physical Review Letters*, **71**, 65–68.

Deane, J.H.B. (1992) Chaos in a current-mode controlled boost DC-DC converter. *IEEE Transactions on Circuits and Systems – I: Fundamental Theory and Applications*, **39**, 680–683.

Deane, J.H.B. and Hamill, D.C. (1990) Instability, subharmonics, and chaos in power electronic systems. *IEEE Transactions on Power Electronics*, **5**, 260–268.

Deane, J.H.B. and Hamill, D.C. (1996) Improvement of power supply EMC by chaos. *Electronics Letters*, **32**, 1045.

Dedieu, H., Kennedy, M.P., and Hasler, M. (1993) Chaos shift keying: modulation and demodulation of a chaotic carrier using self-synchronizing Chua's circuits. *IEEE Transactions on Circuits and Systems – II*, **40**, 634–642.

Di Bernardo, M. and Tse, C.K. (2002) Chaos in power electronics: an overview, in *Chaos in Circuits and Systems* (eds G. Chen and T. Ueta) World Scientific, Hong Kong.

Di Bernardo, M., Garofalo, F., Glielmo, L., and Vasca, F. (1998) Switchings, bifurcations, and chaos in DC/DC converters. *IEEE Transactions on Circuits and Systems – I: Fundamental Theory and Applications*, **45**, 133–141.

Ditto, W. and Munakata, T. (1995) Principles and applications of chaotic systems. *Communications of the ACM*, **38**, 96–102.

Endo, T. and Chua, L.O. (1988) Chaos from phase-locked loops. *IEEE Transactions on Circuits and Systems*, **35**, 987–1003.

Endo, T., Chua, L.O., and Narita, T. (1989) Chaos from phase-locked loops: High-dissipation case. *IEEE Transactions on Circuits and Systems*, **36**, 255–263.

Feigenbaum, M.J. (1978) Quantitative universality for a class of nonlinear transformations. *Journal of Statistical Physics*, **19**, 25–52.

Feng, J.C. and Tse, C.K. (2001) On-line adaptive chaotic demodulator based on radial-basis-function neural networks. *Physical Review E*, **63**, 026202:1–026202:10.

Fossas, E. and Olivar, G. (1996) Study of chaos in the buck converter. *IEEE Transactions on Circuits and Systems – I: Fundamental Theory and Applications*, **43**, 13–25.

Galias, Z. and Maggio, G.M. (2001) Quadrature chaos-shift keying: theory and performance analysis. *IEEE Transactions on Circuits and Systems I: Fundamental Theory and Applications*, **48**, 1510–1519.

Gandhi, G., Cserey, G., Zbrozek, J., and Roska, T. (2009) Anyone can build Chua's circuit: hands-on-experience with chaos theory for high school students. *International Journal of Bifurcation and Chaos*, **19**, 1113–1125.

Gao, Y. and Chau, K.T. (2002) Chaotification of permanent-magnet synchronous motor drives using time-delay feedback. IEEE Industrial Electronics Conference, pp. 762–766.

Gao, Y. and Chau, K.T. (2003a) Chaotification of induction motor drives under periodic speed command. *Electric Power Components and Systems*, **31**, 1083–1099.

Gao, Y. and Chau, K.T. (2003b) Design of permanent magnets to avoid chaos in PM synchronous machines. *IEEE Transactions on Magnetics*, **39**, 2995–2997.

Gao, Y. and Chau, K.T. (2004a) Hopf bifurcation and chaos in synchronous reluctance motor drives. *IEEE Transactions on Energy Conversion*, **19**, 296–302.

Gao, Y. and Chau, K.T. (2004b) Design of permanent magnets to avoid chaos in doubly salient PM machines. *IEEE Transactions on Magnetics*, **40**, 3048–3050.

Gao, Y., Chau, K.T., and Ye, S. (2005) A novel chaotic-speed single-phase induction motor drive for cooling fans. Proceedings of IEEE Industry Applications Society Annual Meeting, pp. 1337–1341.

Hamdan, A.M.A., Nayfeh, M.A., and Nayfeh, A.H. (1990) Nonlinear analysis of a single-machine-quasi-infinite-busbar system. Proceedings of IEEE Southeastcon, pp. 140–144.

Hamill, D.C. and Jefferies, D.J. (1988) Subharmonics and chaos in a controlled switched-mode power converter. *IEEE Transactions on Circuits and Systems*, **35**, 1059–1061.

Hamill, D.C., Deane, J.H.B., and Jefferies, D.J. (1992) Modeling of chaotic DC-DC converters by iterated nonlinear mappings. *IEEE Transactions on Power Electronics*, **7**, 25–36.

Hemati, N. (1994) Strange attractors in brushless DC motors. *IEEE Transactions on Circuits and Systems I: Fundamental Theory and Applications*, **41**, 40–45.

Hilborn, R.C. (1994) *Chaos and Nonlinear Dynamics*, Oxford University Press, Oxford, UK.

Huang, J.F., Chen, J.H., and Chau, K.T. (2005) Chaoization of switched reluctance motor drives. Proceedings of International Conference on Electrical Machines and Systems, pp. 504–508.

Ito, S. and Narikiyo, T. (1998) Abrasive machining under wet condition and constant pressure using chaotic rotation (in Japanese). *Journal of the Japan Society for Precision Engineering*, **64**, 748–752.

Itoh, M. and Murakami, H. (1995) New communication systems via chaotic synchronizations and modulations. *IEICE Transactions on Fundamentals of Electronics, Communications and Computer Sciences*, **E78-A**, 285–290.

Ivan, C. and Serbanescu, A. (2009) Applications of nonlinear time-series analysis in unstable periodic orbits identification – Chaos control in buck converter. Proceedings of International Symposium on Signals, Circuits and Systems, pp. 1–4.

Kavitha, A. and Uma, G. (2008) Control of chaos by resonant parametric perturbation in a current mode controlled buck-boost DC-DC converter. Proceedings of IEEE Applied Power Electronics Conference and Exposition, pp. 323–327.

Kocarev, L. and Chua, L.O. (1993) On chaos in digital filters: Case $b=-1$. *IEEE Transactions on Circuits and Systems II: Analog and Digital Signal Processing*, **40**, 404–407.

Kocarev, L. and Parlitz, U. (1993) General approach for chaotic synchronization with applications to communication. *Physical Review Letters*, **74**, 5028–5031.

Kocarev, L., Halle, K.S., Eckert, K. *et al.* (1992) Experimental demonstration of secure communications via chaotic synchronization. *International Journal of Bifurcation and Chaos*, **2**, 709–713.

Kolumbán, G. and Kennedy, M.P. (2000) Digital communications using chaos, in *Controlling Chaos and Bifurcation in Engineering Systems* (ed. G. Chen) CRC Press, Boca Raton, FL.

Kolumbán, G., Jako, Z., and Kennedy, M.P. (1999) Enhanced versions of DCSK and FM-DCSK data transmission systems. Proceedings of IEEE International Symposium on Circuits and Systems, pp. 475–478.

Kolumbán, G., Kennedy, M.P., and Chua, L.O. (1998) The role of synchronization in digital communications using chaos — Part II: Chaotic modulation and chaotic synchronization. *IEEE Transactions on Circuits and Systems – I*, **45**, 1129–1140.

Kolumbán, G., Kennedy, M.P., and Kis, G. (1997) Performance improvement of chaotic communications systems. Proceedings of European Conference on Circuit Theory and Design, pp. 284–289.

Kolumbán, G., Vizvari, B., Schwarz, W., and Abel, A. (1996) Differential chaos shift keying: A robust coding for chaos communications. Proceedings of International Specialist Workshop on Nonlinear Dynamics of Electronics Systems, pp. 87–92.

Krein, P.T. and Bass, R.M. (1990) Types of instability encountered in simple power electronic circuits: unboundedness, chattering, and chaos. Proceedings of IEEE Applied Power Electronics Conference, pp. 191–194.

Kuroe, Y. and Hayashi, S. (1989) Analysis of bifurcation in power electronic induction motor drive systems. Proceedings of IEEE Power Electronics Specialists Conference, pp. 923–930.

Lau, F.C.M. and Tse, C.K. (2003) *Chaos-Based Digital Communication Systems*, Springer-Verlag, NY.

Lee, B. and Ajjarapu, V. (1993) Period-doubling route to chaos in an electrical power system. *IEE Proceedings C – Generation, Transmission and Distribution*, **140**, 490–496.

Li, T.Y. and Yorke, J.A. (1975). Period three implies chaos. *American Mathematical Monthly*, **82**, 985–92.

Li, Z., Park, J.B., Joo, Y.H. *et al.* (2002) Bifurcations and chaos in a permanent-magnet synchronous motor. *IEEE Transactions on Circuits and Systems I: Fundamental Theory and Applications*, **49**, 383–387.

Liao, G.C. and Tsao, T.P. (2004) A novel GA-based and meta-heuristics method for short-term unit commitment problem. Proceedings of IEEE Power Engineering Society General Meeting, pp. 1088–1093.

Liu, H. (1998) *The Semantics and Philosophy of Chaos (in Chinese)*, Hunan Education Press, Changsha, China.

Liu, S., Wang, M., and Hou, Z. (2003) Hybrid algorithm of chaos optimisation and SLP for optimal power flow problems with multimodal characteristic. *IEE Proceedings – Generation, Transmission and Distribution*, **150**, 543–547.

Liu, Y. and Tang, W.K.S. (2008) Cryptanalysis of chaotic masking secure communication systems using an adaptive observer. *IEEE Transactions on Circuits and Systems – II: Express Briefs*, **55**, 1183–1187.

Long, Y.J. (2001) Chaotic dynamics and compaction engineering. Proceedings of International Conference on Soft Soil Engineering, pp. 143–147.

Lorenz, E.N. (1963) Deterministic nonperiodic flow. *Journal of the Atmospheric Sciences*, **20**, 130–141.

Loría, A. (2009) Robust linear control of (chaotic) permanent-magnet synchronous motors with uncertainties. *IEEE Transactions on Circuits and Systems I: Fundamental Theory and Applications*, **56**, 2109–2122.

Macchi, O. and Jaidane-Saidane, M. (1989) Adaptive IIR filtering and chaotic dynamics: application to audiofrequency coding. *IEEE Transactions on Circuits and Systems*, **36**, 591–599.

Mamishev, A.V., Russell, B.D., and Benner, C.L. (1996) Analysis of high impedance faults using fractal techniques. *IEEE Transactions on Power Systems*, **11**, 435–440.

Mandelbrot, B.B. (1982) *The Fractal Geometry of Nature*, Freeman, San Francisco.

Matsumoto, T. (1984) A chaotic attractor from Chua's circuit. *IEEE Transaction on Circuits and Systems*, **31**, 1055–1058.

Matsumoto, T., Chua, L., and Kobayashi, K. (1986) Hyperchaos: Laboratory experiment and numerical confirmation. *IEEE Transactions on Circuits and Systems*, **33**, 1143–114.

May, R.M. (1974) Biological populations with non-overlapping generations: stable points, stable cycles, and chaos. *Science*, **186**, 645–647.

May, R.M. (1976) Simple mathematical models with very complicated dynamics. *Nature*, **261**, 459–467.

Milanović, V. and Zaghloul, M.E. (1996) Improved masking algorithm for chaotic communications systems. *Electronics Letters*, **32**, 11–12.

Min, X., Xu, W., Wang, L., and Chen, G. (2010) Promising performance of a frequency-modulated differential chaos shift keying ultra-wideband system under indoor environments. *IET Communications*, **4**, 125–134.

Mishina, T., Kohmoto, T., and Hashi, T. (1985) Simple electronic circuit for the demonstration of chaotic phenomena. *American Journal of Physics*, **53**, 332–334.

Mori, H. and Urano, S. (1996) Short-term load forecasting with chaos time series analysis. Proceedings of International Conference on Intelligent Systems Applications to Power Systems, pp. 133–137.

Nagashima, H. and Baba, Y. (1999) *Introduction to Chaos: Physics and Mathematics of Chaotic Phenomena*, Institute of Physics Publishing, Bristol, UK.

Nagy, I. (1994) Tolerance band based current control of induction machines highlighted with the theory of chaos. Proceedings of International Power Electronics Congress, pp. 155–160.

Niu, X. and Qiu, J. (2002) Investigation of torsional instability, bifurcation, and chaos of a generator set. *IEEE Transactions on Energy Conversion*, **17**, 164–168.

Okuno, H., Takeshita, M., and Kanari, Y. (2002) OGY control by asymptotically transition method in power system. Proceedings of SICE Annual Conference, pp. 3163–3168.

Ott, E., Grebogi, C., and Yorke, J.A. (1990) Controlling chaos. *Physical Review Letter*, **64**, 1196–1199.

Parlitz, U., Chua, L.O., Kocarev, L. *et al.* (1992) Transmission of digital signals by chaotic synchronization. *International Journal of Bifurcation and Chaos*, **2**, 973–977.

Poddar, G., Chakrabarty, K., and Banerjee, S. (1998) Control of chaos in DC-DC converters. *IEEE Transactions on Circuits and Systems – I: Fundamental Theory and Applications*, **45**, 672–676.

Poincaré, J.H. (1996) *Science and Method*, (Reprint of 1914 Edition), Routledge/Thoemmes Press, London, UK.

Ren, H. and Liu, D. (2006) Nonlinear feedback control of chaos in permanent magnet synchronous motor. *IEEE Transactions on Circuits and Systems II: Express Briefs*, **53**, 45–50.

Rodríguez-Vázquez, A., Huertas, J.L., and Chua, L.O. (1985) Chaos in a switched-capacitor circuit. *IEEE Transactions on Circuits and Systems*, **32**, 1083–1085.

Saito, T. (1985) A chaos generator based on a quasi-harmonic oscillator. *IEEE Transactions on Circuits and Systems*, **32**, 320–331.

Schiff, S.J., Jerger, K., Duong, D.H. *et al.* (1994) Controlling chaos in the brain. *Nature*, **370**, 615–620.

Shinriki, M., Yamamoto, M., and Mori, S. (1981) Multimode oscillations in a modified van Der Pol oscillator containing a positive nonlinear conductance. *Proceedings of IEEE*, **69**, 394–395.

Simpson, J. (2004) *Oxford English Dictionary Online*, Oxford University Press, Oxford, UK.

Song, Y. and Yu, X. (2000) Multi-parameter modulation for secure communication via Lorenz chaos. Proceedings of IEEE Conference on Decision and Control, pp. 42–45.

Srivastava, K.N. and Srivastava, S.C. (1998). Elimination of dynamic bifurcation and chaos in power systems using FACTS devices. *IEEE Transactions on Circuits and Systems I: Fundamental Theory and Applications*, **45**, 72–78.

Sushchik, M., Tsimring, L.S., and Volkovskii, A.R. (2000) Performance analysis of correlation-based communication schemes utilizing chaos. *IEEE Transactions on Circuits and Systems I: Fundamental Theory and Applications*, **47**, 1684–1691.

Sütő, Z., Nagy, I., and Masada, E. (2000) Period adding route to chaos in a hysteresis current controlled AC drive. Proceedings of International Workshop on Advanced Motion Control, pp. 299–304.

Suzuki, R. and Yasuda, K. (1998) Analysis of chatter vibration in an automotive wiper assembly. *JSME International Journal, Series C*, **41**, 616–620.

Tang, Y., Mees, A., and Chua, L. (1983) Synchronization and chaos. *IEEE Transactions on Circuits and Systems*, **30**, 620–626.

Tao, Y. and Chua, L.O. (1996) Secure communication via chaotic parameter modulation. *IEEE Transactions on Circuits and Systems – II*, **43**, 817–819.

Tse, C.K. (1994) Chaos from a buck switching regulator operating in discontinuous mode. *International Journal of Circuit Theory and Applications*, **22**, 263–278.

Tse, C.K., Fung, S.C., and Kwan, M.W. (1996) Experimental confirmation of chaos in a current-programmed Ćuk converter. *IEEE Transactions on Circuits and Systems – I: Fundamental Theory and Applications*, **43**, 605–608.

Tse, K.K., Ng, R.W.-M., Chung, H.S.-H., and Hui, S.Y.R. (2003) An evaluation of the spectral characteristics of switching converters with chaotic carrier-frequency modulation. *IEEE Transactions on Industrial Electronics*, **50**, 171–182.

Ueda, Y. (1992) *The Road to Chaos*, Aerial Press, Santa Cruz, California.

Van Wyk, M.A. and Steeb, W.-H. (1997) *Chaos in Electronics*, Kluwer Academic Publishers, Netherlands.

Wang, Z. and Chau, K.T. (2006) Stabilization of chaotic vibration in automobile wiper systems. Proceedings of Asia International Symposium on Mechatronics, pp. DS-02:1–DS-02:6.

Wang, Z. and Chau, K.T. (2007) Design and analysis of a chaotic PWM inverter for electric vehicles. Proceedings of IEEE Industry Applications Society Annual Meeting, pp. 1954–1961.

Wang, Z. and Chau, K.T. (2008) Anti-control of chaos of a permanent magnet DC motor system for vibratory compactors. *Chaos, Solitons and Fractals*, **36**, 694–708.

Wang, Z. and Chau, K.T. (2009a) Control of chaotic vibration in automotive wiper systems. *Chaos, Solitons and Fractals*, **39**, 168–181.

Wang, Z. and Chau, K.T. (2009b) Design, analysis and experimentation of chaotic permanent magnet DC motor drives for electric compaction. *IEEE Transactions on Circuits and Systems II: Express Briefs*, **56**, 245–249.

Wang, Z., Chau, K.T., and Jian, L. (2008a) Chaoization of permanent magnet synchronous motors using stator flux regulation. *IEEE Transactions on Magnetics*, **44**, 4151–4154.

Wang, Z., Chau, K.T., and Cheng, M. (2008b) A chaotic PWM motor drive for electric propulsion. Proceedings of IEEE Vehicle Power and Propulsion Conference, pp. H08357:1–H08357:6.

Wang, Z., Chau, K.T., and Liu, C. (2007) Improvement of electromagnetic compatibility of motor drives using chaotic PWM. *IEEE Transactions on Magnetics*, **43**, 2612–2614.

Wildberger, M. (1994) Stability and nonlinear dynamics in power systems. *IEEE Power Engineering Review*, **14**, 16–18.

Wood, J.R. (1989) Chaos: a real phenomenon in power electronics. Proceedings of IEEE Applied Power Electronics Conference, pp. 115–123.

Ye, S. and Chau, K.T. (2005a) Destabilization control of a chaotic motor for industrial mixers. Proceedings of IEEE Industry Applications Society Annual Meeting, pp. 1724–1730.

Ye, S. and Chau, K.T. (2005b) Design of permanent magnets to chaoize PM synchronous motors for industrial mixer. Proceedings of IEEE International Magnetics Conference, pp. 723–724.

Ye, S. and Chau, K.T. (2007) Chaoization of DC motors for industrial mixing. *IEEE Transactions on Industrial Electronics*, **54**, 2024–2032.

Ye, S., Chau, K.T., and Niu, S. (2006) Chaoization of a single-phase induction motor for washing machines. Proceedings of IEEE Industry Applications Society Annual Meeting, pp. 855–860.

Yuan, G.H., Banerjee, S., Ott, E., and Yorke, J.A. (1998) Border collision bifurcation in the buck converter. *IEEE Transactions on Circuits and Systems – I: Fundamental Theory and Applications*, **45**, 707–716.

Zhong, G.Q. and Ayrom, F. (1985) Experimental confirmation of chaos from Chua's circuit. *International Journal of Circuit Theory and Applications*, **13**, 93–98.

Zhou, Y., Tse, C.K., Qiu, S.S., and Lau, F.C.M. (2003) Applying resonant parametric perturbation to control chaos in the buck DC/DC converter with phase shift and frequency mismatch considerations. *International Journal of Bifurcation and Chaos*, **13**, 3459–3471.

Zou, F. and Nossek, J.A. (1991) A chaotic attractor with cellular neural networks. *IEEE Transactions on Circuits and Systems*, **38**, 811–812.

2

Introduction to Chaos Theory and Electric Drive Systems

As this book is a happy marriage between two disciplines – chaos theory and electric drive systems – the coverage of knowledge is so broad that it is desirable to briefly introduce basic chaos theory and the fundamentals of electric drive systems.

In this chapter, the necessary background knowledge of this book, namely the basic theory of chaos and some fundamentals of electric drive systems, are discussed.

2.1 Basic Chaos Theory

This section describes some of the basic principles of chaos, including the concept of dynamical systems, discrete maps, limit sets, attractors, stability, and manifolds. The criteria of chaos are then illustrated using the Lyapunov exponents, fractal dimensions, and entropy. Hence, bifurcations and routes to chaos will both be discussed. Finally, various methods for chaos analysis are introduced, which include waveforms, phase portraits, the Poincaré map, bifurcation diagrams, time-series reconstruction, and the calculations of Lyapunov exponents, embedded unstable periodic orbits, power spectra, fractal dimensions, and entropy.

2.1.1 Basic Principles

2.1.1.1 Dynamical Systems

Autonomous Dynamical Systems

An nth-order autonomous dynamical system is defined by $\dot{x} = f(x)$, $x(t_0) = x_0$. The corresponding solution is $\phi_t(x_0)$, and $\phi_t(x_0)$ is called as a flow. For autonomous continuous systems, the vector field f does not depend on time t.

Nonautonomous Dynamical Systems

An nth-order nonautonomous dynamical system is defined by $\dot{x} = f(x, t)$, $x(t_0) = x_0$. The corresponding underlying flow is $\phi_t(x, t_0)$. For nonautonomous continuous systems, the vector field f depends not only on the state variable x but also on the time t.

Chaos in Electric Drive Systems: Analysis, Control and Application, First Edition. K.T. Chau and Zheng Wang.
© 2011 John Wiley & Sons (Asia) Pte Ltd. Published 2011 by John Wiley & Sons (Asia) Pte Ltd.

2.1.1.2 Discrete Maps

Orbit

A discrete system can be defined by a map $P : \Re^n \rightarrow \Re^n$ with the state equation $x_{k+1} = P(x_k)$, where $x_k \in \Re^n$ are the states at the kth iterative time, and P maps the state x_k to the next state x_{k+1}. Starting from an initial condition x_0, repeated application of P generates a sequence of points $\{x_k\}_{k=0}^{\infty}$ which is known as an orbit (Parker and Chua, 1999).

Poincaré Map

A classical technique for discretization of a dynamical system is the Poincaré map. It replaces the flow of an nth-order continuous system with an $(n-1)$th-order discrete map. The Poincaré map is useful to reduce the system order and bridge the gap between continuous and discrete systems (Parker and Chua, 1999).

For an nth-order nonautonomous system with a minimum period T, the resulting Poincaré map is defined by $P(x) = \phi_{t_0+T}(x_0, t_0)$. The corresponding orbit $P(x_k)$ is a sampling of a single trajectory every T seconds, which is similar to the action of a stroboscope flashing with period T.

For an nth-order autonomous system, the underlying flow $\phi_t(x_0)$ is with the trajectory Γ. When Σ is chosen to be an $(n-1)$-dimensional hyperplane so that the trajectory Γ intersects Σ transversely, and x_0 is a point of Γ on the hyperplane Σ, the trajectory Γ starting from x_0 will intersect Σ at point x_1. Due to the continuity of $\phi_t(x_0)$, trajectories starting on Σ in a sufficiently small neighborhood of x_0 will also intersect Σ in the vicinity of x_1. Thus, Γ and Σ define the Poincaré map P of some neighborhood $U \subset \Sigma$ of x_0 onto a neighborhood $V \subset \Sigma$ of x_1. Figures 2.1(a) and 2.1(b) show the Poincaré map for the autonomous system and nonautonomous system, respectively.

2.1.1.3 Limit Sets and Attractors

If there exists a point y under the condition that the trajectory $\phi_t(x_0)$ repeatedly enters every neighborhood U of y for $t \rightarrow \infty$, y is known as the limit point of x_0 (Parker and Chua, 1999). The set of such points y is defined as a limit set of x_0, which can be represented as a function of $Y(x_0)$. If there exists an open neighborhood U of a limit set Y, and for all $x_0 \in U$, $Y(x_0) = Y$, the limit set Y is attracting. The attracting limit set is also called an attractor (Ott, 1993). The union of all U is defined as the attracting basin of the limit set Y.

Equilibrium Point

If there exists a point x_{eq} which satisfies $x_{eq} = \phi_t(x_{eq})$ for all t, such a point x_{eq} is called an equilibrium point. The attractor of an equilibrium point is the equilibrium point itself (Parker and Chua, 1999). For the discrete system, there is no limit set corresponding to an equilibrium point.

Limit Cycle

If there exists a trajectory $\phi_t(x_0)$ which satisfies $\phi_t(x_0) = \phi_{t+T}(x_0)$ with a minimum period T for all t, such a trajectory is known as a periodic behavior. The attractor of a periodic behavior is a closed trajectory $\phi_t(x_0)$, also called a limit cycle. If, in an nth-order periodic nonautonomous system with forcing period T_f, the period T is an integer multiple k of T_f, then the underlying flow is known as both a period-k behavior and a kth-order subharmonic (Parker and Chua, 1999). For the discrete system, the limit sets of the periodic behavior are fixed points.

Torus

If the solution trajectory of an autonomous system is the sum of a countable number of periodic trajectories whose frequencies are incommensurate, then the whole solution trajectory exhibits a quasiperiodic

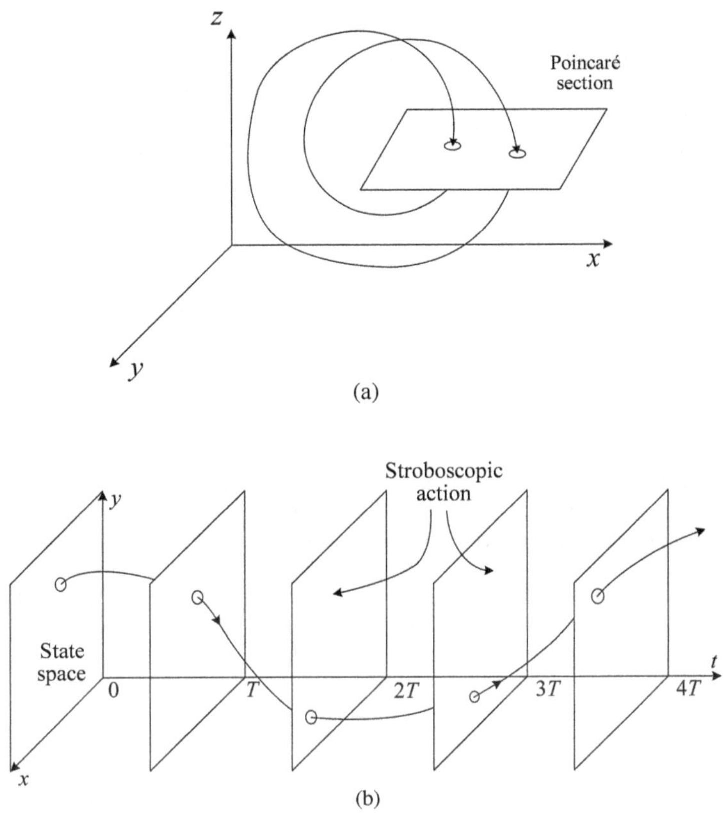

Figure 2.1 Poincaré map. (a) Autonomous system. (b) Nonautonomous system

behavior. If the countable number is p, the quasiperiodic behavior is known as p-periodic behavior (Parker and Chua, 1999). The attractor of the p-periodic behavior is the p-torus. For the discrete system, the attractor of the p-periodic behavior is one or more embedded $p-1$ tori.

Strange Attractor

A limit set is said to be chaotic if the corresponding solution trajectory is exponentially sensitive to the initial conditions (Ott, 1993). The trajectory of chaos exhibits randomly mixed behaviors. Generally, the strange attractor is used to describe the geometrical attracting limit set for chaos. For a discrete system, the limit set of chaos is a strange geometry, which is different from the simple geometry of periodic and quasiperiodic behavior. It is a fine-layered structure (Sprott, 2003).

For most cases involving differential equations, chaos commonly occurs together with geometrical strangeness (Ditto *et al.*, 1990). The strange attractors are not a finite set of points, nor a smooth curve or surface, nor a volume bounded by a piecewise smooth closed surface (Ding, Grebogi, and Ott, 1989). However, it is possible for a chaotic attractor not to be strange. For example, the logistic map under a certain parameter has a chaotic attractor with a positive Lyapunov exponent, but this attractor is not strange since it is a simple interval within [0, 1]. For Hamiltonian systems, the dynamics can be chaotic. However, these are conservative systems that have no attractors at all (Ott, 1993).

On the other hand, it has been identified that strange nonchaotic attractors are indeed typical in systems that are driven by a two-frequency quasiperiodic force (Grebogi *et al.*, 1984). This phenomenon can be observed experimentally by Lyapunov exponents, information dimensions, Fourier amplitude spectra, and a phase portrait (Yang and Bilimgut, 1997). The power spectrum of strange chaotic attractors is comparatively broader and has a much richer harmonic content than periodic attractors or strange nonchaotic attractors (Zhou and Moss, 1992).

2.1.1.4 Stability

Eigenvalues can be used to judge the stability of the equilibrium points. If f is a vector field describing an autonomous dynamic system, the equation governing the time evolution of a perturbation δx_0 in a neighborhood of the equilibrium point x_{eq} can be represented as:

$$\delta \dot{x} = Df(x_{eq})\delta x, \delta x(0) = \delta x_0 \tag{2.1}$$

where $Df(\cdot)$ is the differential equation of f. If λ_i are the eigenvalues of $Df(x_{eq})$, the real part of λ_i, namely $\mathrm{Re}[\lambda_i]$, are used to judge the stability of x_{eq}. If $\mathrm{Re}[\lambda_i] < 0$ for all λ_i, x_{eq} is asymptotically stable. If any $\mathrm{Re}[\lambda_i] > 0$, x_{eq} is unstable. The eigenvalue distributions of different equilibrium points are shown in Figure 2.2.

The stability of a limit cycle is judged by its characteristic multipliers, which are also known as Floquet multipliers. The limit cycle corresponds to a fixed point x^* on the Poincaré map. The local behavior of the map near x^* is determined by the differential equations of the map P at x^* as:

$$\delta x_{k+1} = DP(x^*)\delta x_k \tag{2.2}$$

The eigenvalues of $DP(x^*)$ are defined to be m_i, which are known as the Floquet multipliers of the limit cycle. They determine the stability of the fixed point x^*, and the corresponding limit cycle. If $|m_i| < 1$ for all m_i, x^* is asymptotically stable. If any $|m_i| > 1$, x^* is unstable. Actually, there are relationships between

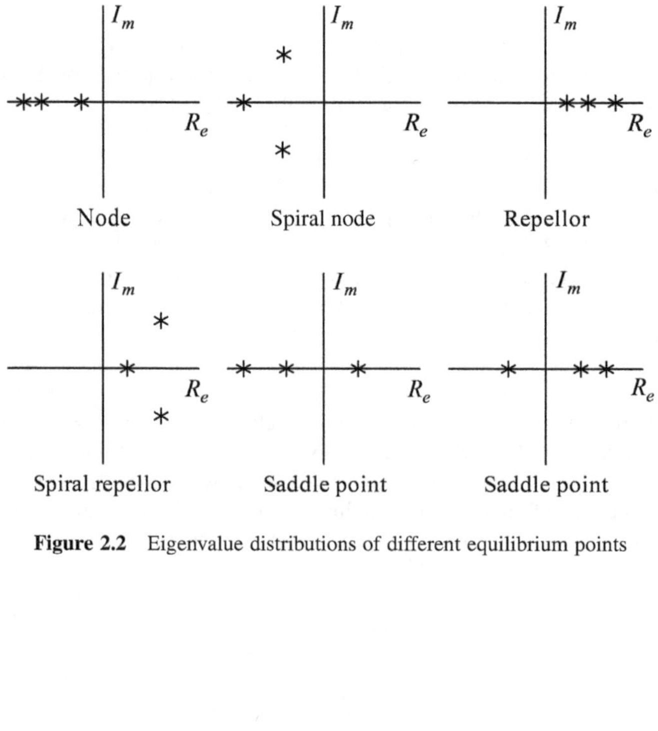

Figure 2.2 Eigenvalue distributions of different equilibrium points

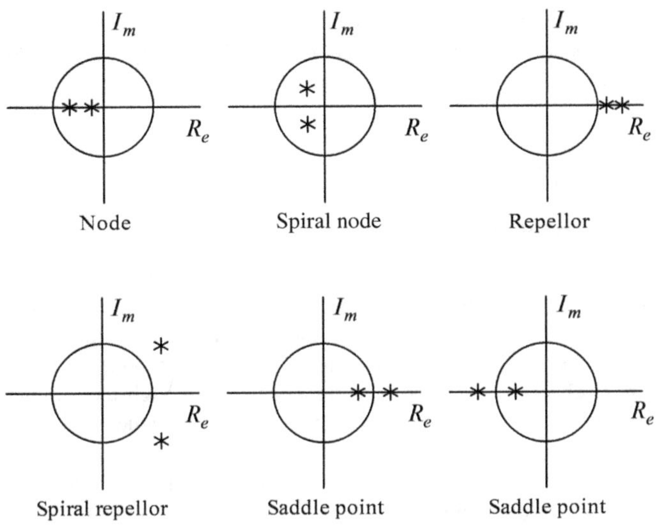

Figure 2.3 Floquet multiplier distributions of different limit cycles

λ_i and m_i which can be represented by $m_i = e^{\lambda_i T}$, where T is the period of the limit cycle. Figure 2.3 shows the Floquet multiplier distributions of different limit cycles.

2.1.1.5 Stable and Unstable Manifolds

Outside the near neighborhood of a fixed point, the curved lines passing through the fixed point are used to evaluate the stability of the dynamical system. If the initial condition on the curves remains on the curves after several iterations of the map, the curves are said to be invariable manifolds (Banerjee and Verghese, 2001).

Stable and unstable manifolds exist for nonchaotic limit sets Y. The stable manifold is the set of all points x_0 whose trajectory $\phi_t(x_0)$ approaches Y as $t \to \infty$. The unstable manifold of the limit set Y is the set of all points x_0 whose trajectory $\phi_t(x_0)$ approaches $Y(x_0)$ as $t \to -\infty$ (Parker and Chua, 1999). If the initial condition is not on the invariable manifolds, the iterative states will move away from the stable manifolds and move closer to the unstable manifolds, as illustrated in Figure 2.4. As an unstable manifold

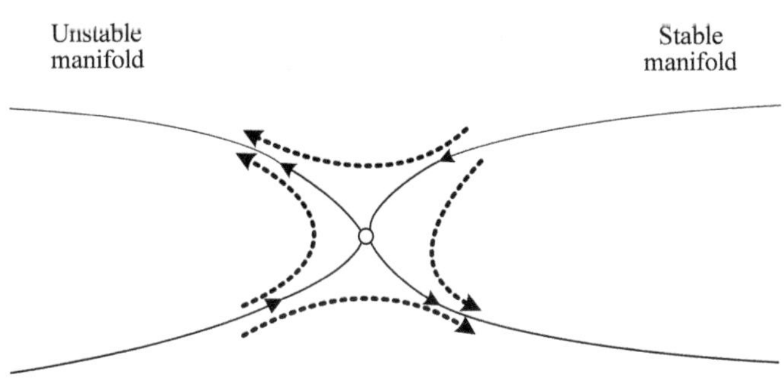

Figure 2.4 Stable and unstable manifolds of a saddle point

attracts the points in the state space, and a stable manifold repels the points, the stable manifold of a saddle point often acts as the boundary separating the attracting basins of different attractors in the state space (Banerjee and Verghese, 2001).

2.1.2 Criteria for Chaos

2.1.2.1 Lyapunov Exponents

The Lyapunov exponent can be used to determine the stability of quasiperiodic and chaotic behavior, and also the stability of equilibrium points and periodic behaviors. It provides a way to quantify the rates of stretching and squeezing of the attractor in the state space. The Lyapunov exponent is the exponential rate of this divergence and convergence. If the maximum Lyapunov exponent of a dynamical system is positive, this system is chaotic; otherwise, it is nonchaotic. Although the Lyapunov exponents are a generation of eigenvalues, there are many differences: the eigenvalue is a local quantity while the Lyapunov exponent is a global quantity; the eigenvalue is a constant value while the Lyapunov exponent is an average value; the eigenvalue is a complex number while the Lyapunov exponent is a real number; and the eigenvalues are not usually orthogonal while the Lyapunov exponents are mutually orthogonal (Sprott, 2003).

The Lyapunov exponents are directly related to the criteria of chaos. Therefore, it is significant to calculate the Lyapunov exponents. If the analytical model of the dynamical system is known, the process to compute the Lyapunov exponents is as follows. The solution flow of the system state variables is expressed as:

$$X(t) = T^t X_0 \tag{2.3}$$

where T^t is the map describing the time-t evolution of X, and the solution flow of their deviation δX is given by:

$$\delta X(t) = U^t_{X_0} \delta X_0 \tag{2.4}$$

where $U^t_{X_0}$ is the map describing the time-t evolution of δX. Then, the Lyapunov exponents λ_i of the d-dimensional system can be computed as (Shimada and Nagashima, 1979):

$$\lambda_i(i = 1 \sim \mathrm{d}) = \lim_{h \to \infty} \frac{1}{h\Delta t} \sum_{j=0}^{h-1} \log \frac{\left\| U^{\Delta t}_{X_j} e^j_i \right\|}{\left\| e^j_i \right\|} \tag{2.5}$$

where Δt is the evolution time, and e^j_i is the ith base vector of the d-dimensional state space at the jth step. It should be noted that e^j_i should be orthogonalized and normalized at each iterative step. Under the restrictions of $\Delta t \ll 1$ and $\|e^j_i\| \ll 1$, λ_i can be approximated as (Benettin, Galgani, and Strelcyn, 1976):

$$\lambda_i(i = 1 \sim \mathrm{d}) = \lim_{h \to \infty} \frac{1}{h\Delta t} \sum_{j=0}^{h-1} \log \frac{\left\| T^{\Delta t}(X_j + e^j_i) - T^{\Delta t}(X_j) \right\|}{\left\| e^j_i \right\|} \tag{2.6}$$

Some calculation approaches have been developed for the special dynamical systems such as the dynamical system with discontinuities and the finite-dimensional time-delay system. For a dynamical system with discontinuities, including the widely used power electronic circuits and motor drives, certain transitions are supplemented to the instants of discontinuities, thus enabling the flow of δX to be computed

(Müller, 1995). For a dynamical system with a time delay, X on the interval $[t, t-\tau]$ can be approximated by N samples taken at intervals $\Delta t = \tau/(N-1)$. These N samples can equivalently be considered as the N variables of an N-dimensional discrete map. With this approximated map, the Lyapunov exponents of a system with a time delay can be computed (Farmer, 1982).

2.1.2.2 Fractal Dimensions

Identifying chaos includes searching for a strange attractor in the state-space dynamics, which can be characterized by its fractal structure. The dimension of an attractor is a measure of the number of active variables and the complexity of the equations required to model the system dynamics (Sprott, 2003). Typical characteristics of the fractal structure are: (a) they have a fine structure and no characteristic scale length; (b) they are too irregular to be described by ordinary geometry, both locally and globally; and (c) they have some degree of self-similarity, which means that small pieces of the object resemble the whole in some respects. Fractal structures can be either deterministic where they are exactly self-similar, or random where they are only statistically self-similar. The self-similarity implies that the structure is scale invariant.

Geometrical objects with fractional dimensions are called fractals and, in dynamical systems, it has been found that chaotic attractors are fractal objects. The determination of the fractal dimensions is thus one of the methods of characterizing a chaotic attractor (Banerjee and Verghese, 2001). The dimension is a kind of quantifier which describes the attractor from the geometric aspects. The dimensionality of an attractor gives us an estimate of the number of active degrees of freedom in the system. If the attractor's dimension is not an integer, the attractor is a strange attractor. The capacity dimension D_b – which is a measure of fractal dimensions – of a dynamical system is defined as (Kolmogorov, 1958):

$$D_b = \lim_{R \to 0} \left\{ -\frac{\log N(R)}{\log R} + \frac{\log k}{\log R} \right\} \tag{2.7}$$

where R is the side length of the constructed "box," $N(R)$ is the number of boxes required to contain all the points on the attractor, and k is a proportionality constant.

2.1.2.3 Entropy

The sum of the positive Lyapunov exponents is the Kolmogorov–Sinai (K–S) entropy (Pesin, 1977). This K–S entropy has a similarity to the usual thermodynamic entropy since it measures the expansion of nearby trajectories into new regions of state space. At the same time, unlike the thermodynamic entropy, the K–S entropy has units of inverse time, or inverse iterations for maps. It is a measure of the average rate at which predictability is lost. Its inverse is a rough estimate of the time for which a reasonable prediction is expected (Sprott, 2003). A purely random system has infinite entropy, and a periodic system has zero entropy; therefore, the K–S entropy is a positive constant for a chaotic system, and the chaotic degree increases with the value of the K–S entropy.

2.1.3 *Bifurcations and Routes to Chaos*

2.1.3.1 Bifurcations

Pitchfork Bifurcation

The pitchfork bifurcation is also known as the cusp bifurcation. For the supercritical pitchfork bifurcation, there exists a fixed point when $\kappa < 0$. When $\kappa > 0$, the stable fixed point becomes unstable and gives birth to two new stable fixed points. The supercritical pitchfork points in the direction of positive κ. For the subcritical pitchfork bifurcation, the stable fixed point at $\kappa < 0$ also becomes unstable when $\kappa > 0$. However, the pitchfork points in the direction of negative κ. Consequently, there exist two other unstable branches when $\kappa < 0$. The principle of supercritical pitchfork bifurcation is shown in Figure 2.5(a).

Saddle-Node Bifurcation

The saddle-node bifurcation is also called a fold bifurcation or tangent bifurcation. When $\kappa < 0$, there is no fixed point in the system. At $\kappa = 0$, two fixed points are newly formed but separate when $\kappa > 0$. One fixed point is stable and the other is unstable. The principle of the saddle-node bifurcation is shown in Figure 2.5(b).

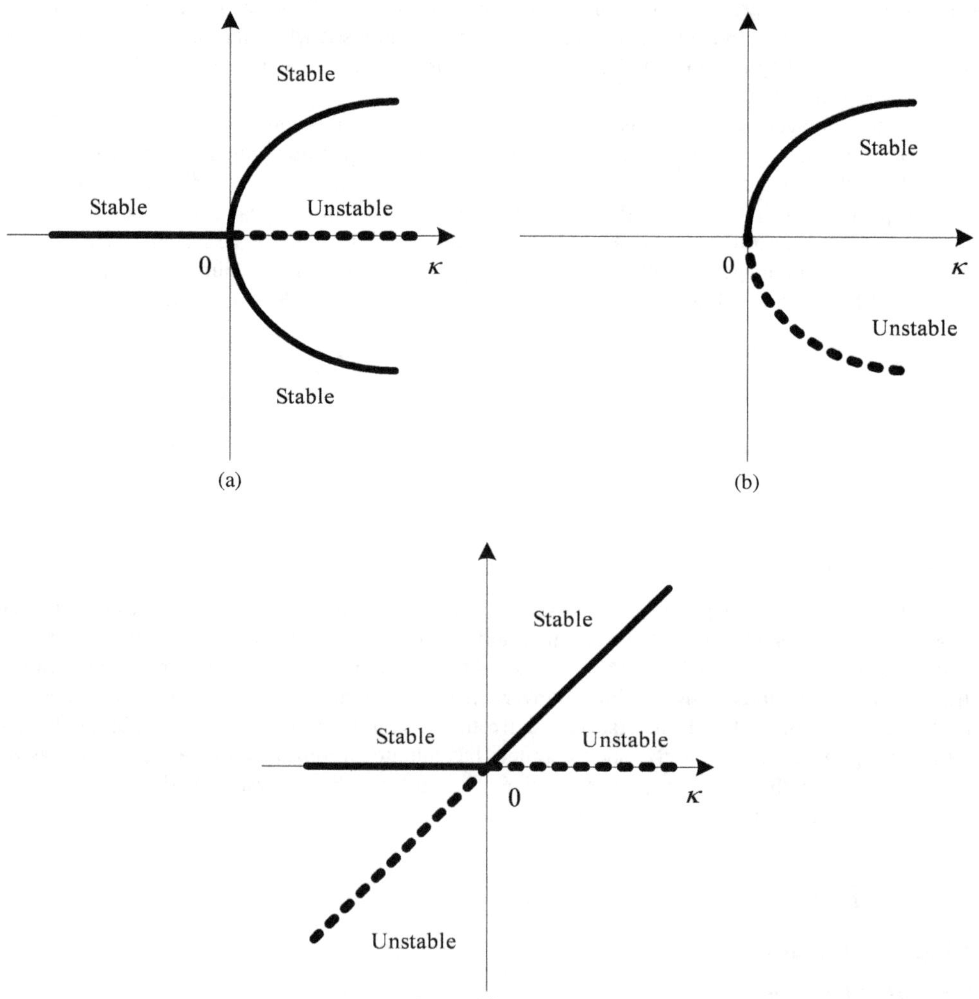

Figure 2.5 Basic bifurcations. (a) Pitchfork. (b) Saddle-node. (c) Transcritical

Transcritical Bifurcation

For a transcritical bifurcation, there are two equilibrium points $x_{eq} = 0$ and $x_{eq} = \kappa$, and their stability switches at $\kappa = 0$. For $\kappa < 0$, the stable equilibrium is at $x_{eq} = 0$. For $\kappa > 0$ the stable equilibrium is at $x_{eq} = \kappa$. Figure 2.5(c) also shows the principle of a transcritical bifurcation.

Period-Doubling Bifurcation

A period-doubling bifurcation is also known as a flip bifurcation. In the first iterative map of the system, there is a stable fixed point when $\kappa < 0$. When $\kappa > 0$, this fixed point becomes unstable. In the second iterative map of the system, the stable fixed point at $\kappa < 0$ becomes unstable when $\kappa > 0$ and two new stable fixed points are produced, causing a period-2 dynamics. This type of bifurcation is known as a supercritical period-doubling bifurcation. For a subcritical period-doubling bifurcation, the period-2 dynamics at $\kappa < 0$ is unstable. Actually, the period-doubling bifurcation is in analogy with the pitchfork bifurcation.

Hopf Bifurcation

A Hopf bifurcation is similar to a pitchfork bifurcation in a system with a dimension higher than 1. For a supercritical Hopf bifurcation, there exists a stable focus when $\kappa < 0$. At $\kappa = 0$, the eigenvalues, which are a complex conjugate pair, cross the imaginary axis. When $\kappa > 0$, the original focus becomes unstable, and a stable limit cycle is born. For a subcritical Hopf bifurcation, there exist a stable focus and an unstable limit cycle when $\kappa < 0$. At $\kappa = 0$, they coalesce and annihilate each other. Figure 2.6 shows the supercritical Hopf bifurcation.

Neimark–Sacker Bifurcation

A Neimark–Sacker bifurcation occurs when the limit cycle becomes unstable, and gives birth to a torus, which produces a quasiperiodic flow. When a torus is born in the flow, the eigenvalues of the corresponding map are a complex conjugate pair with a magnitude of 1. By analogy with the Hopf bifurcation, the birth of a torus in a flow is called a secondary Hopf.

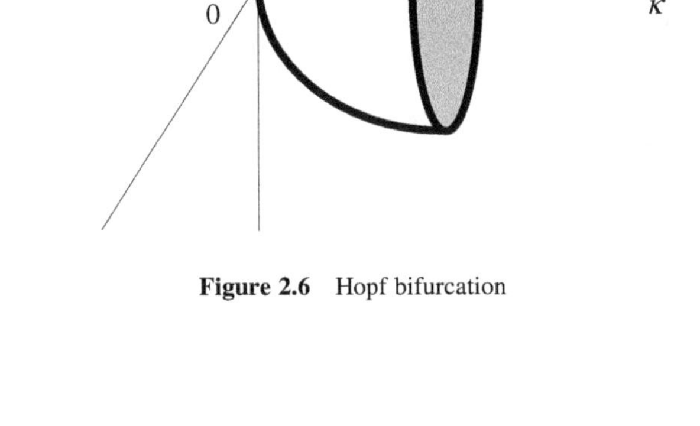

Figure 2.6 Hopf bifurcation

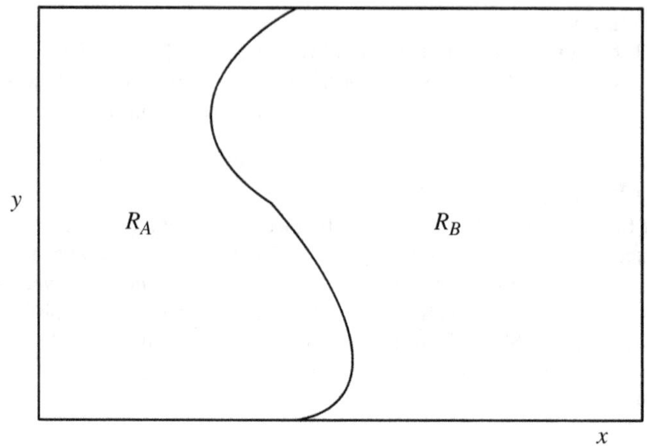

Figure 2.7 Schematic representation of piecewise-smooth map

Border Collision Bifurcation
For a flow or a map that is continuous but its derivative is discontinuous at the hyperplane M or the line χ, the M or the χ is called the border, which separates the phase space into two regions R_A and R_B, as shown in Figure 2.7. Unlike the bifurcations that occur in one of the smooth regions, some bifurcations occur when the equilibrium or fixed point collides with the border, and there is a discontinuous jump in the eigenvalues of the Jacobian matrix. This type of bifurcation is known as a border collision bifurcation (Banerjee and Verghese, 2001). There are two kinds of border collision bifurcation. The first is known as a border collision pair bifurcation, where there is no equilibrium or fixed point for $\kappa < 0$ and there are two equilibrium or fixed points for $\kappa > 0$ – one on region R_A and one on region R_B. The second kind is known as a border-crossing bifurcation, where the equilibrium or fixed point crosses the border when κ goes through zero.

Nonstandard Bifurcations in Discontinuous Systems
In discontinuous systems, some nonstandard bifurcation phenomena occur. In continuous systems, equilibrium or fixed points appear or disappear only in pairs while, in discontinuous systems, a single equilibrium or fixed point may appear or disappear. Furthermore, the eigenvalues of an equilibrium or fixed point of discontinuous systems do not signal the occurrence of a bifurcation. For discontinuous systems, the basin boundary of two attracting equilibrium or fixed points can be formed from points of discontinuity, as depicted in Figure 2.8(a). On the other hand, the basin boundary in continuous systems can be the stable manifold of a saddle point or an unstable periodic orbit, as depicted in Figure 2.8(b) (Banerjee and Verghese, 2001).

Homoclinic and Heteroclinic Bifurcations
If x_0 and y_0 are two different equilibrium or fixed points, $W^s(x_0)$ and $W^u(x_0)$ are respectively the stable and unstable manifolds of x_0, while $W^s(y_0)$ and $W^u(y_0)$ are respectively the stable and unstable manifolds of y_0. If a trajectory or orbit exists in $W^s(x_0) \cap W^u(x_0)$, it is called a homoclinic trajectory or orbit. Figure 2.9(a) shows the homoclinic trajectory of a saddle point x_0. If a trajectory or orbit exists in

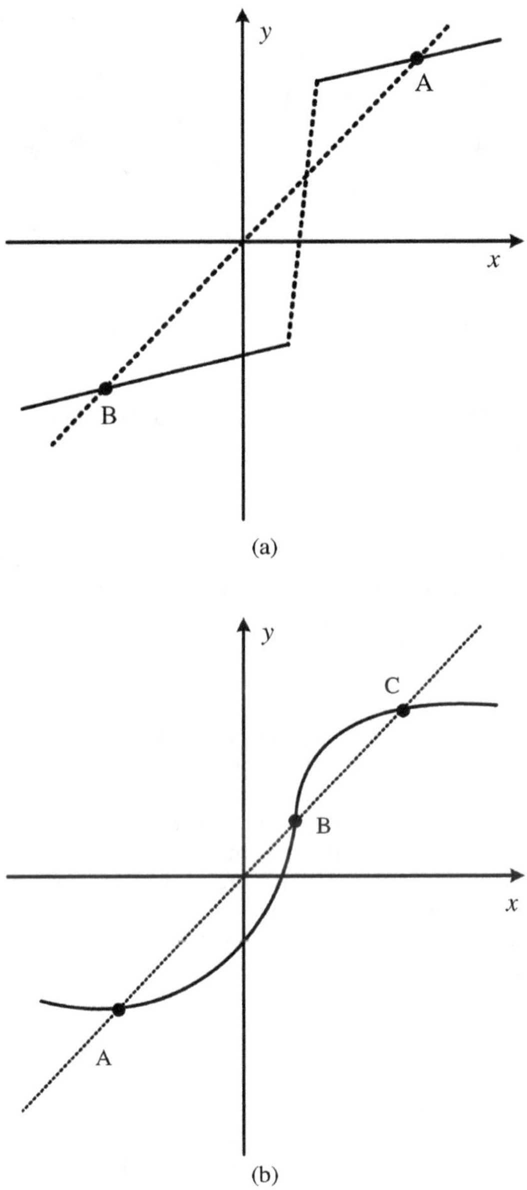

Figure 2.8 Maps with two stable fixed points. (a) Discontinuous. (b) Continuous

$W^s(x_0) \cap W^u(y_0)$, it is called a heteroclinic trajectory or orbit. Figure 2.9(b) shows the structure of a heteroclinic trajectory.

A homoclinic bifurcation occurs when the stable and unstable manifolds of an equilibrium or fixed point touch one another, while a heteroclinic bifurcation occurs when the stable manifold of one equilibrium or

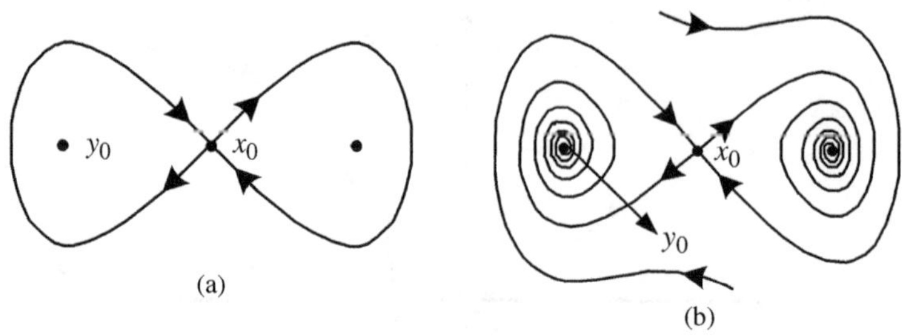

Figure 2.9 Homoclinic and heteroclinic trajectories. (a) Homoclinic. (b) Heteroclinic

fixed point touches the unstable manifold of another (Parker and Chua, 1999). Homoclinic and heteroclinic bifurcations are both global bifurcations.

Crisis

The crisis is another class of global bifurcation that occurs when a chaotic attractor collides with an unstable periodic orbit or its attracting basin. There are three types of crises. The first is the boundary crisis where the chaotic attractor touches the basin boundary that separates it from another coexisting attractor. After this crisis, the attractor is destroyed and the trajectories in this region are first transiently chaotic and then asymptotically approach the other attractor (Sprott, 2003). The second crisis is the interior crisis when the chaotic attractor collides with a periodic trajectory or orbit within its basin. The chaotic attractor suddenly expands in size but remains bounded after the collision. After the crisis, the trajectory or orbit on the attractor spends a long time in the original smaller attractor, and intermittently jumps from this small attractor to the newly created large attractor (Banerjee and Verghese, 2001), resulting in a crisis-induced intermittency. The third crisis is the attractor-merging crisis where two or more chaotic attractors simultaneously touch a periodic trajectory or orbit on the basin boundary that separates them. The two or more attractors then merge into one multipiece chaotic attractor. Intermittency occurs when the trajectory or orbit moves among different pieces of the attractor at random intervals. Figure 2.10 depicts different types of crises (Banerjee and Verghese, 2001).

2.1.3.2 Routes to Chaos

Period-Doubling Cascade Route to Chaos

The first kind of route to chaos through local bifurcation is the period-doubling cascade route, where stable fixed points become unstable in a series of period-doubling bifurcations, and subharmonic behavior occurs. The well-known Logistic map $\xi_{n+1} = A\xi_n(1-\xi_n)$ is identified to be a typical case for illustrating the period-doubling cascade route to chaos, as shown in Figure 2.11. As the parameter A is increased, there is a period doubling cascade. After an infinite number of period doublings, chaos and periodic "windows" are finely intermixed (Ott, 1993). For the Logistic map, chaos onsets when A is beyond the accumulating point, around 3.56994.

Intermittency Transition to Chaos

Another route to chaos resulting from local bifurcations is the intermittency transition, which can be classified into three types. The first type emanates from a saddle-node bifurcation, which often occurs in

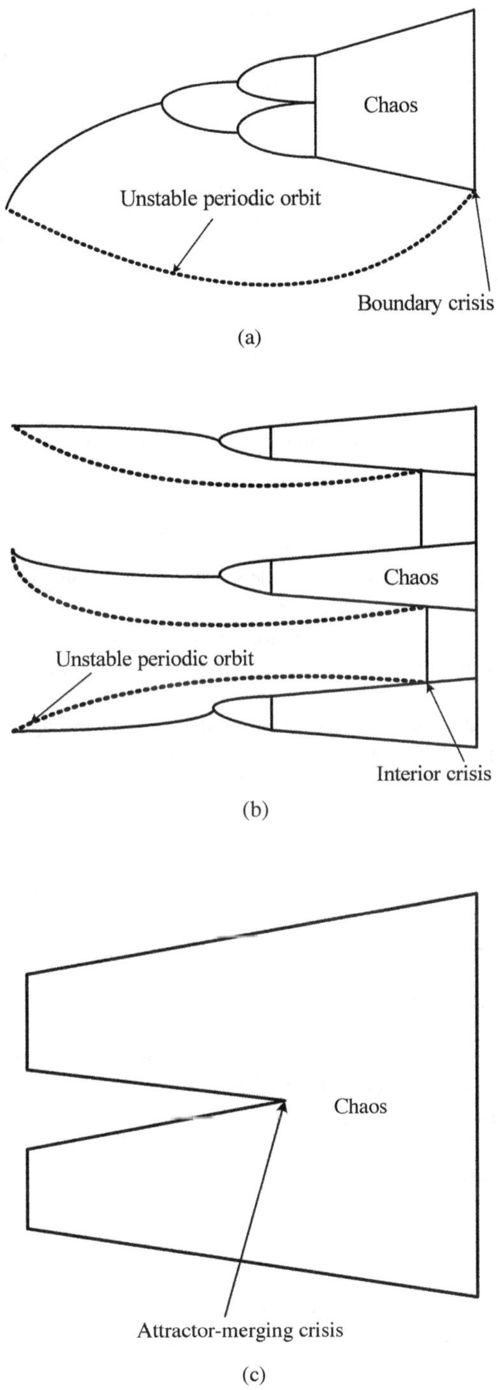

Figure 2.10 Different types of crises. (a) Boundary crisis. (b) Interior crisis. (c) Attractor-merging crisis

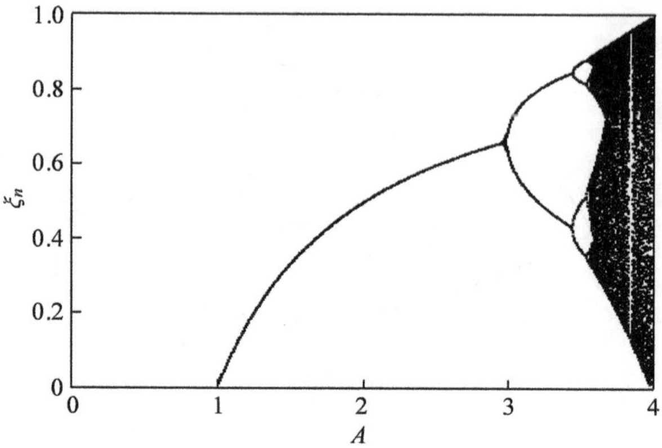

Figure 2.11 Period doubling cascade route to chaos of Logistic map

maps and flows where a stable point and an unstable point come together and coalesce. The second type is the quasiperiodic route that often occurs through subcritical Hopf bifurcations. After a secondary Hopf bifurcation – namely a Neimark–Sacker bifurcation – a new torus is generated which induces chaos. The third type is the inverse period-doubling bifurcation, where an unstable periodic orbit collapses onto a stable periodic orbit of one half its period, and they are both replaced by an unstable periodic orbit of a lower period (Ott, 1993). In the intermittency transition to chaos, the attractor is a periodic orbit for $\kappa < 0$. When κ is slightly larger than 0, there is a long duration during which the trajectory or orbit appears to be periodic and closely resembles the orbit for $\kappa < 0$. However, this regular behavior is intermittently interrupted by a finite bursting duration. Figure 2.12 illustrates these three types of intermittency transition to chaos (Ott, 1993).

Chaos from Manifold Tangle

If x_0 is a point with a stable manifold $W^s(x_0)$ and an unstable manifold $W^u(x_0)$, and the two manifolds intersect transversely (with nonzero angle) at another point x^*, then x^* is called a homoclinic point. The action of intersection of two manifolds is called a manifold tangle. Actually, when the manifolds intersect transversely once, they intersect an infinite number of times. This results in stretching and folding actions, giving an embedded horseshoe map which leads to chaos. The stable manifold cannot intersect with the stable manifold and the unstable manifold cannot intersect with the unstable manifold. Figure 2.13 shows a homoclinic tangle where the stable and unstable manifolds of a saddle point intersect with each other infinitely, where only a portion of each manifold is depicted.

Chaos from Crisis

As mentioned above, chaotic attractors can be changed discontinuously in a crisis. That is, the boundary crisis can destroy the chaotic attractors, the interior crisis can suddenly expand the size of the chaotic attractors, and the attractor-merging crisis can combine different chaotic attractors together. On the other hand, the inverse process can create chaotic attractors, shrink chaotic attractors, and split chaotic attractors with a parameter change in the opposite direction.

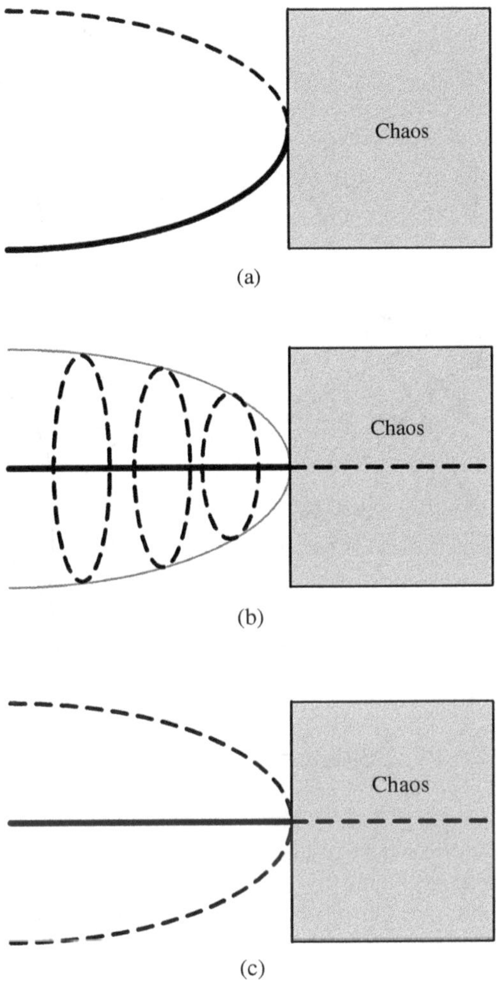

Figure 2.12 Intermittency transitions to chaos. (a) Saddle-node bifurcation. (b) Subcritical Hopf bifurcation. (c) Inverse period doubling bifurcation

2.1.4 Analysis Methods

2.1.4.1 Waveforms and Phase Portraits

The waveforms of equilibrium or periodic behaviors are regular while the waveform of a chaotic behavior is irregular. The phase portraits of the equilibrium points and periodic behaviors are points and close curves respectively, while the phase portrait of chaos is randomly distributed in a bounded region. So, the waveforms and phase portraits can be observed to discern chaotic behaviors from regular behaviors. The oscilloscope can record the waveforms at any instant, and the irregular behavior of a chaotic waveform can be observed. The phase portrait can be displayed directly by using the X–Y display mode of

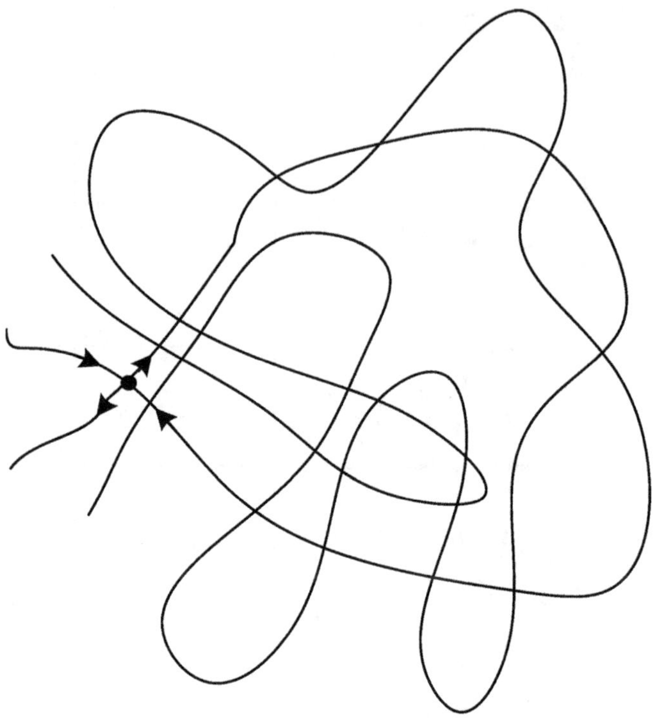

Figure 2.13 Homoclinic tangle

oscilloscope. Both the waveform and the phase portrait can also be plotted when based on the measured data sampled by the oscilloscope. Figure 2.14(a) shows the measured chaotic speed waveform of a DC motor, and Figure 2.14(b) plots the phase portrait of the measured speed versus the measured armature current of the DC motor.

2.1.4.2 Poincaré map

For the autonomous system, the Poincaré map can be obtained by using a Poincaré section to cut the attractor, which is illustrated in Figure 2.15 (Banerjee and Verghese, 2001). The Poincaré section can be chosen by fixing one system state z to be constant, and the projection of the attractor is obtained on the $x-y$ plane. Thus, by using the X–Y mode of the oscilloscope, a 2-D projection of the attractor can be displayed. For the nonautonomous system, the Poincaré map can also be obtained by using the Poincaré section to cut the attractor. The main difference is that the nonautonomous system uses the time t to be the z state, and the Poincaré section is obtained by enabling z to be many times the sampling period.

2.1.4.3 Bifurcation Diagrams

Bifurcation diagrams function to illustrate the bifurcation behaviors in the system when changing the parameters. When changing one parameter in the system for a bifurcation diagram, other parameters

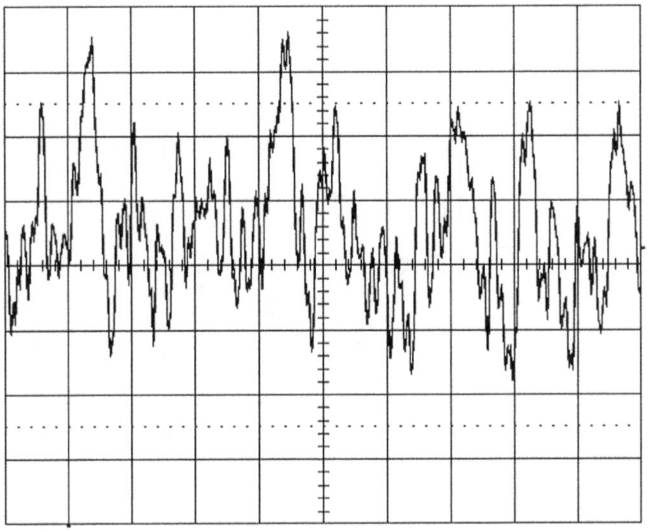

(a)

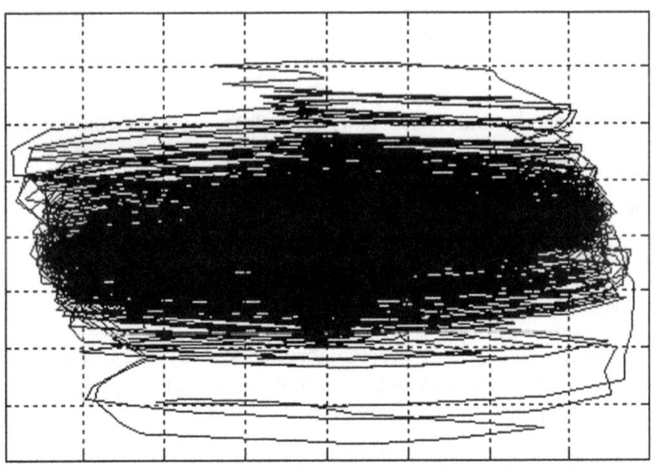

(b)

Figure 2.14 Measured chaotic behavior of DC motor. (a) Waveform of motor speed. (b) Phase portrait of motor speed versus armature current

should remain constant. The variable parameter is along the X-axis, while the system states are along the Y-axis of the bifurcation diagram.

Based on the Poincaré map, the system states are obtained by applying the Poincaré section for autonomous systems, or by stroboscopic sampling for nonautonomous systems. The system states should be chosen arbitrarily, but should make the bifurcation behavior of the system apparent. The variable parameter is changed step by step, and the system states are observed at each value of this parameter.

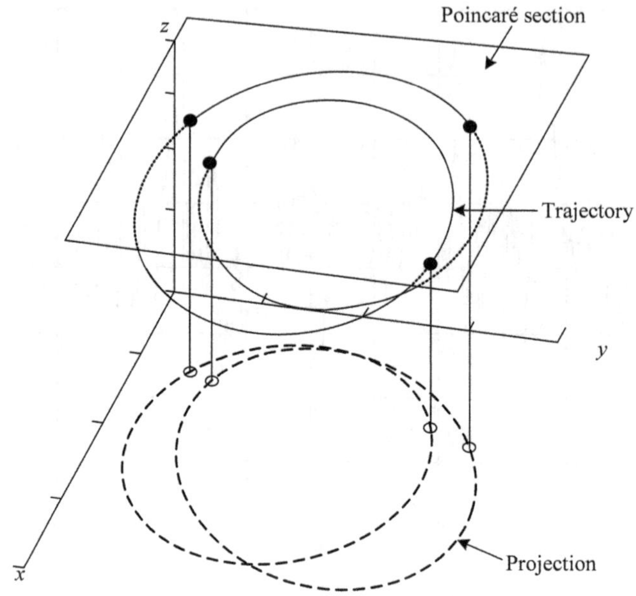

Figure 2.15 Poincaré map using projection

In a dynamical system where multiple attractors exist, different initial conditions should be attempted for each value of the variable parameter so that different attractors can be observed. Also, the iteration times for each bifurcation diagram should be large enough to avoid some transient system behaviors such as chaotic transients.

2.1.4.4 Time Series Reconstruction

The time series reconstruction is another useful tool for chaotic analysis, especially for experimentation. It allows the attractor to be reconstructed even in an infinite-dimensional system or a system where one or more of the state variables cannot be measured directly. For a system that has unknown dynamics, and only one state variable can be observed, the time series of this variable and its time lags can be used to create a multidimensional embedding space. This embedding space has the same geometric and dynamical characteristics as the actual system (Takens, 1980). Beginning with a time series $\{x_1, x_2, \cdots\}$, vectors of $X_i = [x_i, x_{i+L}, x_{i+2L}, \cdots, x_{i+(m-1)L}]$ are constructed, where m is the embedding dimension and L is the time lag. Various ways are put forward to choose m and L properly. Normally, m is chosen to be at least $2D+1$, where D is the actual dimension of the system and L is chosen to be equal to the autocorrelation time, which is the time required for the autocorrelation function to drop to $1/e$ of its initial value (Cellucci *et al.*, 1997).

2.1.4.5 Lyapunov Exponent Calculation

In some practical systems, the equations of motion are usually unknown so that the analytical calculation of Lyapunov exponents is not applicable. Thus, one can employ a technique to estimate the Lyapunov exponents on the basis of the experimental data of a time series (Wolf *et al.*, 1985). An attractive method to calculate the maximum Lyapunov exponent has been especially designed for small data sets (Rosenstein, Collins, and De Luca, 1993). The method is accurate because it takes advantage of all the available data.

Firstly, the reconstructed trajectory **X** is expressed as a matrix, with each row being a state-space vector:

$$\mathbf{X} = [X_1\ X_2 \cdots X_M]^T \tag{2.8}$$

where $\mathbf{X}_i$ is the state of the system at some discrete time i. For an N-point time series $\{x_1, x_2, \cdots, x_N\}$, each $\mathbf{X}_i$ is given by:

$$\mathbf{X}_i = [x_i\ x_{i+L} \cdots x_{i+(m-1)L}] \tag{2.9}$$

Thus, **X** is an $M \times m$ matrix, and the constants m, M, L, and N have a relationship of $M = N - (m-1)L$. After reconstructing the dynamics, the algorithm locates the nearest neighbor of each point on the trajectory. The nearest neighbor $\mathbf{X}_{\hat{j}}$ is obtained by searching for the point that minimizes the distance to the point $\mathbf{X}_j$, which can be expressed as:

$$\mathrm{d}_j(0) = \min_{X_{\hat{j}}} \left\| X_j - X_{\hat{j}} \right\| \tag{2.10}$$

where $\mathrm{d}j(0)$ is the initial distance from the jth point to its nearest neighbor, and $\|\cdots\|$ denotes the Euclidean norm. An additional constraint is imposed that the nearest neighbors have a temporal separation greater than the mean period of the time series, which can be estimated as the reciprocal of the mean frequency of the power spectrum:

$$\left\| j - \hat{j} \right\| > \text{mean period} \tag{2.11}$$

This allows each pair of neighbors to be considered as nearby initial conditions for different trajectories. It is expected that two random vectors of initial conditions will diverge exponentially at a rate given by the maximum Lyapunov exponent $\lambda_{\max}$ (Eckmann and Ruelle, 1985). Therefore, $\lambda_{\max}$ is estimated to be the mean rate of exponential divergence of the nearest neighbors:

$$\mathrm{d}(t) = Ce^{\lambda_{\max} t} \tag{2.12}$$

where $d(t)$ is the average divergence at time t, and C is a constant that normalizes the initial separation. For each jth pair of nearest neighbors, they diverge approximately at the rate given by $\lambda_{\max}$:

$$\mathrm{d}_j(i) - C_j e^{\lambda_{\max}(i\Delta t)} \tag{2.13}$$

where C_j is the initial separation. Taking the logarithm of both sides of (2.13) yields:

$$\ln \mathrm{d}_j(i) = \ln C_j + \lambda_{\max}(i\Delta t) \tag{2.14}$$

which represents a set of approximately parallel lines (for $j = 1, 2, \cdots, M$), each with a slope roughly proportional to $\lambda_{\max}$. Hence, $\lambda_{\max}$ is easily and accurately calculated by using a least-squares fitting to the average line, defined by:

$$y(i) = \frac{1}{\Delta t}\langle \ln \mathrm{d}_j(i) \rangle \tag{2.15}$$

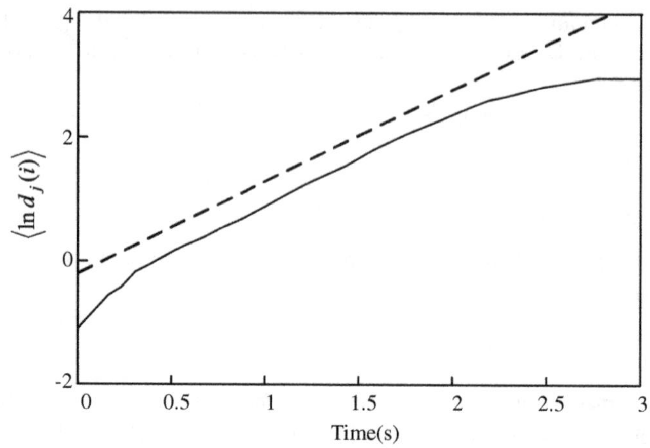

Figure 2.16 Calculated $\langle \ln d_j(i)\rangle$ versus time for Lorenz attractor

where $\langle\cdots\rangle$ denotes the average overall values of j. This process of averaging is the key to calculating accurate values of $\lambda_{\max}$ using small, noisy data sets. It should be noted that C_j in (2.13) performs the function of normalizing the separation of the neighbors. However, as shown in (2.14), this normalization is not necessary for estimating $\lambda_{\max}$.

This method is easy to implement and fast because it uses a simple measure of exponential divergence that circumvents the need to approximate the tangent map. By avoiding the normalization, it also gains a computational advantage. Figure 2.16 shows the typical plot of $\langle \ln d_j(i)\rangle$ versus time for the Lorenz attractor, where the solid curve indicates the calculated results and the slope of the dashed line is the result of $\lambda_{\max}$ (Rosenstein, Collins, and De Luca, 1993).

2.1.4.6 Embedded Unstable Periodic Orbit

In order to achieve the stabilization of chaotic behavior with small perturbations, it is necessary to use the unstable periodic orbit (UPO) embedded in the strange attractor. The period of the UPO τ thus needs to be estimated. Basically, the measured time series can be used to reconstruct the attractor (Takens, 1980), and hence to find the period of the UPO.

There is a well-accepted method of estimating the period of the UPO in the strange attractor (Pawelzik and Schuster, 1991). Firstly, $\{Z_i\}, Z_i \in \Re^m$ are the reconstructed m-dimensional state series with a prerecorded time series, and $\varepsilon_p(t)$ is the distance ε for which there are p returns in the reconstructed series. By sorting the distances $\varepsilon_i(t) = |Z_i - Z_{i+t}|$ for every t, the pth smallest distance of $\varepsilon_i(t)$ is $\varepsilon_p(t)$. The peak values of $\ln(\varepsilon_p^{-2}(t)/2m)$ for different t can approximately predict the periods of the UPOs. Figure 2.17 shows the distribution of $\ln(\varepsilon_p^{-2}(t)/2m)$ against t based on the measured time series of armature current of a wiper motor, which will be described in Chaper 8. Hence, the time for the first peak value at $t = 0.1066$ s is evaluated to be the period of the period-one UPO.

2.1.4.7 Power Spectrum Calculation

Power spectrum is also an important character that can be used to identify chaos. For the equilibrium point, there exist spectral peaks at zero frequency. For periodic behavior, there exist spectral peaks at the fundamental frequency and their multiple frequencies. For quasiperiodic behavior, there exist spectral

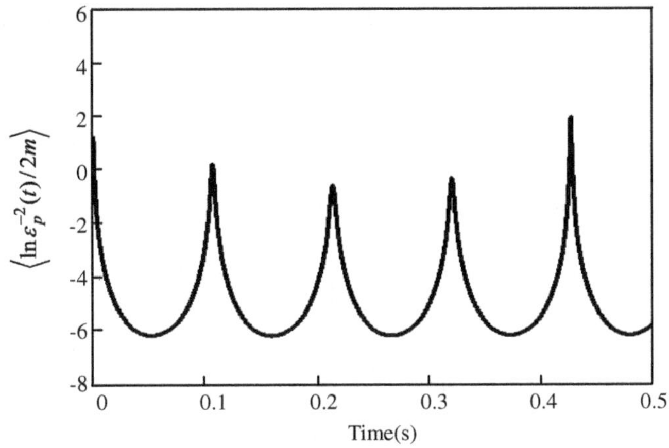

Figure 2.17 Distribution of $\ln(\varepsilon_p^{-2}(t)/2m)$

peaks at the incommensurable frequencies and their common multiple frequencies. For random behavior, there is no spectral peak, and the spectrum is totally continuous. For chaotic behavior, the spectrum also exhibits continuous and broadband spectrum while there exist some spectral peaks which correspond to the average periods of the orbits to travel along the strange attractors. The power spectrum can be computed by using the Fourier transform of the autocorrelation function $R(t)$, which is defined as:

$$P(f) = \int_{-\infty}^{\infty} R(t) e^{i2\pi f t} \mathrm{d}t \tag{2.16}$$

where $R(t) = \lim_{T\to\infty} \frac{1}{T} \int_0^{T-t} x(z)x(z+t)dz$. Equivalently, it can also be defined as the modulus square of its Fourier amplitude per unit time, which is given by:

$$P(f) - \lim \frac{1}{T} \left| \int_0^T x(t) e^{i2\pi f t} \mathrm{d}t \right|^2 \tag{2.17}$$

There is a useful approach for calculating the power spectrum on the basis of the time series (Valsakumar, Satyanarayana, and Sridhar, 1997). The time series is given by sampling $x(t)$ with a nonzero sampling time τ, and with a finite length N. Thus, it yields $\{x_j = x(t = j\tau); j = 0, \cdots, N-1\}$. The discrete version of the autocorrelation is defined as:

$$R_j(N) = \left\langle \frac{1}{N} \sum_{l=0}^{N-1-|j|} x_l x_{l+|j|} \right\rangle \tag{2.18}$$

where $\langle \cdots \rangle$ represents the average over several initial conditions and trajectories. This averaging is performed in order to ensure that the autocorrelation function of the discrete time series $R_j(N)$ is identical to that of the continuous time process $R(t)$ evaluated at $t = j\tau$ in the limit $N \to \infty$. Then, $P(f, N, \tau)$ is defined as:

$$P(f, N, \tau) = \tau \sum_{j=-(N-1)}^{(N-1)} R_j(N) e^{i2\pi f \tau j} \tag{2.19}$$

which can also be represented in a form analogous to (2.17). The discrete Fourier transform of $\{x_j\}$ is expressed as:

$$X(f,N,\tau) = \tau \sum_{j=0}^{N-1} x_j e^{i2\pi f \tau j} \tag{2.20}$$

Hence, the corresponding power spectrum is given by:

$$P(f,N,\tau) = \frac{\tau}{N}\left\langle |X(f,N,\tau)|^2 \right\rangle \tag{2.21}$$

The computed power spectrum $P(f,N,\tau)$ is equal to the true power spectrum $P(f)$ under the following limit:

$$P(f) = \lim_{\tau\to 0}\lim_{N\to\infty} P(f,N,\tau) \tag{2.22}$$

2.1.4.8 Fractal Dimension Calculation

Compared with the aforementioned capacity dimension to measure the fractal dimension, the correlation dimension D_c has a computational advantage because it uses trajectory points $\{X_i\}$ and does not require a separate partition of the state space. The correlation dimension is defined as (Grassberger and Procaccia, 1983):

$$P(f) = \lim_{\tau\to 0}\lim_{N\to\infty} P(f,N,\tau) \tag{2.22}$$

$$D_c = \lim_{R\to 0}\lim_{N\to\infty} \frac{\log C(R)}{\log R} \tag{2.23}$$

$$C(R) = \frac{2}{N(N-1)} \sum_{i=1}^{N} \sum_{j=i+1}^{N} \Theta\left(R - |X_i - X_j|\right) \tag{2.24}$$

where $\Theta(\cdot)$ is the Heaviside step function, which is given by:

$$\Theta(x) = \begin{cases} 0 & \text{if} \quad x < 0 \\ 1 & \text{if} \quad x \geq 0 \end{cases} \tag{2.25}$$

For the m-dimensional time-delay embedding, this yields:

$$|X_i - X_j| = \sqrt{\sum_{k=0}^{m-1} (x_{i-k} - x_{j-k})^2} \tag{2.26}$$

When analyzing a scalar time series in which the optimal embedding is unknown, D_c should be calculated in increasing embeddings until it ceases to change. The correlation dimension may be more physically meaningful since it emphasizes regions of the attractor visited most frequently by the orbit, rather than being a purely geometric quantity like the capacity dimension (Sprott, 2003).

2.1.4.9 Entropy Calculation

The correlation entropy is a close lower bound on the aforementioned K–S entropy, which is actually similar to the case where the correlation dimension is a close lower bound on the capacity dimension (Grassberger and Procaccia, 1983b). The correlation entropy can also be computed with trajectory points, and is given by:

$$K_c = \lim_{m\to\infty}\lim_{r\to 0}\lim_{N\to\infty} \log\frac{C(m,R)}{C(m+1,R)} \tag{2.27}$$

where $C(m,R)$ is the correlation sum in (2.24)–(2.26) for an embedding dimension m.

2.2 Fundamentals of Electric Drive Systems

An electric drive system is depicted in Figure 2.18, which can be divided into two parts – electrical and mechanical. The electrical part consists of the subsystems of electric motor, power converter, electronic controller, and sensor, whereas the mechanical part includes the subsystems of mechanical transmission and mechanical load. The boundary between the electrical and mechanical parts is the air-gap of the motor, where electromechanical energy conversion takes place. The electronic controller can be further divided into two functional units: interface circuitry and processor. The sensor is used to translate the measurable quantities, such as current, voltage, temperature, speed, torque, and flux, into electronic signals. Through the interface circuitry, these signals are conditioned to the appropriate level before being fed into the processor. The processor output signals are usually amplified via the interface circuitry to drive power devices of the power converter. The converter acts as a power conditioner that regulates the power flow between the energy source and the electric motor for motoring and regeneration. Finally, the motor interfaces with the mechanical load via the mechanical transmission.

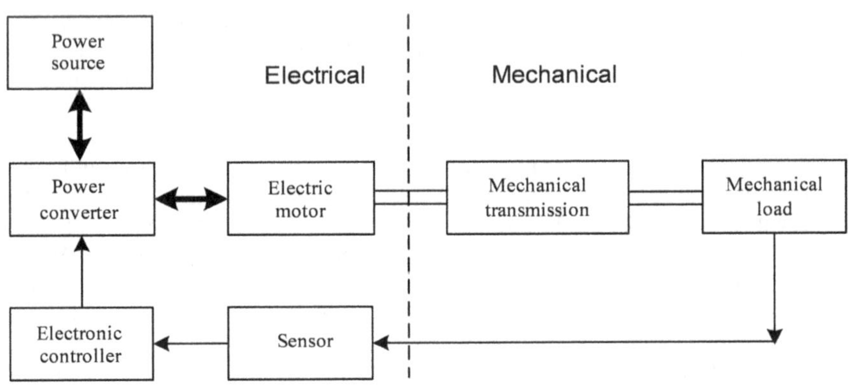

Figure 2.18 Functional block diagram of electric drive system

2.2.1 General Considerations

The development of electric drive systems has been based on the growth of various technologies, especially electric motors, power converters, and control strategies.

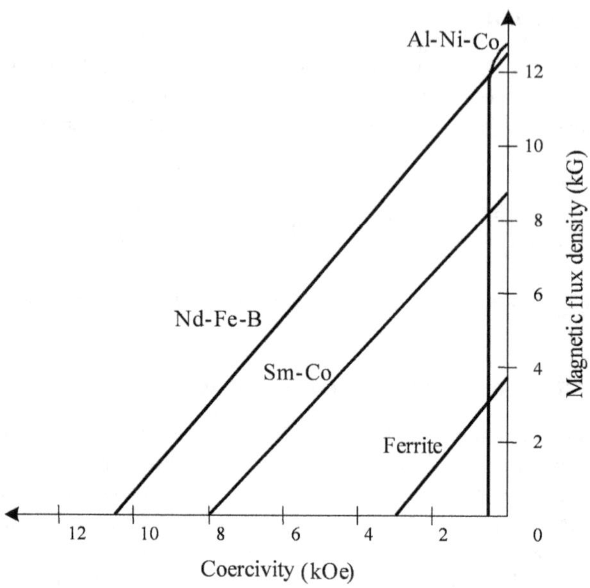

Figure 2.19 Demagnetization curves of PM materials

Electric motors have been available for over a century. The evolution of motors, unlike that of electronics and computer science, has been long and relatively slow. Nevertheless, the development of motors is continually fueled by new materials and sophisticated topologies.

The permanent magnet (PM) is one of the most influential materials for the development of motors, which can provide electric motors with life-long excitation. Figure 2.19 shows the typical characteristics of major PM materials for motors. Notice that Gauss (G) and Oersted (Oe) are non-SI units of magnetic flux density and coercivity, respectively, which are widely adopted in the field of magnetics. They are related to the SI units by $1\,\mathrm{G} = 10^{-4}\,\mathrm{T}$ and $1\,\mathrm{Oe} = 10^3/(4\pi)\mathrm{A/m}$. The ferrite PM is the lowest in cost and has a straight demagnetization characteristic. They were once widely applicable to motors, but suffered from being bulky in size because of their low remanence. The aluminum–nickel–cobalt (Al–Ni–Co) PM has very high remanence but very low coercivity, thus its application to motors is normally limited by the demagnetizing field that it can withstand. Nevertheless, this property can be positively utilized to perform online demagnetization and remagnetization for the newly invented memory motor. Although the samarium–cobalt (Sm–Co) PM has both a high remanence and a high coercivity, its high initial cost restricts its widespread application to motors. The neodymium–iron–boron (Nd–Fe–B) PM has very high remanence and coercivity, and because of its reasonable cost, it has been widely adopted in recent PM motors. Other important parameters of PM materials are the maximum energy product, which is a measure of the maximum stored energy, and the temperature coefficient, which is a measure of the variation of magnetic characteristics with respect to temperature. In general, PMs lose remanence as temperature increases, but within a limited temperature range, the changes are reversible. When exposed to a temperature known as the Curie temperature, the magnetization of a PM is reduced to zero. In general, the relative permeability of PMs is similar to that of air, and much lower than that of iron. A brief summary of typical PM properties is given in Table 2.1.

There are various topologies of electric motors, which create various classifications of electric motors and hence various classifications of electric drive systems. Electric motors were classified into two groups: DC and AC. The DC motors are fed by DC voltage, whereas the AC motors are fed by AC voltage. The AC motors can be further split into the two subgroups of induction motors and synchronous motors. In recent

Table 2.1 Properties of PM materials

	Nd–Fe–B	Sm–Co	Al–Ni–Co	Ferrite
Remanence (kG)	12.5	8.7	12.8	3.8
Coercivity (kOe)	10.5	8.0	0.6	3.0
Max. energy product (MGOe)	36.0	18.3	5.5	3.5
Temperature coefficient (%/°C)	−0.13	−0.04	−0.03	−0.19
Curie temperature (°C)	310	720	800	310
Relative permeability	1.8	1.0	4.0	1.0

years, there have been other motors that cannot be simply grouped into DC or AC. As shown in Figure 2.20, the latest classification of electric motors is to split them into commutator motors and commutatorless motors. The former simply denote that they have a commutator and carbon brushes, while the latter have neither a commutator nor carbon brushes.

All DC motors belong to the commutator motor group, which can be further split into two subgroups – self-excited and separately excited. The self-excited subgroup can be either series excited or shunt excited; the separately excited subgroup can be excited by either field winding or PMs. On the other hand, the commutatorless motors can be further split into three subgroups – induction, synchronous, and doubly salient. Sometimes, the induction motors and synchronous motors can be regrouped as AC motors. Conventionally, the induction motors are split into wound-rotor and cage-rotor types. Similarly, the synchronous motors are split into wound-rotor, PM-rotor, and reluctance-rotor types – the so-called wound-rotor synchronous motor, PM brushless motor, and synchronous reluctance motor. The doubly salient motors refer to those motors having salient poles in both the stator and the rotor. The switched reluctance motors are doubly salient motors with a solid iron rotor. By incorporating PMs into the stator structure of doubly salient motors, doubly salient PM motors have recently been developed, which can be split into two subgroups – yoke-PM and tooth-PM types. The tooth-PM motors can be further split into two types – with PMs mounted on the teeth, and with PMs buried inside the teeth. These are termed flux-reversal PM motors and flux-switching PM motors, respectively.

In order to evaluate the aforementioned motors, some viable types, namely the conventional DC motor with field winding, the PMDC motor, the cage-rotor induction motor, the synchronous reluctance (SynR) motor, the PM brushless (PMBL) motor, the switched reluctance (SR) motor, and the doubly salient PM (DSPM) motor, are compared in terms of their power density, efficiency, controllability, reliability, maturity, and cost. A point grading system (1 to 5) is used, in which 1 is the worst and 5 is the best. As listed in Table 2.2, this evaluation indicates that the commutatorless motors are preferred to the commutator types. Among those commutatorless motors, the cage-type induction motor and the PMBL motor are the most attractive. Among those commutator motors, the PMDC motor is more acceptable.

The evolution of power converters aims to achieve high power density, high efficiency, high controllability, and high reliability. Power converters may be AC–DC, AC–AC, DC–DC or DC–AC. Loosely, DC–DC converters are known as DC choppers while DC–AC converters are known as inverters, and are respectively used for DC and AC motors for electric drive systems. Initially, DC choppers were introduced in the early 1960s using force-commutated thyristors that were constrained to operate at low switching frequency. Due to the advent of fast-switching power devices, they can now be operated at tens or hundreds of kilohertz. Inverters are generally classified into voltage-fed and current-fed types. Because of the need of a large series inductance to emulate a current source, current-fed inverters are seldom used for electric drive systems. In fact, voltage-fed inverters are almost exclusively used because they are very simple and can have a power flow in either direction. Their output waveforms may be rectangular, six-step or pulse width modulation (PWM), depending on the switching strategy for different applications. For example, a rectangular output waveform is produced for a PM brushless DC motor, while a six-step or

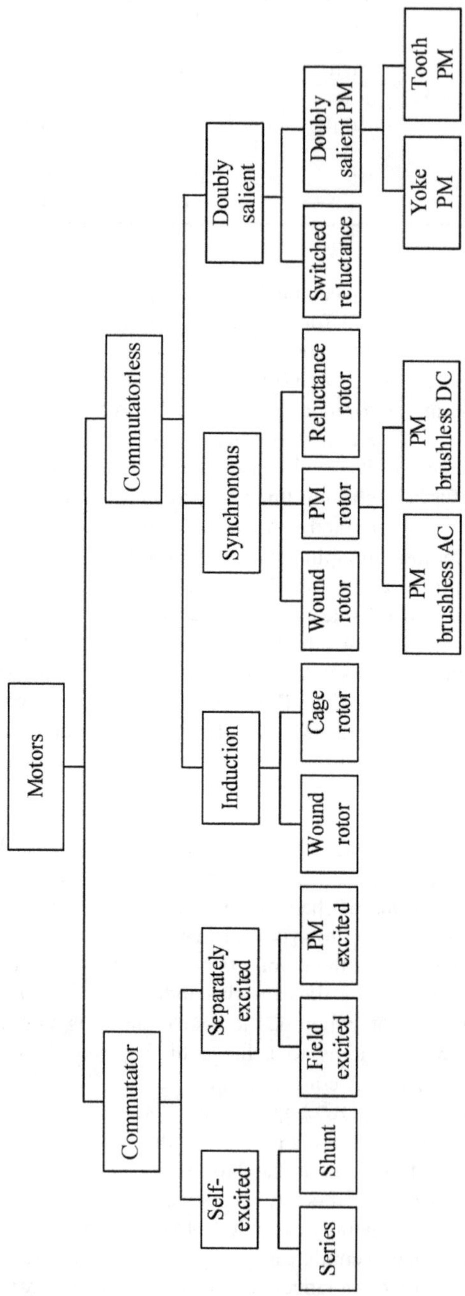

Figure 2.20 Classification of electric motors

Table 2.2 Evaluation of viable motors

	DC motor	PMDC motor	Ind. motor	SynR motor	PMBL motor	SR motor	DSPM motor
Power density	2	3	3.5	3.5	5	3.5	4
Efficiency	2	3	3.5	3.5	5	3.5	5
Controllability	5	5	4	4	4	3	4
Reliability	3	3	5	5	4	5	4
Maturity	5	5	5	3	5	4	3
Cost	4	3	5	4	3	4	3
Total	**21**	**22**	**26**	**23**	**26**	**23**	**23**

PWM output waveform is produced for an induction motor. It should be noted that the six-step output is becoming obsolete because its amplitude cannot be directly controlled and its harmonics are rich. On the other hand, the PWM waveform is harmonically optimal and its fundamental magnitude and frequency can be smoothly varied for speed control.

Since the last two decades, numerous PWM switching schemes have been developed for voltage-fed inverters, focusing on the harmonic suppression, better utilization of DC voltage, tolerance of DC voltage fluctuation as well as suitability for real-time and microcontroller-based implementation. These schemes can be classified as voltage-controlled and current-controlled PWM. The state-of-the-art voltage-controlled PWM schemes are sinusoidal PWM, regular PWM, optimal PWM, delta PWM, and random PWM. On the other hand, the use of current control for voltage-fed inverters is particularly attractive for high-performance drive systems because the motor torque and flux are directly related to the controlled current. The state-of-the-art current-controlled PWM schemes are hysteresis-band PWM and space vector PWM.

Instead of using hard switching, power converters can adopt soft switching. The key of soft switching is to employ a resonant circuit to shape the current or voltage waveform such that the power device switches at zero-current or zero-voltage condition. In general, the use of soft-switching converters possesses the following advantages:

- Due to zero-current or zero-voltage switching condition, the device switching loss is almost zero, thus giving high efficiency.
- Because of low heat sinking requirement and snubberless operation, the converter size and weight are reduced, thus giving high power density.
- The device reliability is improved because of minimum switching stress during soft switching.
- The electromagnetic interference (EMI) problem is less severe and the machine insulation is less stressed because of lower dv/dt resonant voltage pulses.
- The acoustic noise is very small because of high frequency operation.

On the other hand, their key drawbacks are the additional cost of the resonant circuit and the increased complexity.

Although there have been many soft-switching DC–DC converters developed for switched-mode power supplies, these converters cannot be directly applied to DC drive systems. Apart from suffering excessive voltage and current stresses, they generally cannot handle backward power flow during regenerative braking. It should be noted that the capability of regenerative braking is very essential for electric drive systems. Nevertheless, a soft-switching DC–DC converter, having the capability of bidirectional power flow for motoring and regenerative braking as well as the minimum hardware count, was developed for DC drive systems (Chau *et al.*, 1997).

The development of soft-switching inverters for AC motors (including the induction motor and PM brushless AC motors) has become a research direction in power electronics. The milestone of soft-switching inverters, namely the three-phase voltage-fed resonant DC link inverter, was developed in

1986 (Divan, 1986). Consequently, many improved soft-switching topologies were proposed, such as the quasiresonant DC link, series resonant DC link, parallel resonant DC link, synchronized resonant DC link, resonant transition, auxiliary resonant commutated pole, and auxiliary resonant snubber inverters.

Compared with the development of soft-switching inverters for AC motors, there has been little development with SR motors (Cho *et al.*, 1997). A soft-switching converter, the so-called zero-voltage-transition version, was particularly developed for SR motors (Ching, Chau, and Chan, 1998). This converter possesses the advantages that all main switches and diodes can operate at zero-voltage condition, unity device voltage, and current stresses, as well as a wide operating range. Moreover, it offers simple circuit topology, minimum hardware count, and low cost, leading to achieve high switching frequency, high power density and high efficiency.

Conventional linear control such as a PID (proportional-integral-derivative) can no longer satisfy the stringent requirement placed on high-performance electric drive systems. In recent years, many modern control strategies have been proposed. The state-of-the-art control strategies that have been proposed for electric drive systems are adaptive control, variable structure control, fuzzy control, and neural network control.

Adaptive control includes self-tuning control (STC) and model-referencing adaptive control (MRAC). Using STC, the controller parameters are tuned to adapt to variations in the system parameters. The key is to employ an identification block to track changes in system parameters and to update the controller parameters through controller adaptation in such a way that a desired closed-loop performance can be obtained. Using MRAC, the output response is forced to track the response of a reference model irrespective of any variations in parameters of the system. Based on an adaptation algorithm that utilizes the difference between the reference model and system outputs, the controller parameters are adjusted to give a desired closed-loop performance.

Variable structure control (VSC) has also been applied for electric drive systems to compete with adaptive control. Using VSC, the system can be designed to provide parameter-insensitive features, prescribed error dynamics, and simplicity in implementation. Based on a set of switching control laws, the system is forced to follow a predefined trajectory in the phase plane irrespective of system parameter variations.

Emerging technologies such as fuzzy logic and neural networks have been introduced into the field of electric drive systems. Fuzzy control is essentially a linguistic process which is based on the prior experience and heuristic rules used by human operators. Making use of neural network control (NNC), the controller can possibly interpret the behavior of system dynamics, then self-learn and self-adjust accordingly. Furthermore, these state-of-the-art control strategies can incorporate one another, such as adaptive fuzzy control, fuzzy NNC, and fuzzy VSC. In near future, controllers incorporating artificial intelligence (AI) can permit diagnosis of systems and correction of faults to supplant the need of human intervention.

2.2.2 DC Drive Systems

A DC drive system is composed of the DC motor, power converter, electronic controller, sensor, mechanical transmission, and mechanical load. The classification of various DC drive systems depends on the location of field excitation in the DC motor. Namely, they can be grouped as a self-excited DC drive system or a separately excited DC drive system. Based on the source of field excitation in the DC motor, they can also be grouped as a wound-field DC drive system or a PMDC drive system. The former has a field winding so that any field excitation can be controlled by the DC current, whereas the latter has no field winding and the PM excitation is uncontrollable.

The name applied to those wound-field DC drives is usually determined by the mutual interconnection between the field winding and armature winding. As shown in Figure 2.21, common wound-field DC

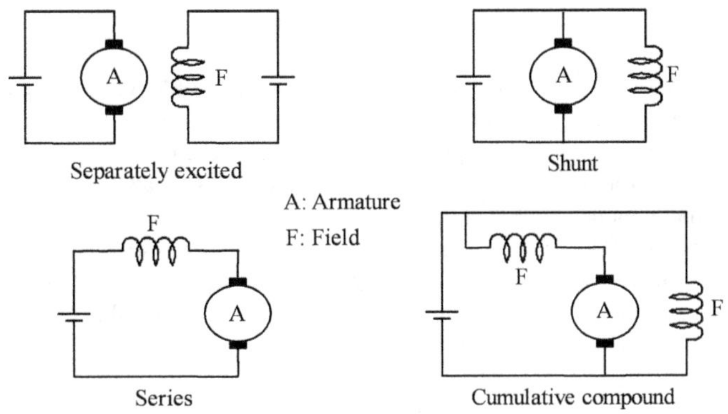

Figure 2.21 Wound-field DC motors

drives are the separately excited, series, shunt, and cumulative compound types. Without external control, their torque-speed characteristics at the rated voltage are shown in Figure 2.22. In the separately excited DC drive, the field and armature voltages can be controlled independently of each other. The torque-speed characteristic is linearly related so that speed decreases as torque increases and speed regulation depends on the armature circuit resistance. In the series DC drive, the field current is the same as the armature current. An increase in torque is accompanied by an increase in the armature current and hence an increase in flux. The speed must drop to maintain the balance between the supply and the induced voltages. The torque-speed characteristic has an inverse relationship. In the shunt DC drive, the field and armature are connected to a common voltage source. The corresponding characteristic is similar to that of the separately excited DC drive. In a cumulative compound DC drive, the MMF of the series field is in the same direction as the MMF of the shunt field. The characteristic lies between those of the series DC and shunt DC drives, depending on the relative strength of the series and shunt fields.

By replacing the field winding and pole structure with PMs, the PMDC drive can readily be generated from the separately excited DC drive. Compared with the separately excited DC drive, the PMDC drive has relatively higher power density and higher efficiency because of the space-saving benefit by PMs and the absence of field losses. Owing to the low permeability of PMs (similar to that of air), the corresponding armature reaction is usually reduced and commutation is improved. However, since the field excitation in

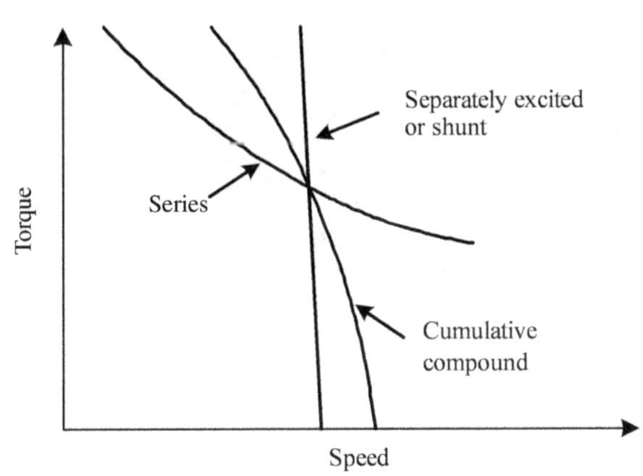

Figure 2.22 Torque-speed characteristics of wound-field DC motors

the PMDC drive is uncontrollable, it cannot readily attain the same operating characteristics as the separately excited DC drive.

Both wound-field DC and PMDC drives suffer from the same problem due to the use of commutators and brushes in their motors. Commutators cause torque ripples and limit the motor speed, while brushes are responsible for friction and radio-frequency interference. Moreover, due to wear and tear, periodic maintenance of commutators and brushes is always required. These drawbacks make them less reliable and unsuitable for maintenance-free operation.

The major advantages of DC drive systems are their maturity and simplicity. The simplicity is mainly due to their simple control because the air-gap flux Φ and the armature current I_a – hence the motor speed ω_m and torque T – can be independently controlled. Irrespective of whether the motors are wound-field DC or PMDC, they are governed by the following basic equations:

$$E = K_e \Phi \omega_m \tag{2.28}$$

$$V_a = E + R_a I_a \tag{2.29}$$

$$T = K_e \Phi I_a \tag{2.30}$$

where E is the back EMF, V_a is the armature voltage, R_a is the armature resistance, and K_e is named as the back EMF constant or torque constant. For those wound-field DC motors, Φ is linearly related to the field current I_f, which may be independently controlled, dependent on I_a, dependent on V_a, or dependent on both I_a and V_a, respectively, for those separately excited, series, shunt, or cumulative compound types. In contrast, Φ is essentially uncontrollable for those PMDC motors.

The basic topology of DC motors is shown in Figure 2.23. The corresponding design consideration includes the electric loading, magnetic loading, stator outside and inside diameters, rotor outside and inside diameters, core length, air-gap length, number of poles, number of armature slots, armature tooth width and slot depth, number of turns per coil, slot fill factor, number of commutator bars, commutation

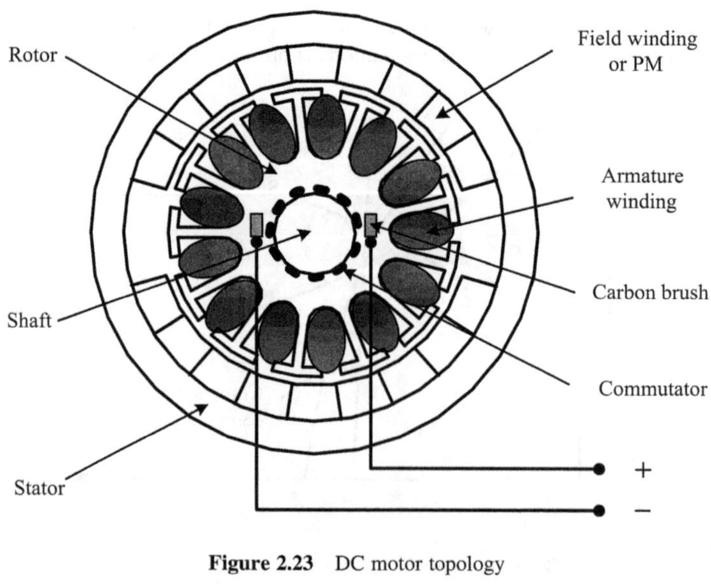

Figure 2.23 DC motor topology

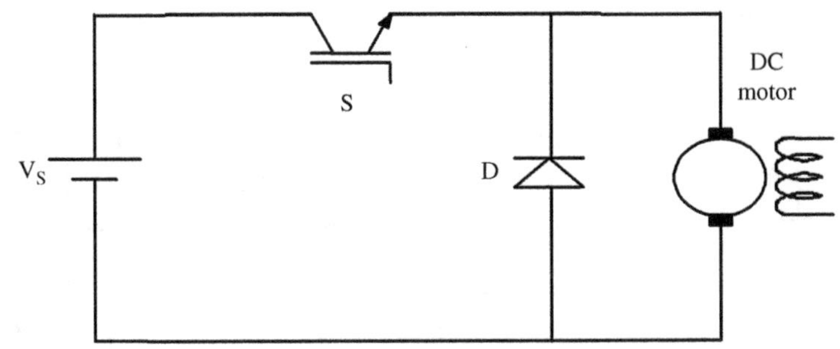

Figure 2.24 First-quadrant DC chopper

arrangement, field winding or PM excitation arrangement, thermal arrangement, speed, torque, power, efficiency, torque density, and power density.

In order to perform motion control for the DC drive systems, the use of power converters is mandatory. There are two major power converters suitable for DC drive systems, namely the DC–DC converter and AC–DC converter. The former has been used extensively for various applications, whereas the latter, also termed the controlled rectifier, is rarely used for DC drive systems due to its low input power factor. Actually, even when the supply is from the grid, the AC voltage is first converted to DC voltage by using a simple diode rectifier and then controlled by the DC–DC converter, rather than directly using the AC–DC converter.

When DC–DC converters adopt the chopping mode of operation, they are usually known as DC choppers. These DC choppers are classified as first-quadrant, second-quadrant, two-quadrant, and four-quadrant versions. The first-quadrant DC chopper (shown in Figure 2.24) is suitable for motoring, and the corresponding power flow is from the source to the load, whereas the second-quadrant is for regenerative braking and the power flow is out from the load into the source. Incorporating both of the motoring and regenerative braking, the two-quadrant DC chopper is shown in Figure 2.25. Moreover, instead of using mechanical contactors to achieve reversible operation, the four-quadrant DC chopper shown in Figure 2.26 can be employed so that motoring and regenerative braking in both forward and reversible operations are controlled electronically.

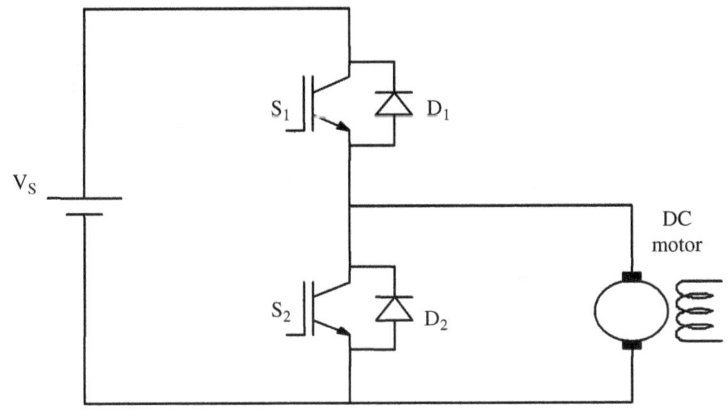

Figure 2.25 Two-quadrant DC chopper

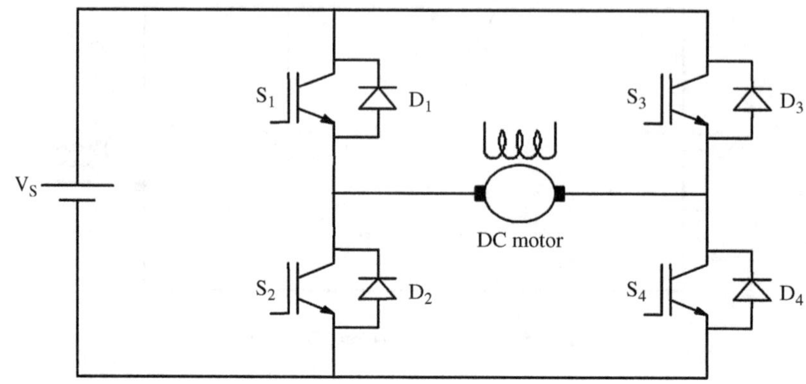

Figure 2.26 Four-quadrant DC chopper

As shown in Figure 2.27, there are three ways in which the chopper output voltage can be varied, namely PWM control, frequency-modulated control, and current-limit control. In the first method, the chopper frequency is kept constant and the pulse width is varied. The second method has a constant pulse width and a variable chopping frequency. In the third method, both the pulse width and frequency are varied to control the load current between certain specified maximum and minimum limits. For conventional DC drive systems, PWM control of the two-quadrant DC chopper is generally adopted. The corresponding

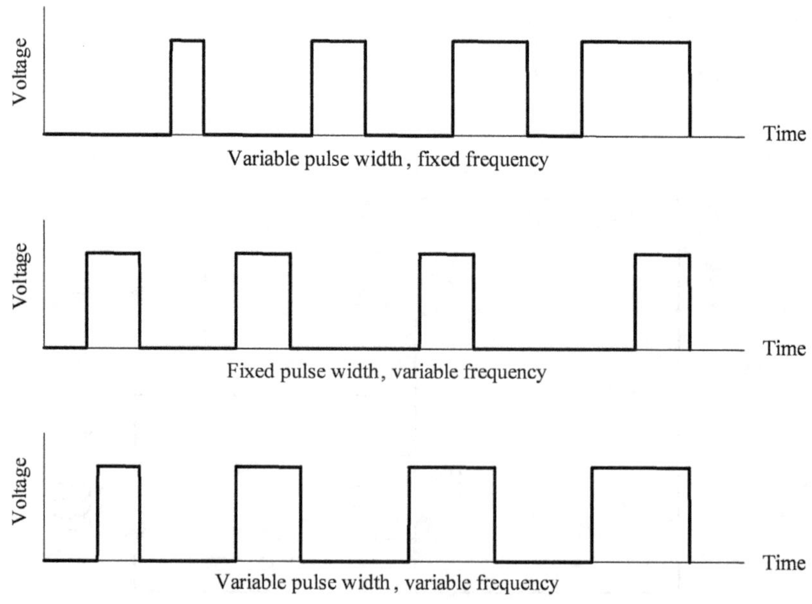

Figure 2.27 DC chopper controlled output voltages

control is based on the variation of duty cycle δ:

$$V_a = \delta V_s \tag{2.31}$$

$$I_a = \frac{V_a - E}{R_a} \tag{2.32}$$

where V_s is the DC supply voltage. Hence, the motoring operation ($I_a > 0$) occurs when $\delta > (E/V_s)$, and regenerative braking ($I_a < 0$) occurs when $\delta < (E/V_s)$. The no-load operation is obtained when $\delta = (E/V_s)$. As the current is always flowing, a discontinuous conduction mode does not occur.

In general, speed control of DC drive systems can be accomplished by two methods – armature control and field control. When the armature voltage of the DC motor is reduced, the armature current (and hence the motor torque) decrease, causing the motor speed to decrease. In contrast, when the armature voltage is increased, the motor torque increases, causing the motor speed to increase. Since the maximum allowable armature current remains constant and the field is fixed, this armature voltage control has the advantage of retaining the maximum torque capability at all speeds. However, since the armature voltage cannot be further increased beyond the rated value, this control is used only when the DC drive system operates below its base speed. On the other hand, when the field voltage of the DC motor is weakened while the armature voltage is fixed, the motor induced EMF decreases. Because of low armature resistance, the armature current will increase by an amount much larger than the decrease in the field. Thus, the motor torque is increased, causing the motor speed to increase. Since the maximum allowable armature current is constant, the induced EMF remains constant for all speeds when the armature voltage is fixed. Hence, the maximum allowable motor power becomes constant so that the maximum allowable torque varies inversely with the motor speed. Therefore, in order to achieve wide-range speed control of DC drive systems, the armature control has to be combined with the field control. By maintaining the field constant at the rated value, armature control is employed for speeds from standstill to the base speed. Then, by keeping the armature voltage at the rated value, field control is used for speeds beyond the base speed. The corresponding maximum allowable torque and power in the combined armature and field control are shown in Figure 2.28.

It should be noted that only the separately excited DC drive system can perform the combined armature and field control. The corresponding torque-speed characteristics during motoring and regenerative

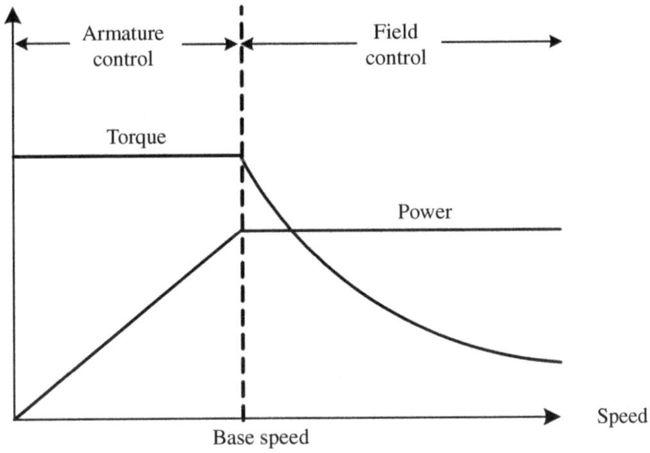

Figure 2.28 Combined armature and field control for DC drive systems

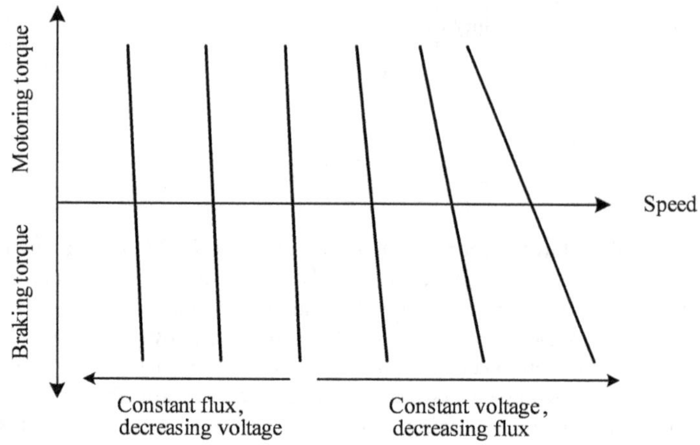

Figure 2.29 Separately excited DC drive system characteristics

braking are depicted in Figure 2.29. For the shunt, series and cumulative compound DC drive systems, both of armature voltage and field voltage are varied simultaneously, whereas the PM drive system can only perform armature control.

2.2.3 *Induction Drive Systems*

At present, the induction drive system is the most mature technology among various commutatorless drive systems. There are two types of induction motors, namely wound-rotor and cage-rotor. Because of high cost, need of maintenance, and lack of sturdiness, the wound-rotor induction motor is less attractive than its cage-rotor counterpart. Apart from the common advantages of commutatorless drive systems, the induction drive systems possess the definite advantages of low cost and ruggedness. These advantages can generally outweigh their major disadvantage of control complexity, and facilitate them to be widely accepted.

The most common induction drive system is composed of the cage-rotor induction motor, PWM inverter, electronic controller, sensor, mechanical transmission and mechanical load. Reasonable high-voltage low-current motor design should be employed to reduce the copper loss of motor windings as well as the cost and size of the PWM inverter. High-speed operation should also be adopted to minimize the motor size and weight, although the maximum speed of the motor is limited by the bearing friction and windage losses as well as the transaxle tolerance. Low stray reactance is also necessary to favor flux-weakening operation.

The basic topology of induction motors is shown in Figure 2.30. The corresponding design consideration includes the electric loading, magnetic loading, stator outside and inside diameters, rotor outside and inside diameters, core length, air-gap length, number of poles, number of stator slots, number of rotor slots, stator tooth width and slot depth, rotor tooth width and slot depth, number of turns per phase, slot fill factor, thermal arrangement, speed, torque, power, efficiency, torque density, and power density.

The three-phase voltage-fed PWM inverter shown in Figure 2.31 is almost exclusively used for induction drive systems. The inverter design highly depends on the technology of power devices. At present, the IGBT-based inverter is most attractive. Since the hard-switching inverter topology is almost fixed, the inverter design generally depends on the selection of power devices and PWM switching schemes. The selection of power devices is based on the criteria that (1) the voltage rating is at least twice the nominal supply voltage because of the voltage surge during switching, (2) the current rating is large enough so that there is no need to connect several power devices in parallel, and (3) the switching speed is sufficiently high to suppress motor harmonics and acoustic noise levels. The power

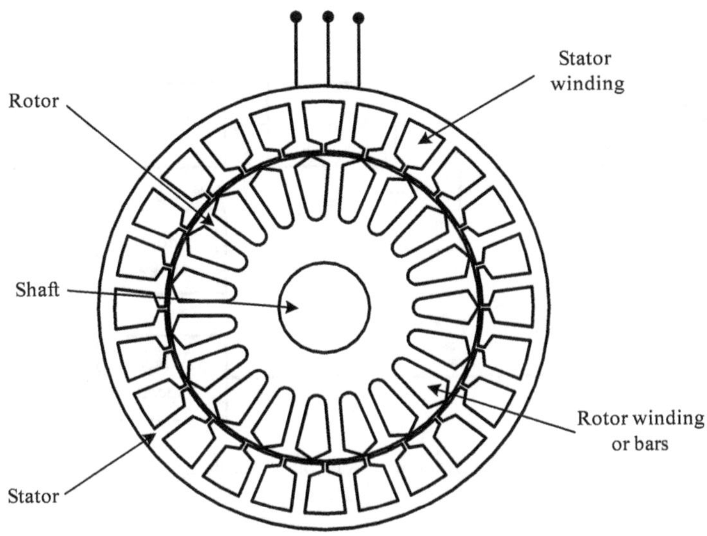

Figure 2.30 Induction motor topology

module is normally a two-in-one or even six-in-one type, namely two or six devices are internally connected with an antiparallel diode across each device, to minimize wiring and stray impedance. On the other hand, the selection of PWM switching schemes is based on the criteria that the magnitude and frequency of the fundamental component of the output waveform can be smoothly varied, the harmonic distortion of the output waveform is minimum, the switching algorithm can be real-time implemented with minimum hardware and compact software, and the variation of input voltage can be handled. There are numerous PWM schemes that have been available, such as the sinusoidal PWM, regular PWM, optimal PWM, delta PWM, random PWM, hysteresis-band PWM, and space vector PWM. Currently, both the sinusoidal PWM and space vector PWM are widely used for induction drive systems.

Speed control in induction drive systems is considerably more complex than that of DC drive systems because the induction motors suffer from nonlinearity of the dynamic model with coupling between direct and quadrature axes. There are two representative control strategies, namely variable-voltage

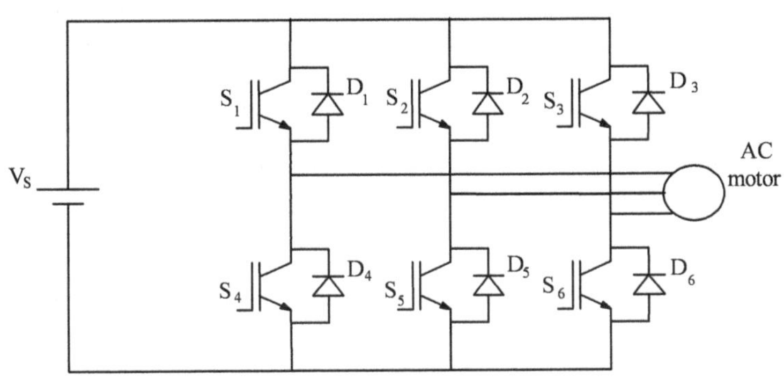

Figure 2.31 Three-phase voltage-fed PWM inverter for induction drive systems

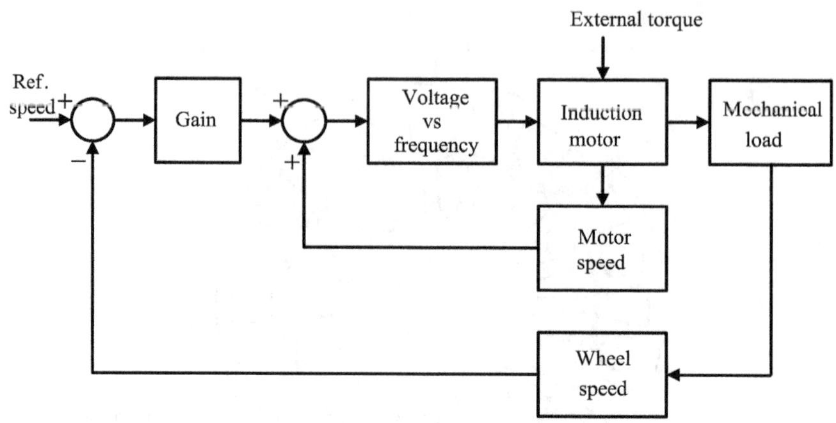

Figure 2.32 Block diagram of VVVF control for induction drive systems

variable-frequency (VVVF) control and field-oriented control (FOC) which is also known as vector control or decoupling control. The basic equation of speed control is governed by:

$$n = n_s(1-s) = \frac{f}{p}(1-s) \tag{2.33}$$

where n is the motor speed in rev/s or rps, n_s is the rotating-field synchronous speed in rev/s, s is the slip, p is the number of pole pairs, and f is the supply frequency in Hz. Thus, the motor speed can be controlled by variations in f, p, and/or s. In general, more than one control variable is adopted. On top of these control strategies, sophisticated control algorithms such as adaptive control and optimal control have also been employed to achieve faster response, higher efficiency, and wider operating ranges.

Figure 2.32 shows the functional block diagram of VVVF control of induction drive systems. This strategy is based on constant volts/hertz control for frequencies below the motor rated frequency, whereas variable frequency control with constant rated voltage for frequencies beyond the rated frequency. For very low frequencies, voltage boosting is applied to compensate the difference between the applied voltage and the induced EMF due to the stator resistance drop. As shown in Figure 2.33, the induction drive system characteristics can be divided into three operating regions. The first region is called the constant torque region in which the motor can deliver its rated torque for frequencies below the rated frequency. In the second region, called the constant power region, the slip is increased to the maximum value in a preprogrammed manner so that the stator current remains constant and the motor can maintain its rated power capability. In the high-speed region, the slip remains constant while the stator current decreases. Thus, the torque capability declines with the square of the speed. Because of the disadvantages of air-gap flux drifting and sluggish response, the VVVF control strategy is becoming less attractive for high-performance induction drive systems.

In order to improve the dynamic performance of induction drive systems, FOC is preferred to VVVF control. Figure 2.34 shows the functional block diagram of FOC of induction drive systems. By using FOC, the mathematical model of induction motors is transformed from the stationary reference frame (α–β frame) to the general synchronously rotating frame (x–y frame) as shown in Figure 2.35. Thus, at steady state, all the motor variables such as supply voltage v_s, stator current i_s, rotor current i_r, and rotor flux linkage λ_r can be represented by DC quantities. When the x-axis is purposely selected to be coincident with the rotor flux linkage vector, the reference frame ($d-q$ frame) rotates synchronously with the rotor

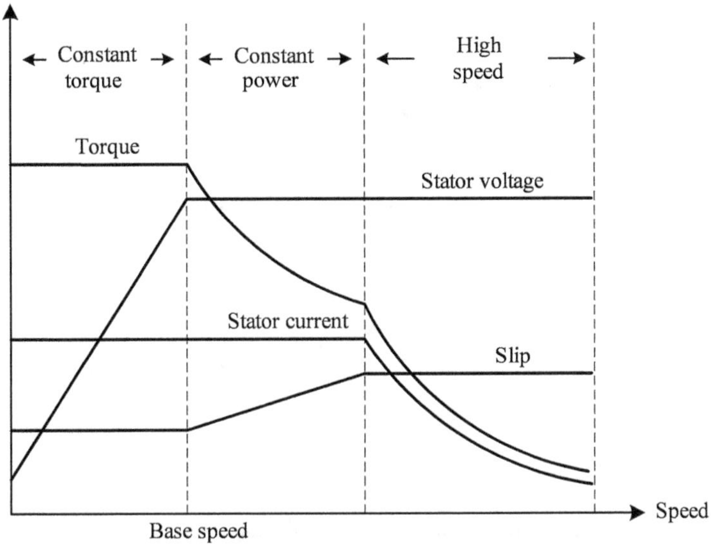

Figure 2.33 Characteristics of induction drive systems

flux, as shown in Figure 2.36, where i_{sd} and i_{sq} are the d-axis and q-axis components of stator current, respectively. Hence, the motor torque T can be obtained as:

$$T = \frac{3}{2} p \frac{M}{L_r} \lambda_r i_{sq} \tag{2.34}$$

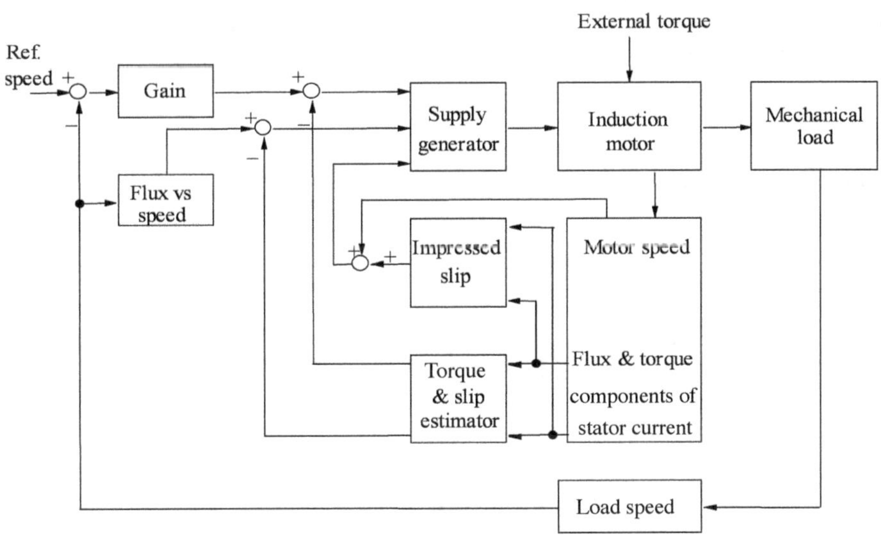

Figure 2.34 FOC of induction drive systems

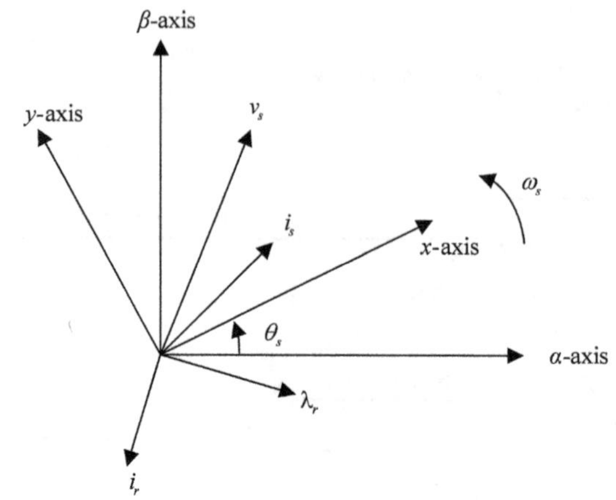

Figure 2.35 x–y frame rotating synchronously in general

where M and L_r are, respectively, the mutual and rotor inductances per phase. Since λ_r can be written as Mi_{sd}, the torque equation can be rewritten as:

$$T = \frac{3}{2}p\frac{M^2}{L_r}i_{sd}i_{sq} \tag{2.35}$$

This torque equation is very similar to that of separately excited DC motors. Namely, i_{sd} resembles the field current I_f while i_{sq} resembles the armature current I_a. Thus, i_{sd} can be considered as the field component of i_s, which is responsible for establishing the air-gap flux. On the other hand, i_{sq} can be considered as the torque component of i_s, which produces the desired motor torque. Therefore, by means of this FOC, the motor torque can be effectively controlled by adjusting the torque component as long as the field component remains constant. Hence, induction drive systems can offer the same fast transient response as separately excited DC drive systems. In order to attain the above FOC, the rotor flux linkage vector is always aligned with the d-axis. This criterion, the so-called decoupling condition, can be attained through slip frequency ω_{slip} control as given by:

$$\omega_{slip} = \frac{R_r i_{sq}}{L_r i_{sd}} \tag{2.36}$$

where R_r is the rotor resistance per phase.

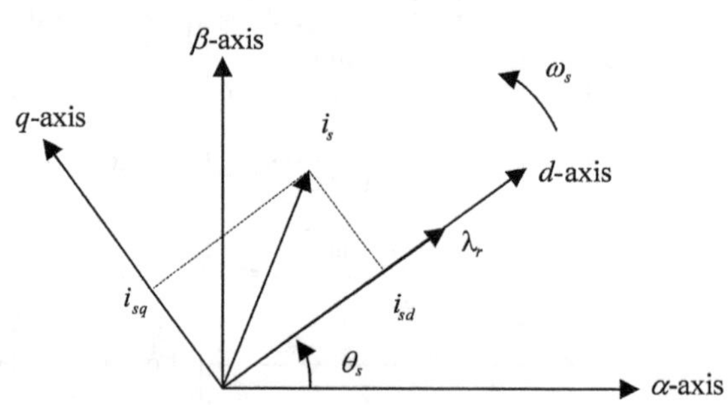

Figure 2.36 $d-q$ frame rotating synchronously with rotor flux

Since the advent of FOC, a number of methods have been proposed for implementation. Basically, these methods can be classified into two groups, namely direct FOC and indirect FOC. The direct FOC requires the direct measurement of the rotor flux, which not only increases the complexity in implementation, but also suffers from unreliable measurement at low speeds. In contrast, the indirect FOC determines the rotor flux by calculation, instead of by direct measurement. This method has the definite advantage of being easier to implement than the direct FOC. Therefore, the indirect FOC is widely adopted for the high-performance motion control of induction drive systems.

Although the indirect FOC has been widely used for high-performance induction drive systems, it still suffers from some drawbacks. In particular, the rotor time constant L_r/R_r (which has a dominant effect on the decoupling condition) changes severely, depending on operating temperature and magnetic saturation, and can lead to deterioration in the desired FOC. In general, there are two ways to solve this problem: one is to perform an online identification of the rotor time constant and, accordingly, update the parameters used in the FOC controller; the other is to adopt a sophisticated control algorithm to render the FOC controller insensitive to variations in the parameters of the motor.

The model reference adaptive control (MRAC) algorithm has been widely used for the FOC control of induction drive systems. Firstly, a reference model – which is designed to be an optimal system in general – is made to satisfy the desired dynamic performance of the drive system. An adaptive mechanism is then adopted which aims to force the drive system to follow the reference model even after some variation in the parameters of the system, such as a change of L_r/R_r due to prolonged operation. The main criterion of the adaptive mechanism is to assure robustness with asymptotically zero error between the outputs of the reference model and the drive system. The definite advantage of this MRAC scheme is that there is no need to carry out an explicit parameter identification or estimation in the synthesis of the drive system control input. In fact, only the command input, the controlled drive system output, and the reference model are required to establish this control scheme.

2.2.4 Synchronous Drive Systems

As mentioned above, AC drive systems are composed of two major groups – induction drive systems and synchronous drive systems. Previously, the induction motor drives almost monopolized the whole market of AC drive systems, but no longer, due to the fact that the wound-rotor synchronous motor needs an external DC current to excite the field windings in the rotor via slip rings and carbon brushes, which causes many problems such as maintenance requirement, bulky accessories, and safety issues. With the advent of high-energy PM materials, the PM brushless motors (including the PM synchronous motor) are becoming attractive and can directly compete with the induction motors for the market of AC drive systems. Also, with the introduction of new magnetic material and structure, the synchronous reluctance motor can provide a decent performance without the use of field windings or PMs in the rotor.

2.2.4.1 PM Brushless Drive Systems

Among those viable electric drive systems, the PM brushless drive systems are most capable of competing with the mature induction drive systems. Their advantages are summarized below:

- Since the magnetic field is excited by high-energy PMs, the overall weight and volume can be significantly reduced for a given power output, leading to high power density.
- Because of the absence of rotor copper losses, their efficiency is inherently high.
- Since the heat mainly arises in the stator, it can be more efficiently dissipated to surroundings.
- Since PM excitation suffers from no risk of manufacturing defects, overheating, or mechanical damage, their reliability is inherently high.
- Because of the lower electromechanical time constant of the rotor, the rotor acceleration at a given input power can be increased.

Nevertheless, the PM brushless drive systems suffer from the drawbacks of relatively high PM material cost and uncontrollable PM flux.

The system configuration of PM brushless drive systems is similar to that of induction drive systems – namely, it consists of a PM brushless motor, power inverter, electronic controller, sensor, mechanical transmission, and mechanical load. Based on the waveforms feeding into the motor terminals, PM brushless motors can be divided into two types – PM brushless AC (BLAC) and PM brushless DC (BLDC). PM BLAC motors are fed by sinusoidal or near-sinusoidal AC waves. Actually, they are usually known as PM synchronous motors or, sometimes, sinusoidal-fed PM brushless motors. On the other hand, PM BLDC motors are fed by rectangular AC waves, and are sometimes known as rectangular-fed PM brushless motors. Since the interaction between a rectangular field and a rectangular current in the motor can produce a higher torque than that produced by a sinusoidal field and sinusoidal current, the PM BLDC drive system possesses a higher power density than the PM BLAC drive system. Meanwhile, the PM BLDC motor has a significant torque pulsation, whereas the PM BLAC motor produces an essentially constant instantaneous torque, or so-called smooth torque like a wound-rotor synchronous drive system.

The basic consideration of the PM brushless motor design includes the electric loading, magnetic loading, stator outside and inside diameters, rotor outside and inside diameters, core length, air-gap length, PM material characteristics, PM topology and dimensions, number of poles, number of stator slots, stator tooth width and slot depth, number of turns per phase, slot fill factor, thermal arrangement, speed, torque, power, efficiency, torque density, and power density.

On the basis of the placement of PMs, PM brushless motors can be classified as surface-mounted, surface-inset, interior-radial, and interior-circumferential types (Gan *et al.*, 2000; Zhu and Howe, 2007). For the surface-mounted PM brushless motor topology as shown in Figure 2.37, the PMs are simply mounted on the rotor surface by using epoxy adhesives. Since the permeability of PMs is near to that of air, the effective air-gap is the sum of the actual air-gap length and the radial thickness of the PMs. Hence, the

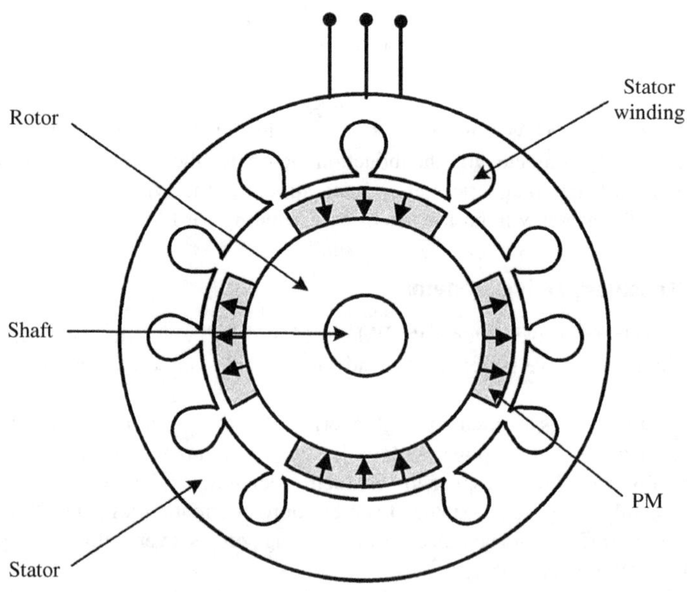

Figure 2.37 Surface-mounted PM brushless motor topology

corresponding armature reaction field is small and the stator winding inductance is low. Also, since the d-axis and q-axis stator winding inductances are nearly the same, its reluctance torque is almost zero. For the surface-inset PM brushless motor topology, as shown in Figure 2.38, the PMs are inset or buried in the rotor surface. Thus, the q-axis inductance becomes higher than the d-axis inductance, hence producing the additional reluctance torque. Also, since the PMs are inside the rotor, it can withstand the centrifugal force at high-speed operation, thus offering good mechanical integrity. For the interior-radial PM brushless motor topology, as shown in Figure 2.39, the PMs are radially magnetized and buried inside the rotor. Similar to the surface-inset type, the PMs are mechanically protected, hence allowing for high-speed operation. Also, because of its d–q saliency, an additional reluctance torque is generated. Contrary to the surface-inset type, this interior-radial topology adopts linear PMs which are more easily inserted and can be easily machinable. For the interior-circumferential PM brushless motor topology shown in Figure 2.40, the PMs are circumferentially magnetized and buried inside the rotor. This gives the definite advantage that the air-gap flux density can be higher than the PM remanent flux density – which is known as flux focusing. Also, this holds the merits of good mechanical integrity and additional reluctance torque. However, because of significant flux leakage at the inner ends of the PMs, a nonmagnetic shaft or collar is generally required.

As mentioned above, PM brushless drive systems have two basic operations, namely BLAC and BLDC, as shown in Figure 2.41. Actually, each PM brushless motor can operate at both modes if the torque density, torque smoothness, and efficiency are not of great concern. As PM BLAC drive systems operate with a sinusoidal current and a sinusoidal air-gap flux, they need a high-resolution position signal for closed-loop control, hence desiring a costly position encoder or resolver. In contrast, the PM BLDC drive systems operate with a rectangular current and a trapezoidal air-gap flux, and only require a low-cost sensor for phase-current commutation. Nevertheless, the PM BLAC drive systems allow for open-loop operation, whereas the position feedback is mandatory for the PM BLDC drive systems.

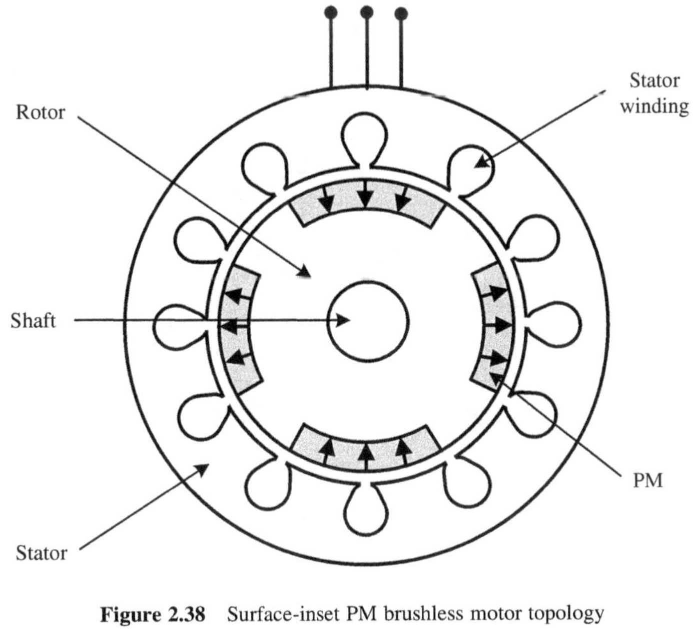

Figure 2.38 Surface-inset PM brushless motor topology

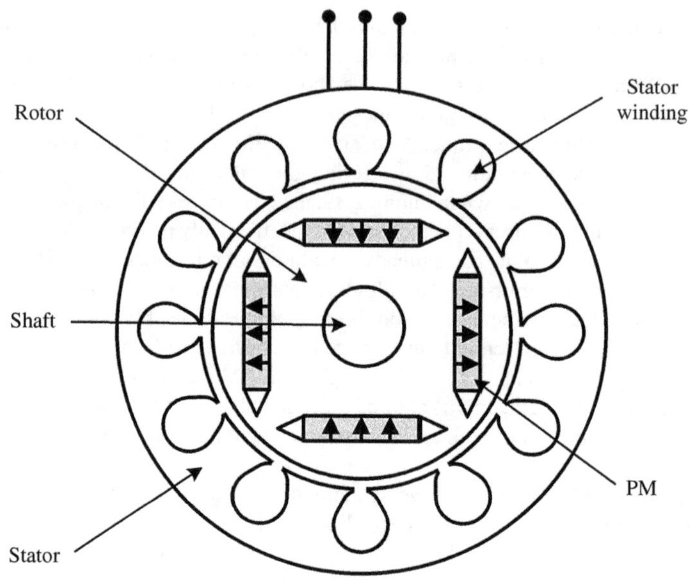

Figure 2.39 Interior-radial PM brushless motor topology

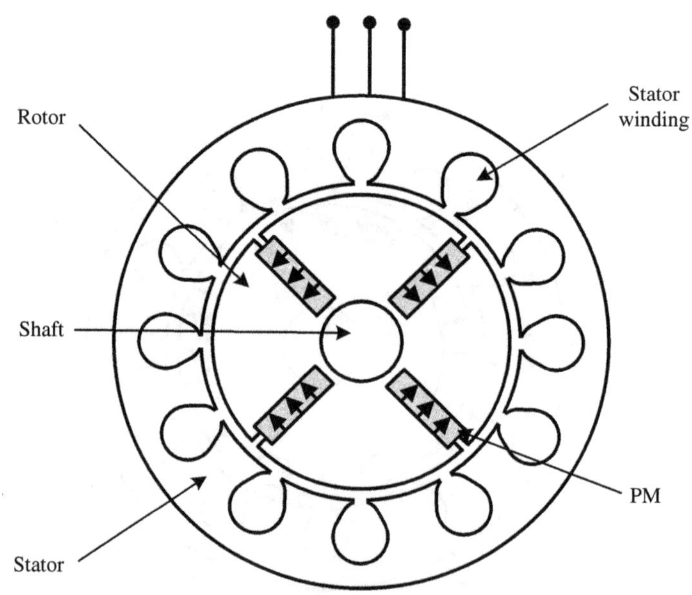

Figure 2.40 Interior-circumferential PM brushless motor topology

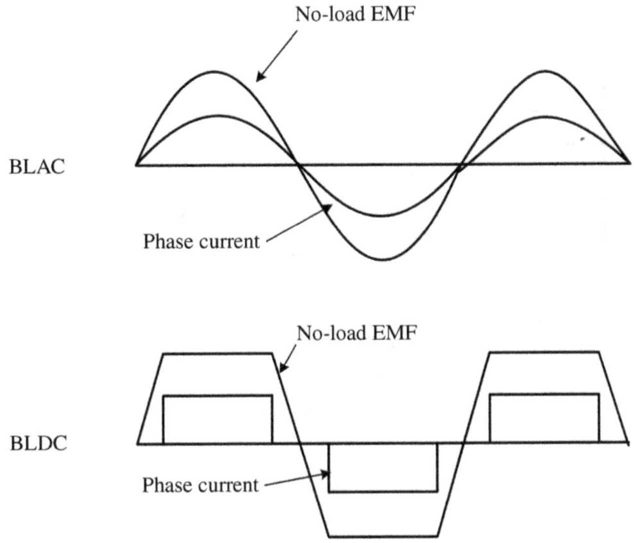

Figure 2.41 Operation modes of PM brushless drive systems

Control of PM BLAC motors is similar to that of induction motors. So, the control strategies for induction drive systems, such as VVVF and FOC, are applicable to PM BLAC drive systems. Based on FOC, the generated torque can be expressed as:

$$T = \frac{3}{2} p \left[\psi_m i_q - \left(L_q - L_d \right) i_d i_q \right] \tag{2.37}$$

where p is the number of pole-pairs, ψ_m is the stator winding flux linkage due to the PMs, L_d, L_q are respectively the d-axis and q-axis stator winding inductances, and i_d, i_q are respectively the d-axis and q-axis currents. Moreover, by incorporating the well-known d–q axis transformation, the well-developed flux-weakening control technique can readily be applied to PM BLAC drive systems for constant power operation. The maximum flux-weakening capability is achieved when the motor is designed to have unity per-unit d-axis inductance (Soong and Ertugrul, 2002):

$$\frac{L_d I_r}{\psi_m} = 1 \tag{2.38}$$

where ψ_m is the PM flux linkage, L_d is the d-axis winding inductance, and I_r is the rated current. In general, the ratio of $L_d I_r / \psi_m$ is less than unity, therefore the higher the ratio, the higher will be the flux-weakening capability. The flux-weakening control has been comprehensively studied in various PM BLAC drive systems (Zhu, Chen, and Howe, 2000; Uddin and Rahman, 2007). The constant-power operation for PM BLDC drive systems is more complex. Since the operating waveforms are no longer sinusoidal, d–q axis transformation and, hence, flux-weakening control are ill-suited. Nevertheless, the corresponding constant-power operation can be offered by using advanced conduction angle control (Chan *et al.*, 1995; Kim, Kook, and Ko, 1997).

Figure 2.42 shows the torque-speed characteristics of the PM brushless drive systems without control, and with either flux-weakening control for the BLAC or advanced conduction angle control for the BLDC. It illustrates that the speed range of constant-power operation can be significantly extended. On the other hand, Figure 2.43 gives a comparison of the torque-speed characteristics of PM

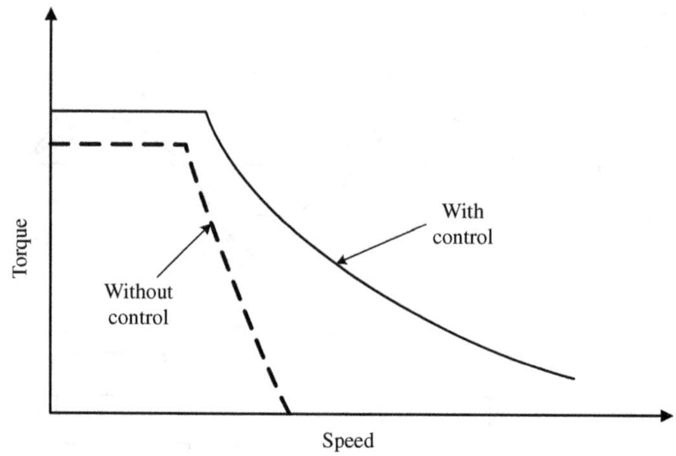

Figure 2.42 Torque-speed characteristics of PM brushless drive systems with and without control

BLAC and PM BLDC drive systems. It can be seen that the BLAC drive systems offer higher torque and higher power capabilities than the BLDC motor drives employing two-phase 120° conduction. Nevertheless, the BLDC motor drives employing three-phase 180° conduction can offer a better high-speed power capability, but with the sacrifice of low-speed torque capability (Zhu and Howe, 2007). Moreover, for PM BLDC drive systems with multiphase polygonal windings (Wang *et al.*, 2002), the corresponding back EMF, rather than the air-gap flux, can be directly varied to enable constant-power operation.

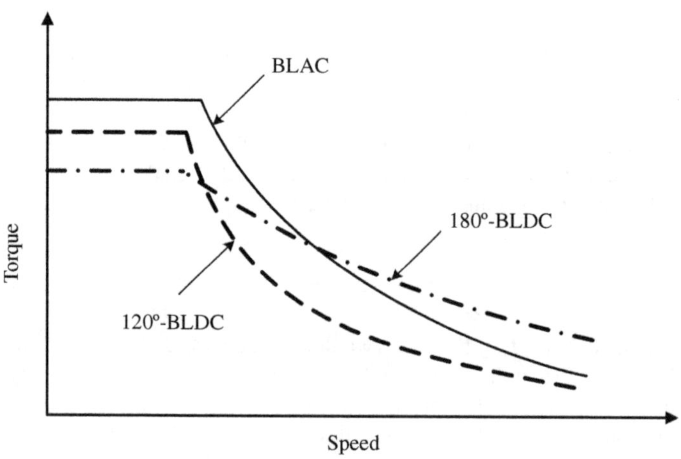

Figure 2.43 Torque-speed characteristics of PM brushless drive systems using BLAC and BLDC operations

2.2.4.2 Synchronous Reluctance Drive Systems

PM brushless motors have been accepted to provide the highest efficiency and highest power density. However, they suffer from the drawbacks of high PM material cost, the accidental demagnetization of

PMs, and the thermal instability of PMs. The synchronous reluctance (SynR) motor relies on reluctance torque, rather than the reaction torque which is dominant in the cylindrical wound-rotor synchronous motor and the surface-mounted PM brushless motor.

The system configuration of SynR drive systems is similar to that of PM BLAC drive systems. It consists of the SynR motor, PWM inverter, electronic controller, sensor, mechanical transmission, and mechanical load. There are a number of advantages associated with this system:

- The SynR motor does not require field windings or PMs in the rotor, hence offering high mechanical integrity to withstand high-speed operation.
- Since the SynR motor does not need expensive PM material, it is much less costly than PM brushless motors.
- As the SynR motor eliminates the problem of accidental demagnetization of PMs, the drive system can offer a very high-current operation.
- As the SynR motor also eliminates the problem of thermal instability of PMs, the drive system can allow for working in a high-temperature environment.

The SynR motor is actually one of the earliest types of electric motors. The first generation was to adopt a cylindrical rotor with multiple slits along the lines of the direct-axis flux (Kostko, 1923). This rotor configuration could not offer a high saliency ratio, leading to the creation of a low reluctance torque. The second generation was to adopt a segmental rotor (Lawrenson and Gupta, 1967). Its saliency ratio could be five or even larger, hence creating a higher reluctance torque. The latest third generation is to adopt an axially laminated rotor, which has steel sheets bent into a U-shape and stacked in a radial direction (Cruickshank, Anderson, and Menzies, 1971). Its saliency can achieve seven or more, which enables the SynR motor to compete favorably with an induction motor. Nevertheless, this rotor involves higher manufacturing cost which, perhaps, can be solved by mass production.

A modern axially laminated SynR motor is shown in Figure 2.44. Its rotor is constructed of thin laminations which are bent in a semicircular shape. These iron segments are separated by insulating material, such as air or plastic. By selecting the ratio between the width of each iron segment and the width of each insulating material, the saliency of this SynR motor can be optimized, enabling ten or more to be achieved (Matsuo and Lipo, 1994).

Since the stator winding of the SynR motor is sinusoidally distributed, the control of SynR motors is similar to that of PM BLAC motors. Therefore, the control strategies for PM BLAC drive systems, such as VVVF and FOC, are also applicable to SynR drive systems (Xu *et al.*, 1991). Based on the d–q axis transformation, the SynR motor can be modeled as:

$$v_d = R_s i_d + \frac{\mathrm{d}}{\mathrm{d}t}(L_d i_d) - \omega_r L_q I_q \tag{2.39}$$

$$v_q = R_s i_q + \frac{\mathrm{d}}{\mathrm{d}t}\left(L_q i_q\right) + \omega_r L_d I_d \tag{2.40}$$

where v_d, v_q are the d–q components of applied voltage, i_d, i_q are the d–q components of stator winding current, L_d, L_q are the d–q components of inductance, R_s is the stator winding resistance, and ω_r is the rotor speed. The saliency is defined as:

$$\kappa = \frac{L_d}{L_q} \tag{2.41}$$

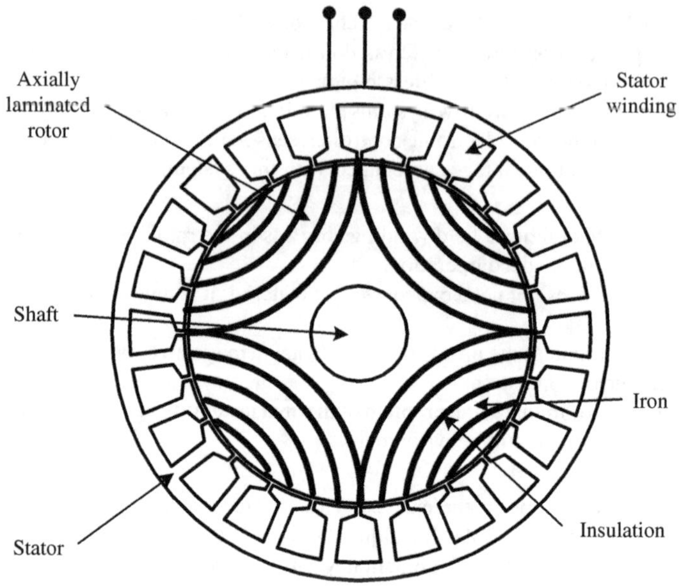

Figure 2.44 Synchronous reluctance motor topology

Consequently, the generated torque can be expressed as:

$$T = \frac{3}{2} p \left(L_d - L_q \right) i_d i_q \tag{2.42}$$

where p is the number of pole-pairs. It is obvious that the larger the value of κ, the higher will be the generated torque. Moreover, the saliency of seven can enable the SynR motor to operate at a power factor of about 0.8, which is comparable with the induction motors for electric drive systems.

2.2.5 *Doubly Salient Drive Systems*

A doubly salient drive system adopts a doubly salient motor, which means that boththe stator and the rotor have salient poles. The two most common doubly salient motors are the switched reluctance (SR) and doubly salient PM (DSPM) types, both of which adopt a simple solid-iron rotor with salient poles. The major difference between their structures is that the SR motor installs only armature windings in the stator, whereas the DSPM motor incorporates both armature windings and PMs in the stator. In terms of torque generation, they are very different – that is, the torque generated in the SR motor is solely a reluctance torque, whereas the torque generated in the DSPM motor is mainly a PM torque.

The DSPM motors, and hence their drive systems, are sometimes classified as a subclass of PM brushless drive systems, since they adopt PMs in the motor structure and offer the merit of brushless. As their PMs are located in the stator rather than in the rotor, they are known as stator-PM brushless drive systems, in contrast to the conventional rotor-PM brushless drive systems.

2.2.5.1 Switched Reluctance Drive Systems

Although the concept of variable reluctance was adopted for electric motors over a century ago, the SR motor did not reach its full potential until the advent of power electronics. In general, the SR drive system

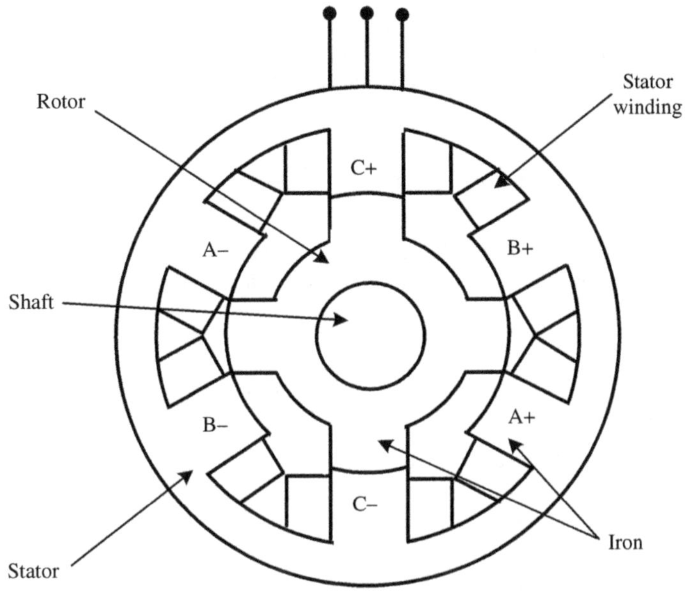

Figure 2.45 Switched reluctance motor topology

consists of an SR motor, power converter, electronic controller, sensor, mechanical transmission, and mechanical load.

Figure 2.45 shows a three-phase 6/4-pole SR motor. Because of the salient nature of both the stator and rotor poles, the inductance L of each phase varies with the rotor position, as shown in Figure 2.46. The operating principle of the SR motor is based on the "minimum reluctance" rule. For instance, when the phase A winding is excited, the rotor tends to rotate clockwise in order to decrease the reluctance of the flux path until the rotor poles align with the stator poles A+ and A−, where the reluctance of the flux path has a minimum value (the inductance has a maximum value). Phase A is then switched off and phase B is switched on so that the reluctance torque tends to make the relevant rotor poles align with the stator poles B+ and B−. The torque direction is always toward the nearest aligned position. Hence, by conducting the

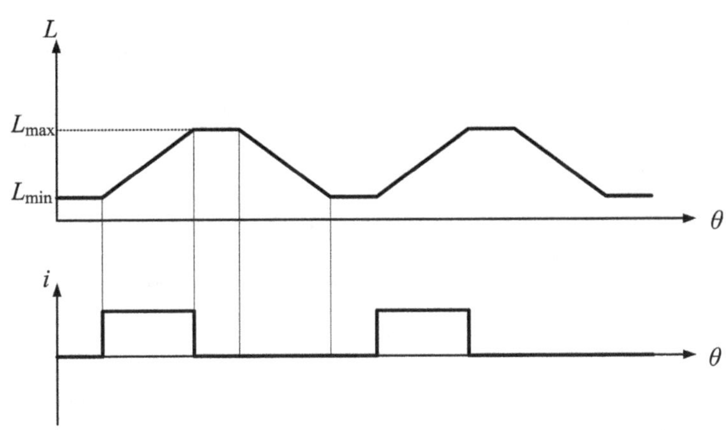

Figure 2.46 Principle of operation of SR drive systems

phase windings in the sequence of A–B–C according to the rotor position feedback from the position sensor, the rotor can continuously rotate clockwise.

According to the co-energy principle, the reluctance torque produced by one phase at any rotor position is given by:

$$T(\theta, i) = \frac{\partial W_c(\theta, i)}{\partial \theta} \tag{2.43}$$

where θ is the rotor position angle, i is the phase current, and $W_c(\theta, i)$ is the co-energy, which is defined as the area below the magnetization curve of flux linkage versus current. This can be expressed as:

$$W_c(\theta, i) = \int_0^i \psi(\theta, i) \mathrm{d}i \tag{2.44}$$

Since the flux linkage $\psi(\theta, I)$ can be written as $\psi(\theta, i) = L(\theta, i)i$, the reluctance torque can be rewritten as:

$$T(\theta, i) = \frac{1}{2} \frac{\partial}{\partial \theta} \int_0^{i^2} L(\theta, i) \mathrm{d}i^2 \tag{2.45}$$

After neglecting magnetic saturation, the inductance is independent of the phase current. The reluctance torque can thus be deduced as:

$$T(\theta, i) = \frac{1}{2} i^2 \frac{\mathrm{d}L}{\mathrm{d}\theta} \tag{2.46}$$

From the above analysis, it can be seen that the SR motor has two significant features. One is that the direction of torque is independent of the polarity of the phase current. The other is that the motoring torque can be produced only in the direction of rising inductance ($\mathrm{d}L/\mathrm{d}\theta > 0$), otherwise, a negative torque (or braking torque) is produced. Each phase is therefore fed with current and hence produces a positive torque only in half a rotor pole-pitch. This is also the reason why the SR motor generally suffers from a large torque ripple. Nevertheless, this torque ripple can be alleviated by increasing the number of phases.

Although SR motors possess a simplicity in construction, that does not imply simplicity in analysis and design. Because of the heavy saturation of pole tips and the fringing effect of poles and slots, the design of SR motors suffers from a great difficulty when using the magnetic circuit approach. In most cases, the electromagnetic finite element analysis is employed to determine the motor parameters and performances. Optimization is then based on the minimization of total losses while taking into account the pole arc constraint, height constraint, and maximum flux density constraint. Nevertheless, there are some basic criteria to initialize the design process of SR motors (Chan *et al.*, 1996).

An appropriate selection of the number of phases and poles is important to satisfy the desired motor performance. In order to allow for starting and running bidirectionally, SR motors should have at least three phases with six stator poles and four rotor poles. The three-phase 6/4-pole SR motor offers the lowest cost and highest efficiency, but its corresponding large torque ripple reduces the smoothness of its operation. On the other hand, the four-phase 8/6-pole SR motor has a relatively higher cost and lower efficiency while possessing the smallest torque ripple. The three-phase 12/8-pole SR motor is a compromise design between the three-phase and four-phase types, and their selection should be based on the operating requirements and cost justification. It should be noted that higher numbers of phases and poles require more power devices and a higher switching frequency, leading to increased costs and switching losses, respectively. The numbers of both the stator and rotor poles, p_s and p_r, are governed by:

$$p_s = 2km \tag{2.47}$$

$$p_r = 2k(m \pm 1) \tag{2.48}$$

where m is the number of phases and k is a positive integer. When the rotor speed is n_r rev/s or rps, the commutating frequency f_{ph} of a particular phase is given by:

$$f_{ph} = n_r p_r \tag{2.49}$$

In order to minimize the switching frequency and to decrease the iron losses in poles and yokes, the number of rotor poles selected should be as small as possible. Thus, the number of rotor poles is usually smaller than the stator poles. Other considerations of SR motor design include the electric loading, magnetic loading, stator outside and inside diameters, rotor outside and inside diameters, stator and rotor pole arcs, stator and pole heights, core length, air-gap length, number of stator windings per phase, thermal arrangement, speed, torque, power, efficiency, torque density, and power density.

Similar to DC drive systems, the chopping frequency of SR drive systems should be above 10 kHz to minimize acoustic noise. Many converter circuits have been developed in attempts to reduce the number of power devices and take full advantage of unipolar operation. However, when the device count is reduced, there is a penalty in the form of lower controllability, lower reliability, lower operating performance, or extra passive components. The converter circuit shown in Figure 2.47 is well suited for three-phase SR drive systems. It utilizes two power devices to independently control the current of each phase and two freewheeling diodes to return any stored magnetic energy to the DC source or link. Since this circuit topology needs two power devices per phase, the converter cost is relatively higher than that with less power devices. However, this bridge arrangement allows control of each phase winding independent of the state of other phase windings. Thus, it is possible to allow for phase overlapping to increase the torque production and extend the constant-power range.

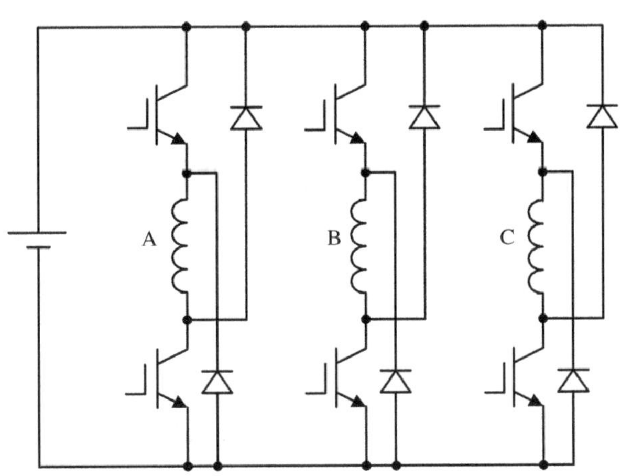

Figure 2.47 Power converter for switched reluctance drive systems

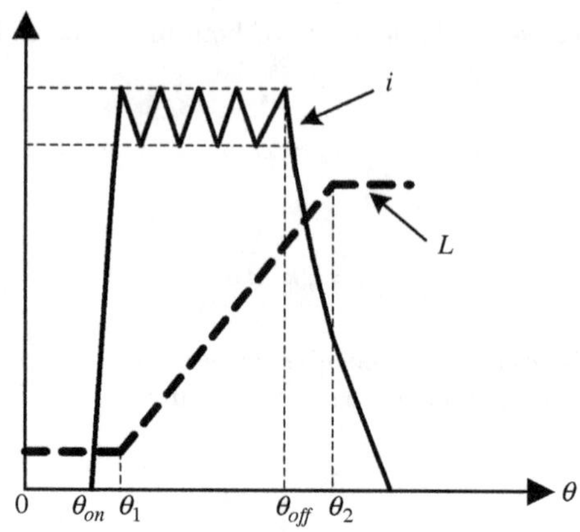

Figure 2.48 Current chopping control for switched reluctance drive systems

SR drive systems have three modes of operation. When the speed is below the base speed ω_b, the current can be limited by chopping, known as current chopping control (CCC). In the CCC mode, as shown in Figure 2.48, the turn-on angle θ_{on} and the turn-off angle θ_{off} are fixed and the firing angle depends only on the speed feedback. The torque can be controlled by changing the current limits, and thus the constant-torque characteristic can be achieved by CCC. During high-speed operation, however, the peak current is limited by the EMF of the phase winding. The corresponding characteristic is essentially controlled by phasing the switching instants relative to the rotor position – known as angular position control (APC). In the APC mode as shown in Figure 2.49, the constant-power characteristic can be achieved. At the critical speed ω_c, both θ_{on} and θ_{off} reach their limit values. Thereafter, the SR drive can no longer keep constant-power operation and thus offer the characteristic similar to that of the DC series drive. The torque-speed characteristics of these three operation modes are shown in Figure 2.50. Moreover, because of the inherently high nonlinearity of SR motors, various advanced control methods have been developed for the SR drive systems, such as the adaptive control, fuzzy logic control, and sliding mode control (Zhan, Chan, and Chau, 1999).

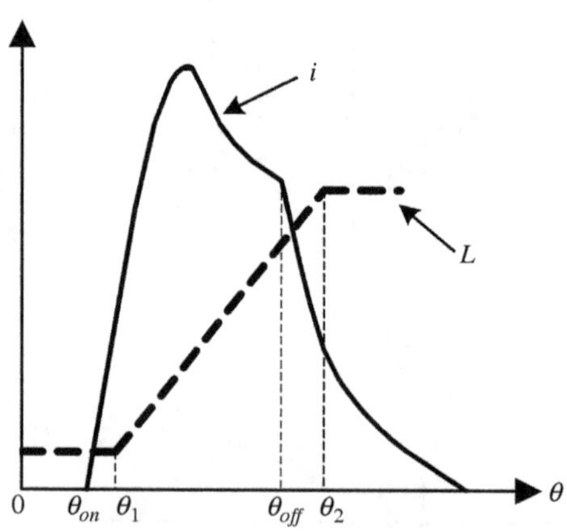

Figure 2.49 Angular position control for switched reluctance drive systems

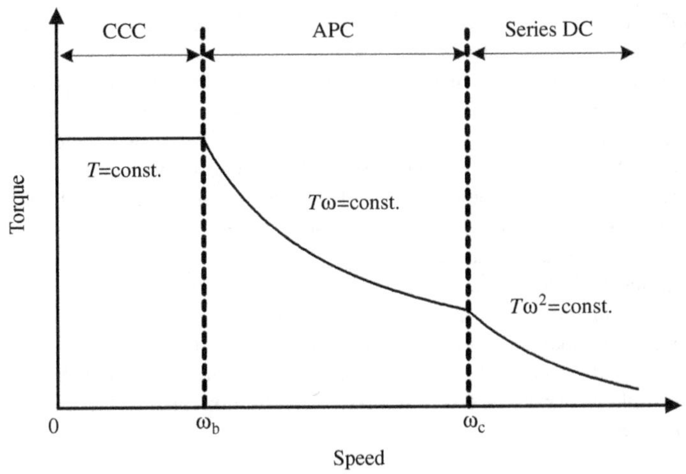

Figure 2.50 Torque-speed characteristics of switched reluctance drive systems

2.2.5.2 Doubly Salient PM Drive Systems

The DSPM motor topologies have salient poles in both of the stator and rotor, and incorporate PMs located in the stator. Similar to the SR motors where their rotor has neither PMs nor windings, these DSPM motors are mechanically simple and robust, and are hence very suitable for high-speed operation. According to the shape and location of the PMs, they can be classified as the yoke-magnet and tooth-magnet types.

Yoke-magnet DSPM motors can be further split into linear-magnet and curved-magnet types. As shown in Figure 2.51, the yoke-linear-magnet type is most common and relatively more mature

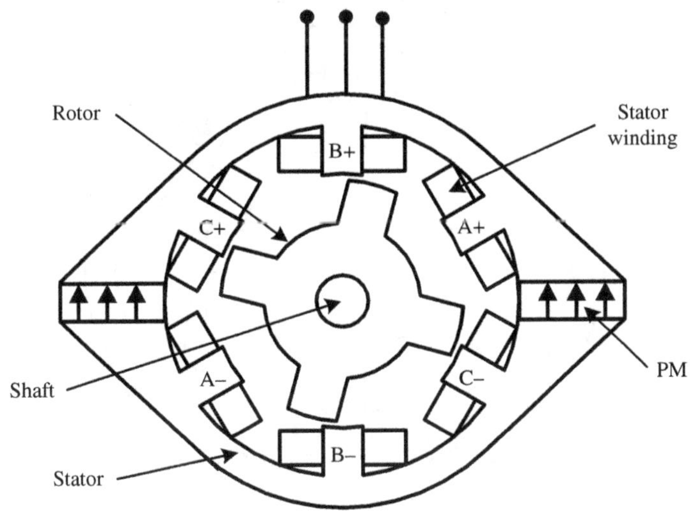

Figure 2.51 Yoke-linear-magnet DSPM motor

(Chau *et al.*, 2005; Cheng, Chau, and Chan, 2001, 2003). Although they have salient poles in the stator and rotor, the PM torque significantly dominates the reluctance torque, hence exhibiting low cogging torque. Since the variation of flux linkage with each coil as the rotor rotates is unipolar, it is very suitable for BLDC operation. On the other hand, when the rotor is skewed, it can offer a BLAC operation. The major disadvantage of this topology is the relatively low torque density, which results from its unipolar flux linkage. As shown in Figure 2.52, the yoke-curved-magnet type is very similar to the previous one, except for the shape of the PMs. As opposed to the yoke-linear-magnet type, the periphery of this topology is essentially circular. Also, since there is more space to accommodate the PMs, this DSPM motor can achieve a higher air-gap flux density. Its major drawback is the difficulty in machining the curved PMs and inserting them into the stator core.

The variation in the air-gap flux of the DSPM motors is induced by the variation in permeance rather than by the rotation of the magnet. Thus, the flux always tends to flow through the shortest path that the stator poles align with the rotor poles from the unaligned position, causing the motor to rotate. The corresponding flux linkage per phase Ψ is composed of the PM flux linkage Ψ_{pm} and the armature reaction flux linkage Li as given by:

$$\Psi = \Psi_{\mathrm{pm}} + Li \tag{2.50}$$

where L is the self-inductance and i is the armature current per phase. The magnetic field co-energy W_c can thus be obtained by subtracting the energy stored in the armature windings from the total input energy:

$$W_c = i\Psi - \frac{1}{2}Li^2 = i\Psi_{\mathrm{pm}} + \frac{1}{2}Li^2 \tag{2.51}$$

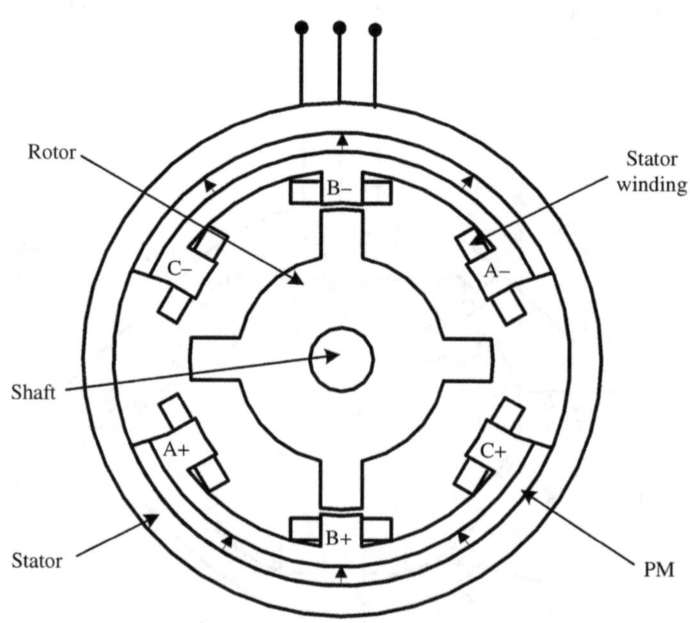

Figure 2.52 Yoke-curved-magnet DSPM motor

Then, by differentiating the co-energy, the generated torque can be obtained, as given by:

$$\begin{aligned} T &= i\frac{d\Psi_{pm}}{d\theta} + \frac{1}{2}i^2\frac{dL}{d\theta} \\ &= T_{pm} + T_r \end{aligned} \tag{2.52}$$

where θ is the rotor position, T_{pm} is the PM torque component which is due to the interaction between the armature current and the PM flux linkage, and T_r is the reluctance torque component, which is due to the variation of self-inductance. The theoretical waveforms of Ψ_{pm} and L are shown in Figure 2.53. In order to produce a continuous unidirectional torque, a bipolar armature current is used, in which a positive current is applied when the flux linkage increases, whereas a negative current is applied when the flux linkage decreases. As a result, the PM torque becomes the dominant torque component, while the reluctance torque is a parasitic pulsating torque with zero average value.

It should be noted that the bipolar armature current operation of DSPM motors is fundamentally different from that of SR motors in which a unipolar armature current is adopted to create the reluctance torque only during the period of increasing inductance. Therefore, the torque density of DSPM drive systems is inherently higher than that of SR drive systems.

Tooth-magnet DSPM motors can be further split into surface-magnet and interior-magnet types. Figure 2.54 shows the tooth-surface-magnet DSPM motor topology, which is commonly known as a flux-reversal PM motor, since the flux linkage with each coil reverses polarity as the rotor rotates (Deodhar *et al.*, 1997; Zhu and Howe, 2007). In this topology, each stator tooth has a pair of PMs of different polarities mounted onto the surface. Hence, the flux linkage variation is bipolar so that the torque density can be higher than that of the yoke-magnet types. However, since the PMs are on the surface of the stator teeth, they are more prone to partial demagnetization. Also, a significant eddy current loss in the PMs may result. On the other hand, Figure 2.55 shows the tooth-interior-magnet motor topology which is commonly known as a flux-switching PM motor (Zhu *et al.*, 2005; Zhu and Howe, 2007). In this topology, each stator

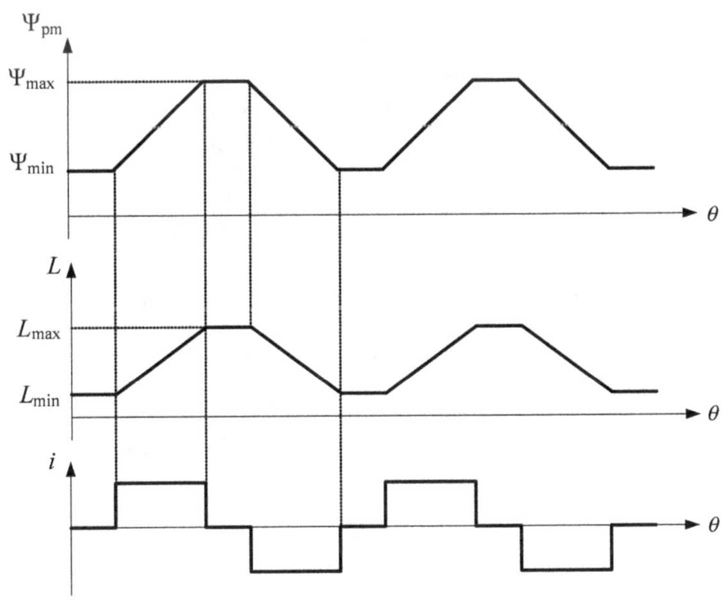

Figure 2.53 Principle of operation of yoke-magnet DSPM drive systems

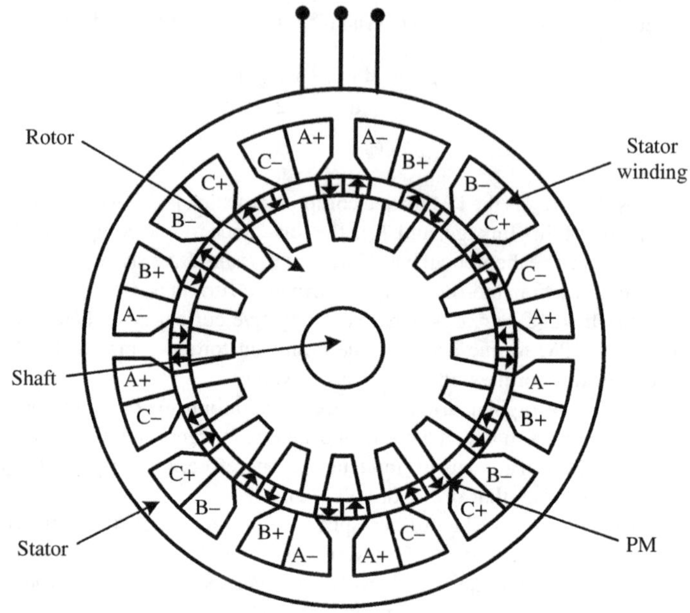

Figure 2.54 Tooth-surface-magnet DSPM motor

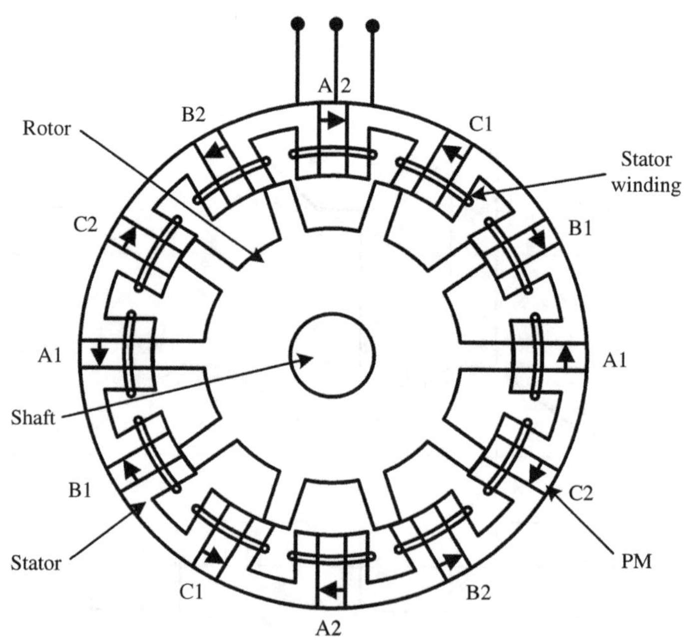

Figure 2.55 Tooth-interior-magnet DSPM motor

tooth consists of two adjacent laminated segments and a PM, and each of these segments is sandwiched by two circumferentially magnetized PMs. Hence, it enables flux focusing. Additionally, this flux-switching PM motor has less armature reaction, hence offering higher electric loading. Since its back EMF waveform is essentially sinusoidal, this motor is more suitable for a BLAC operation.

References

Banerjee, S. and Verghese, G.C. (2001) *Nonlinear Phenomena in Power Electronics*, IEEE Press, New York.

Benettin, G., Galgani, L., and Strelcyn, J.M. (1976) Kolmogorov entropy and numerical experiments. *Physical Review A*, **14**, 2338–2345.

Cellucci, C.J., Albano, A.M., Rapp, P.E. *et al.* (1997) Detecting noise in a time series. *Chaos*, **7**, 414–422.

Chan, C.C., Jiang, J.Z., Xia, W., and Chau, K.T. (1995) Novel wide range speed control of permanent magnet brushless motor drives. *IEEE Transactions on Power Electronics*, **10**, 539–546.

Chan, C.C., Jiang, Q., Zhan, Y.J., and Chau, K.T. (1996) A high-performance switched reluctance drive for P-star EV project. Proceedings of 13th International Electric Vehicle Symposium, pp. 78–83.

Chau, K.T., Ching, T.W., Chan, C.C., and Chan, D.T.W. (1997) A novel two-quadrant zero-voltage transition converter for DC motor drives. Proceedings of IEEE International Conference on Industrial Electronics, pp. 517–522.

Chau, K.T., Sun, Q., Fan, Y., and Cheng, M. (2005) Torque ripple minimization of doubly salient permanent magnet motors. *IEEE Transaction on Energy Conversion*, **20**, 352–358.

Cheng, M., Chau, K.T., and Chan, C.C. (2001) Design and analysis of a new doubly salient permanent magnet motor. *IEEE Transaction on Magnetics*, **37**, 3012–3020.

Cheng, M., Chau, K.T., and Chan, C.C. (2003) New split-winding doubly salient permanent magnet motor drive. *IEEE Transaction on Aerospace Electronic Systems*, **39**, 202–210.

Ching, T.W., Chau, K.T., and Chan, C.C. (1998) A new zero-voltage-transition converter for switched reluctance motor drives. Proceedings of IEEE Power Electronics Specialists Conference, pp. 1295–1301.

Cho, J.G., Kim, W.H., Rim, G.H., and Cho, K.Y. (1997) Novel zero transition PWM converter for switched reluctance motor drives. Proceedings of IEEE Power Electronics Specialists Conference, pp. 887–891.

Cruickshank, A.J.O., Anderson, A.F., and Menzies, R.W. (1971) Theory and performance of reluctance motors with axially laminated anisotropic rotors. *Proceedings of IEE*, **118**, 887–894.

Deodhar, R.P., Andersson, S., Boldea, I., and Miller, T.J.E. (1997) The flux-reversal machine: a new brushless doubly-salient permanent magnet machine. *IEEE Transaction on Industry Applications*, **33**, 925–934.

Ding, M.Z., Grebogi, C., and Ott, E. (1989) Evolution of attractors in quasiperiodically forced systems: from quasiperiodic to strange nonchaotic to chaotic. *Physical Review A*, **39**, 2593–2598.

Ditto, W.L., Spano, M.L., Savage, H.T. *et al.* (1990) Experimental observation of a strange nonchaotic attractor. *Physical Review Letters*, **65**, 533–536.

Divan, D.M. (1986) The resonant DC link converter – a new concept in static power conversion. Proceedings of IEEE Industry Application Society Annual Meeting, pp. 648–656.

Eckmann, J.P. and Ruelle, D. (1985) Ergodic theory of chaos and strange attractors. *Reviews of Modern Physics*, **57**, 617–656.

Farmer, J.D. (1982) Chaotic attractors of an infinite-dimensional dynamical system. *Physica D*, **4**, 366–393.

Gan, J., Chau, K.T., Chan, C.C., and Jiang, J.Z. (2000) A new surface-inset, permanent-magnet, brushless DC motor drive for electric vehicles. *IEEE Transactions on Magnetics*, **36**, 3810–3818.

Grassberger, P. and Procaccia, I. (1983a) Characterization of strange attractors. *Physics Review Letters*, **50**, 346–349.

Grassberger, P. and Procaccia, I. (1983b) Measuring the strangeness of strange attractors. *Physica D*, **9**, 189–208.

Grebogi, C., Ott, E., Pelikan, S., and Yorke, J.A. (1984) Strange attractors that are not chaotic. *Physica D*, **13**, 261–268.

Kim, Y., Kook, Y., and Ko, Y. (1997) A new technique of reducing torque ripples for BDCM drives. *IEEE Transactions on Industrial Electronics*, **44**, 735–739.

Kolmogorov, A.N. (1958) A new invariant for transitive dynamical systems. *Doklady Akademii Nauk SSSR*, **119**, 861–864.

Kostko, J.K. (1923) Polyphase reaction synchronous motors. *Journal of American Institute of Electrical Engineers*, **42**, 1162–1168.

Lawrenson, P.J. and Gupta, S.K. (1967) Developments in the performance and theory of segmental-rotor reluctance motors. *Proceedings of IEE*, **114**, 645–653.

Matsuo, T. and Lipo, T.A. (1994) Rotor design optimization of synchronous reluctance machine. *IEEE Transaction on Energy Conversion*, **9**, 359–365.

Müller, P.C. (1995) Calculation of Lyapunov exponents for dynamic systems with discontinuities. *Chaos, Solitons and Fractals*, **5**, 1671–1681.

Ott, E. (1993) *Chaos in Dynamical Systems*, Cambridge University Press, Cambridge.

Parker, T.S. and Chua, L.O. (1999) *Practical Numerical Algorithms for Chaotic Systems*, Springer Verlag, New York.

Pawelzik, K. and Schuster, H.G. (1991) Unstable periodic orbits and prediction. *Physical Review A*, **43**, 1808–1812.

Pesin, Ya.B. (1977) Characteristic Lyapunov exponents and smooth ergodic theory. *Russian Mathematical Surveys*, **32**, 55–114.

Rosenstein, M.T., Collins, J.J., and De Luca, C.J. (1993) A practical method for calculating largest Lyapunov exponents from small data sets. *Physica D*, **65**, 117–134.

Shimada, I. and Nagashima, T. (1979) A numerical approach to ergodic problem of dissipative dynamical systems. *Progress of Theoretical Physics*, **61**, 1605–1616.

Soong, W.L. and Ertugrul, N. (2002) Field-weakening performance of interior permanent-magnet motors. *IEEE Transactions on Industry Applications*, **38**, 1251–1258.

Sprott, J.C. (2003) *Chaos and Time-Series Analysis*, Oxford University Press, Oxford.

Takens, F. (1980) Detecting strange attractors in turbulence, in *Dynamical Systems and Turbulence, Lecture Notes in Mathematics*, **898** (eds D.A. Rand, and L.-S. Young), Springer-Verlag, pp. 366–381.

Uddin, M.N. and Rahman, M.A. (2007) High-speed control of IPMSM drives using improved fuzzy logic algorithms. *IEEE Transactions on Industrial Electronics*, **54**, 190–199.

Valsakumar, M.C., Satyanarayana, S.V.M., and Sridhar, V. (1997) Signature of chaos in power spectrum. *Pramana Journal of Physics*, **48**, 69–85.

Wang, Y., Chau, K.T., Chan, C.C., and Jiang, J.Z. (2002) Design and analysis of a new multiphase polygonal-winding permanent-magnet brushless dc machine. *IEEE Transactions on Magnetics*, **38**, 3258–3260.

Wolf, A., Swift, J.B., Swinney, H.L., and Vastano, J.A. (1985) Determining Lyapunov exponents from a time series. *Physica D*, **16**, 285–317.

Xu, L., Xu, X., Lipo, T.A., and Novotny, D.W. (1991) Vector control of a synchronous reluctance motor including saturation and iron loss. *IEEE Transactions on Industry Applications*, **27**, 977–985.

Yang, T. and Bilimgut, K. (1997) Experimental results of strange nonchaotic phenomenon in a second-order quasi-periodically forced electronic circuit. *Physics Letters A*, **236**, 494–499.

Zhan, Y.J., Chan, C.C., and Chau, K.T. (1999) A novel sliding-mode observer for indirect position sensing of switched reluctance motor drives. *IEEE Transactions on Industrial Electronics*, **46**, 390–397.

Zhou, T. and Moss, F. (1992) Observation of a strange nonchaotic attractor in a multistable potential. *Physical Review A*, **45**, 5394–5400.

Zhu, Z.Q. and Howe, D. (2007) Electrical machines and drives for electric, hybrid and fuel cell vehicles. *IEEE Proceedings*, **95**, 746–765.

Zhu, Z.Q., Chen, Y.S., and Howe, D. (2000) On-line optimal field weakening control of permanent magnet brushless ac drives. *IEEE Transactions on Industry Applications*, **36**, 1661–1668.

Zhu, Z.Q., Pang, Y., Howe, D. *et al.* (2005) Analysis of electromagnetic performance of flux-switching permanent magnet machines by non-linear adaptive lumped parameter magnetic circuit model. *IEEE Transaction on Magnetics*, **41**, 4277–4287.

Part Two

Analysis of Chaos in Electric Drive Systems

3

Chaos in DC Drive Systems

DC drive systems have been widely used for domestic, industrial and vehicular applications because of their technological maturity and control simplicity. In general, the speed control of DC drives can be accomplished by two methods, namely armature control and field control. In the case of permanent magnet (PM) excitation, the PM field is essentially uncontrollable. During normal operation, the armature circuit is fed by a voltage source while adopting either armature voltage control or armature current control. The voltage control has the definite advantage of simplicity and low cost, whereas the current control has the merit of direct torque control.

In this chapter, chaos in both voltage-controlled DC drive systems and current-controlled DC drive systems are investigated. The corresponding modeling, analysis, simulation, and experimentation are also discussed in detail.

3.1 Voltage-Controlled DC Drive System

The investigation into chaos in power electronic circuits was launched in the late 1980s, focusing on various kinds of switching DC–DC converters (Hamill and Jefferies, 1988; Deane, 1992; Tse, Fung, and Kwan, 1996). By extending the work to DC drive systems that involve a speed-dependent load voltage, chaotic behavior in the voltage-controlled DC drive system was first investigated in 1997 (Chau *et al.*, 1997). Consequently, the corresponding dynamic bifurcation (Chau, Chen, and Chan, 1997) as well as modeling of subharmonics and chaos (Chau *et al.*, 1997) have also been discussed.

3.1.1 Modeling

As shown in Figure 3.1, a voltage-controlled DC chopper-fed PMDC drive system operating in continuous conduction mode is used for exemplification (Chen, Chau, and Chan, 2000a). The corresponding equivalent circuit is shown in Figure 3.2, where the motor speed ω is controlled by constant frequency pulse width modulation (PWM).

Considering that the operational amplifier A_1 has a feedback gain g, the control signal v_c can be expressed as:

$$v_c(t) = g(\omega(t) - \omega_{ref}) \tag{3.1}$$

Chaos in Electric Drive Systems: Analysis, Control and Application, First Edition. K.T. Chau and Zheng Wang.
© 2011 John Wiley & Sons (Asia) Pte Ltd. Published 2011 by John Wiley & Sons (Asia) Pte Ltd.

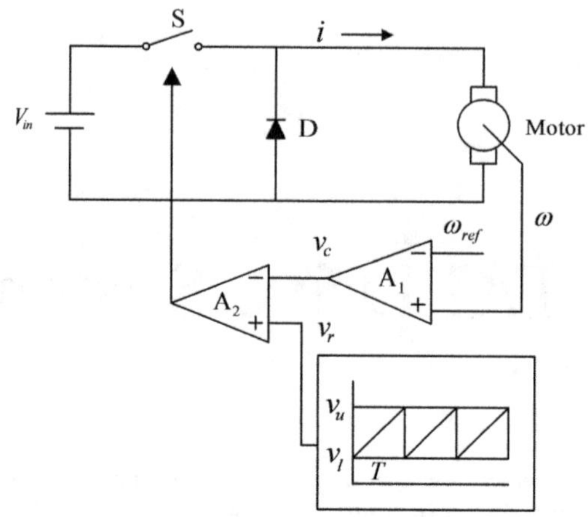

Figure 3.1 Block diagram of voltage-controlled DC drive system

where $\omega(t)$ and ω_{ref} are the instantaneous and reference motor speeds, respectively. The ramp voltage v_r is represented by:

$$v_r(t) = v_l + (v_u - v_l)t/T \tag{3.2}$$

where v_l and v_u are, respectively, the lower and upper voltages of the ramp signal, and T is its period. Then, both v_c and v_r are fed into the comparator A_2 which outputs the signal to turn the power switch S on or off. When the control voltage exceeds the ramp voltage, S is off and hence the diode D comes on; otherwise, S is on and D is off. Thus, the system equation can be divided into two stages as given by:

Stage 1: $v_c \geq v_r$

$$\frac{\mathrm{d}}{\mathrm{d}t}\begin{pmatrix}\omega(t)\\ i(t)\end{pmatrix} = \begin{pmatrix}-B/J & K_T/J\\ -K_E/L & -R/L\end{pmatrix}\begin{pmatrix}\omega(t)\\ i(t)\end{pmatrix} + \begin{pmatrix}-T_l/J\\ 0\end{pmatrix} \tag{3.3}$$

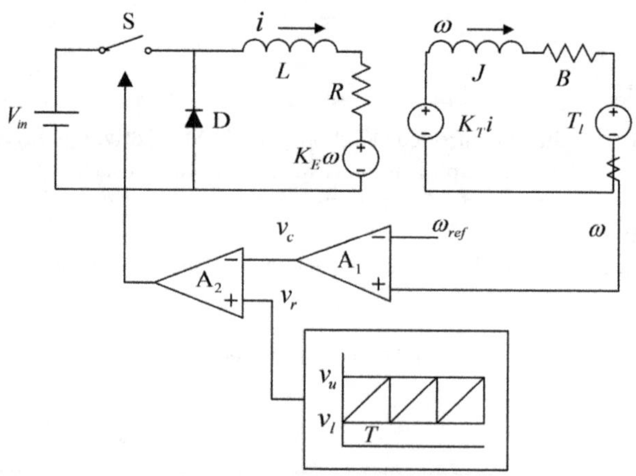

Figure 3.2 Equivalent circuit of voltage-controlled DC drive system

Stage 2: $v_c < v_r$

$$\frac{\mathrm{d}}{\mathrm{d}t}\begin{pmatrix}\omega(t)\\ i(t)\end{pmatrix} = \begin{pmatrix}-B/J & K_T/J\\ -K_E/L & -R/L\end{pmatrix}\begin{pmatrix}\omega(t)\\ i(t)\end{pmatrix} + \begin{pmatrix}-T_l/J\\ V_{in}/L\end{pmatrix} \tag{3.4}$$

where $i(t)$ is the armature current, R is the armature resistance, L is the armature inductance, V_{in} is the DC supply voltage, K_E is the back-EMF constant, K_T is the torque constant, B is the viscous damping, J is the load inertia, and T_l is the load torque.

By defining the state vector $\mathbf{X}(t)$ and the matrices $\mathbf{A}$, $\mathbf{E}_1$, $\mathbf{E}_2$, $\mathbf{E}_3$, $\mathbf{E}_4$ as:

$$\mathbf{X}(t) = \begin{pmatrix}\omega(t)\\ i(t)\end{pmatrix}, \quad \mathbf{A} = \begin{pmatrix}-B/J & K_T/J\\ -K_E/L & -R/L\end{pmatrix} \tag{3.5}$$

$$\mathbf{E}_1 = \begin{pmatrix}-T_l/J\\ 0\end{pmatrix}, \quad \mathbf{E}_2 = \begin{pmatrix}-T_l/J\\ V_{in}/L\end{pmatrix}, \quad \mathbf{E}_3 = \begin{pmatrix}g & 0\end{pmatrix}, \quad \mathbf{E}_4 = \begin{pmatrix}\omega_{ref}\\ 0\end{pmatrix} \tag{3.6}$$

the system equation given by (3.3) and (3.4) can be rewritten as:

$$\dot{\mathbf{X}}(t) = \mathbf{A}\,\mathbf{X}(t) + \mathbf{E}_k \quad (k = 1, 2) \tag{3.7}$$

and the switching condition $v_c = v_r$ can be expressed as:

$$\mathbf{E}_3(\mathbf{X}(t) - \mathbf{E}_4) = 0 \tag{3.8}$$

As k changes value when $v_c = v_r$ while v_c is time dependent, the system given by equation (3.7) is in fact a time-varying state equation. Thus, this DC drive system is a second-order nonautonomous dynamical system.

3.1.2 Analysis

The analysis of system chaotic behavior begins with the solution of (3.7) in a continuous-time domain, namely $\mathbf{X}(t)$. Then, the iterative function that maps this $\mathbf{X}(t)$ at $t = nT$ to its successive one at $t = (n+1)T$ is defined as P: $\Re^2 \rightarrow \Re^2$:

$$\mathbf{X}_{n+1} = P(\mathbf{X}_n) \tag{3.9}$$

which is the so-called Poincaré map. The generalized Poincaré map can fully describe the system behavior using numerical simulation, whereas the specific one can analytically describe the system behavior in terms of periodic orbits and stability.

3.1.2.1 Solution of System Equation

Given an initial value $\mathbf{X}(t_0)$, the continuous-time solution of the system equation given by (3.7) can be expressed as:

$$\begin{aligned}\mathbf{X}(t) &= \Phi(t-t_0)\mathbf{X}(t_0) + \int_{t_0}^{t}\Phi(t-\tau)\mathbf{E}_k d\tau\\ &= -\mathbf{A}^{-1}\mathbf{E}_k + \Phi(t-t_0)(\mathbf{X}(t_0) + \mathbf{A}^{-1}\mathbf{E}_k) \quad (k = 1, 2)\end{aligned} \tag{3.10}$$

where $\Phi(t) = e^{\mathbf{A}t}$ is known as a state transition matrix. By defining the parameters α and Δ as:

$$\alpha = \frac{1}{2}\left(\frac{R}{L} + \frac{B}{J}\right), \; \Delta = \frac{1}{4}\left(\frac{R}{L} - \frac{B}{J}\right)^2 - \frac{K_E K_T}{LJ} \tag{3.11}$$

the eigenvalues λ_1, λ_2 of $\mathbf{A}$ can be expressed as:

$$\Delta = 0, \; \lambda_1 = \lambda_2 = \lambda = -\alpha \tag{3.12}$$

$$\Delta > 0, \; \lambda_1 = -\alpha + \sqrt{\Delta}, \lambda_2 = -\alpha - \sqrt{\Delta} \tag{3.13}$$

$$\Delta < 0, \; \lambda_1 = -\alpha + j\sqrt{-\Delta}, \; \lambda_2 = -\alpha - j\sqrt{-\Delta} \tag{3.14}$$

Hence, the corresponding $\Phi(t)$ can be obtained as:

$$\Delta = 0, \; \Phi(t) = \exp(-\alpha t)[\mathbf{1} - t(2\alpha\mathbf{1} - \mathbf{A})] \tag{3.15}$$

$$\Delta > 0, \; \Phi(t) = \frac{1}{\lambda_2 - \lambda_1}[\exp(\lambda_1 t)(\lambda_2\mathbf{1} - \mathbf{A}) - \exp(\lambda_2 t)(\lambda_1\mathbf{1} - \mathbf{A})] \tag{3.16}$$

$$\Delta < 0, \; \Phi(t) = \exp(-\alpha t)\left[\mathbf{1}\cos(\beta t) + \frac{1}{\beta}(\alpha\mathbf{1} + \mathbf{A})\sin(\beta t)\right] \tag{3.17}$$

where $\mathbf{1}$ is the identity matrix and $\beta = \sqrt{-\Delta}$.

3.1.2.2 Derivation of Generalized Poincaré Map

During each cyclic operation of the drive system, there are two possible situations – either a skipping cycle because of the absence of intersection between v_c and v_r, or an intersecting cycle in which there is at least one intersection between v_c and v_r.

For the skipping cycle, S does not change its state, remaining either on or off. Making use of (3.9), the corresponding Poincaré map P can be easily deduced from (3.10) as given by:

$$\mathbf{X}_{n+1} = -\mathbf{A}^{-1}\mathbf{E}_k + \Phi(T)(\mathbf{X}_n + \mathbf{A}^{-1}\mathbf{E}_k) \quad (k = 1, 2) \tag{3.18}$$

For the intersecting cycle, S may change m times when v_c crosses v_r by m times within T. This is known as multiple pulsing. Thus, the intersections occur at:

$$t = nT + \delta_i T, \; 0 = \delta_0 < \delta_1 <, \cdots, < \delta_m < 1 \quad (i = 1, \cdots, m) \tag{3.19}$$

Hence, the corresponding Poincaré map can be expressed as the following iterative functions:

$$\mathbf{X}(nT + \delta_i T) = \mathbf{f}_i(\mathbf{X}(nT + \delta_{i-1}T), \delta_{i-1}T, \delta_i T) \quad (i = 1, \cdots, m) \tag{3.20}$$

$$\mathbf{X}_{n+1} = \mathbf{f}_{m+1}(\mathbf{X}(nT + \delta_m T), \delta_m T, T) \tag{3.21}$$

Firstly, the derivation is based on $m = 1$ where there is only one intersection within T. When v_c at $t = nT$ is greater than or equal to v_l, the drive operates in Stage 1 from nT to $nT + \delta_1 T$, at which point the switching condition $v_c = v_r$ is satisfied, and then in Stage 2 to $(n+1)T$. Using (3.8) and (3.10), δ_1 can be determined by evaluating the solution of the following transcendental equation:

$$h(\delta_1) = \mathbf{E}_3[-\mathbf{A}^{-1}\mathbf{E}_1 + \Phi(\delta_1 T)(\mathbf{X}_n + \mathbf{A}^{-1}\mathbf{E}_1) - \mathbf{E}_4] - v_r(\delta_1 T) = 0 \quad (i = 1, \cdots, m) \tag{3.22}$$

Hence, the corresponding Poincaré map can be written as:

$$\mathbf{X}(nT+\delta_1 T) = -\mathbf{A}^{-1}\mathbf{E}_1 + \Phi(\delta_1 T)(\mathbf{X}_n + \mathbf{A}^{-1}\mathbf{E}_1) \quad (i = 1,\cdots,m) \tag{3.23}$$

$$\mathbf{X}_{n+1} = -\mathbf{A}^{-1}\mathbf{E}_2 + \Phi(T-\delta_1 T)(\mathbf{X}(nT+\delta_1 T) + \mathbf{A}^{-1}\mathbf{E}_2) \tag{3.24}$$

On the contrary, when v_c at $t=nT$ is lower than or equal to v_l, the system operates in Stage 2 first and then in Stage 1. The corresponding Poincaré map can similarly be obtained as:

$$h(\delta_1) = \mathbf{E}_3[-\mathbf{A}^{-1}\mathbf{E}_2 + \Phi(\delta_1 T)(\mathbf{X}_n + \mathbf{A}^{-1}\mathbf{E}_2) - \mathbf{E}_4] - v_r(\delta_1 T) = 0 \tag{3.25}$$

$$\mathbf{X}(nT+\delta_1 T) = -\mathbf{A}^{-1}\mathbf{E}_2 + \Phi(\delta_1 T)(\mathbf{X}_n + \mathbf{A}^{-1}\mathbf{E}_2) \tag{3.26}$$

$$\mathbf{X}_{n+1} = -\mathbf{A}^{-1}\mathbf{E}_1 + \Phi(T-\delta_1 T)(\mathbf{X}(nT+\delta_1 T) + \mathbf{A}^{-1}\mathbf{E}_1) \tag{3.27}$$

Similar to the derivation for $m=1$, the Poincaré map for $m>1$, in which there are more than one intersection within T, can be determined as:

$$h(\delta_i) = \mathbf{E}_3[-\mathbf{A}^{-1}\mathbf{E}_k + \Phi(\delta_i T - \delta_{i-1}T)(\mathbf{X}(nT+\delta_{i-1}T) + \mathbf{A}^{-1}\mathbf{E}_k) - \mathbf{E}_4] - v_r(\delta_i T) = 0 \tag{3.28}$$

$$\mathbf{X}(nT+\delta_i T) = -\mathbf{A}^{-1}\mathbf{E}_k + \Phi(\delta_i T - \delta_{i-1}T)(\mathbf{X}(nT+\delta_{i-1}T) + \mathbf{A}^{-1}\mathbf{E}_k) \tag{3.29}$$

$$\mathbf{X}_{n+1} = -\mathbf{A}^{-1}\mathbf{E}_k + \Phi(T-\delta_m T)(\mathbf{X}(nT+\delta_m T) + \mathbf{A}^{-1}\mathbf{E}_k) \tag{3.30}$$

where $i = 1,\cdots,m$, and k equals 1 or 2 depending on whether v_c is over v_r or not.

It should be noted that the above Poincaré map relates to generalized mapping which covers all possible solutions, such as real and complex roots due to different system parameters and conditions. Thus, the generalized Poincaré map can be considered as the mapping method for any second-order dynamical system using similar mathematical models. Moreover, the derivation can readily be extended to those higher-order dynamical systems involving power switches.

3.1.2.3 Analysis of Periodic Orbits

The generalized Poincaré map is so general that it includes cycle skipping and multiple pulsing and, owing to the presence of these cases, it is very inconvenient to analyze the steady-state periodic orbits and their stability. Also, the presence of multiple pulsing can greatly increase the switching losses, which should be avoided by using a latch or sample-and-hold. Therefore, instead of using the generalized case, the detailed analysis of periodic orbits is focused on the case in which the orbits cross the ramp signal once per cycle, namely one intersection within T. The corresponding mapping is known as the specific Poincaré map.

The steady-state periodic solution of the DC drive system can be a period-1 orbit $\mathbf{X}^*$, or a period-p orbit $\{\mathbf{X}_1^*,\cdots,\mathbf{X}_p^*\}$ with $p>1$. It should be noted that the period-1 orbit is the fundamental orbit, whereas the period-p orbit means the $(1/p)$th subharmonic orbit. The corresponding specific Poincaré maps are described as:

$$\mathbf{X}^* = P(\mathbf{X}^*) \tag{3.31}$$

$$\mathbf{X}_{k+1}^* = P(\mathbf{X}_k^*)\ (k = 1,\cdots,p-1),\ \mathbf{X}_1^* = P(\mathbf{X}_p^*) \tag{3.32}$$

Firstly, the period-1 orbit is analyzed. Since $m=1$ and the orbit must lie between v_l and v_u, the corresponding Poincaré map can be deduced from (3.23) and (3.24) as given by:

$$\mathbf{X}_{n+1} = -\mathbf{A}^{-1}\mathbf{E}_1 + \Phi(T)(\mathbf{X}_n + \mathbf{A}^{-1}\mathbf{E}_1) + (\mathbf{1} - \Phi(T - \delta T))\mathbf{A}^{-1}(\mathbf{E}_1 - \mathbf{E}_2) \tag{3.33}$$

Substituting (3.33) into the mapping given by (3.31) and taking $\mathbf{X}_{n+1} = \mathbf{X}_n$, the period-1 orbit can be obtained as:

$$\mathbf{X}^* = -\mathbf{A}^{-1}\mathbf{E}_1 + (\mathbf{1} - \Phi(T))^{-1}(\mathbf{1} - \Phi(T - \delta T))\mathbf{A}^{-1}(\mathbf{E}_1 - \mathbf{E}_2) \tag{3.34}$$

After substitution of (3.34) into (3.22), δ_1 can be determined from the corresponding transcendental equation:

$$h(\delta) = \mathbf{E}_3[-\mathbf{A}^{-1}\mathbf{E}_1 + \Phi(\delta T)(\mathbf{1} - \Phi(T))^{-1}(\mathbf{1} - \Phi(T - \delta T))\mathbf{A}^{-1}(\mathbf{E}_1 - \mathbf{E}_2) - \mathbf{E}_4] = 0 \tag{3.35}$$

Hence, provided that $\delta_1 \in (0, 1)$, $\mathbf{X}^*$ can be obtained from (3.34).

Secondly, the period-p orbit $\{\mathbf{X}_1^*, \cdots, \mathbf{X}_p^*\}$ is analyzed. By defining $\delta_1, \cdots, \delta_p$ as the p duty cycles within p periods with $p > 1$, it indicates that v_c crosses p voltage ramps at $(nT + \delta_1 T), \cdots, ((n+p-1)T + \delta_p T)$. Based on the specific Poincaré map derived in (3.33), the p-fold iterative mapping can be formulated as:

$$\begin{aligned}\mathbf{X}_{n+p} &= P^{(p)}(\mathbf{X}_n)\\ &= -\mathbf{A}^{-1}\mathbf{E}_1 + \Phi(pT)(\mathbf{X}_n + \mathbf{A}^{-1}\mathbf{E}_1)\\ &\quad + \sum_{j=1}^{p}[\Phi((p-j)T)(\mathbf{1} - \Phi(T - \delta_j T))]\mathbf{A}^{-1}(\mathbf{E}_1 - \mathbf{E}_2)\end{aligned} \tag{3.36}$$

By using (3.32) and (3.33), the period-p orbit can be obtained as:

$$\mathbf{X}_1^* = -\mathbf{A}^{-1}\mathbf{E}_1 + (\mathbf{1} - \Phi(pT))^{-1}\sum_{j=1}^{p}[\Phi((p-j)T)\,(\mathbf{1} - \Phi(T - \delta_j T))]\mathbf{A}^{-1}(\mathbf{E}_1 - \mathbf{E}_2) \tag{3.37}$$

$$\begin{aligned}\mathbf{X}_i^* &= -\mathbf{A}^{-1}\mathbf{E}_1 + \Phi((i-1)T)(\mathbf{X}_1^* + \mathbf{A}^{-1}\mathbf{E}_1)\\ &\quad + \sum_{j=1}^{i-1}[\Phi((i-j-1)T)(\mathbf{1} - \Phi(T - \delta_j T))]\mathbf{A}^{-1}(\mathbf{E}_1 - \mathbf{E}_2) \quad (i = 2, \cdots, p)\end{aligned} \tag{3.38}$$

$$h_i(\mathbf{d}) = \mathbf{E}_3[-\mathbf{A}^{-1}\mathbf{E}_1 + \Phi(\delta_i T)(\mathbf{X}_i^* + \mathbf{A}^{-1}\mathbf{E}_1) - \mathbf{E}_4] = 0 \quad (i = 1, \cdots, p) \tag{3.39}$$

where $\Phi(0) = \mathbf{1}$ and $\mathbf{d} = (\delta_1, \cdots, \delta_p)$. By substituting (3.37) and (3.38) into (3.39), $\mathbf{d}$ can be determined. Hence, provided that $\delta_i \in (0, 1)$ for $i = 1, \cdots, p$, $\mathbf{X}_1^*, \cdots, \mathbf{X}_p^*$ can then be obtained from (3.37) and (3.38).

Due to the cyclic property of the period-p orbit, $\{\mathbf{X}_2^*, \cdots, \mathbf{X}_p^*, \mathbf{X}_1^*\}, \cdots, \{\mathbf{X}_p^*, \mathbf{X}_1^*, \cdots, \mathbf{X}_{p-1}^*\}$ are other period-p orbits which correspond to the same subharmonic frequency. If $\mathbf{X}_1^* = \cdots = \mathbf{X}_p^*$, the period-$p$ orbit becomes a period-1 orbit, indicating that the period-1 is a particular case of the period-p orbit.

3.1.2.4 Stability and Characteristic Multipliers

Period-1 and period-p orbits may be stable or unstable. It is known that the stability type of a fixed point of mapping corresponds to the stability type of the underlying periodic solution, and the fixed point of mapping is stable if and only if its characteristic multipliers all lie within the unit cycle in the complex plane.

For a period-1 orbit that is a fixed point of the specific Poincaré map given by (3.33), its characteristic multipliers are eigenvalues of the Jacobian matrix of that mapping, which is given by:

$$\begin{aligned} DP(\mathbf{X}^*) &= \Phi(T) - \frac{\partial \Phi(T-\delta T)}{\partial \delta}\mathbf{A}^{-1}(\mathbf{E}_1 - \mathbf{E}_2)\frac{\partial \delta}{\partial \mathbf{X}^*} \\ &= \Phi(T) + \Phi(T-\delta T)(\mathbf{E}_1 - \mathbf{E}_2)T\frac{\partial \delta}{\partial \mathbf{X}^*} \end{aligned} \tag{3.40}$$

where $\frac{\partial \delta}{\partial \mathbf{X}^*} = \left(\frac{\partial \delta}{\partial \omega}, \frac{\partial \delta}{\partial i}\right)$. According to the implicit-function theorems, $\frac{\partial \delta}{\partial \mathbf{X}^*}$ can be deduced from (3.22) as:

$$\begin{aligned} \frac{\partial \delta}{\partial \mathbf{X}^*} &= -\left(\frac{\partial h}{\partial \delta}\right)^{-1}\frac{\partial h}{\partial \mathbf{X}^*} \\ &= -[\mathbf{E}_3\Phi(\delta T)(\mathbf{A}\mathbf{X}^* + \mathbf{E}_1)T]^{-1}\mathbf{E}_3\Phi(\delta T) \end{aligned} \tag{3.41}$$

For a period-p orbit $\{\mathbf{X}_1^*, \cdots, \mathbf{X}_p^*\}$, $\mathbf{X}_1^*$ is a fixed point of the p-fold iterative specific Poincaré map. Therefore, its characteristic multipliers are the eigenvalues of the Jacobian matrix of that mapping, which is given by:

$$DP^{(p)}(\mathbf{X}_1^*) = \Phi(pT) + \sum_{j=1}^{p}\Phi((p-j)T)\Phi(T-\delta_j T)(\mathbf{E}_1 - \mathbf{E}_2)T\frac{\partial \delta_j}{\partial \mathbf{X}_1^*} \tag{3.42}$$

By substituting (3.38) into (3.39)

$$\mathbf{H}(\mathbf{d}, \mathbf{X}_1^*) = [h_1(\mathbf{d}, \mathbf{X}_1^*), \cdots, h_p(\mathbf{d}, \mathbf{X}_1^*)]^T = \mathbf{0} \tag{3.43}$$

Hence, according to the implicit-function theorems, the partial derivative in (3.42) can be expressed as:

$$\frac{\partial \mathbf{d}}{\partial \mathbf{X}_1^*} = \left(\frac{\partial \delta_1}{\partial \mathbf{X}_1^*}, \cdots, \frac{\partial \delta_p}{\partial \mathbf{X}_1^*}\right)^{\mathrm{T}} = -\left[\frac{\partial \mathbf{H}}{\partial \mathbf{d}}\right]^{-1}\left[\frac{\partial \mathbf{H}}{\partial \mathbf{X}_1^*}\right] \tag{3.44}$$

where

$$\begin{aligned} \frac{\partial \mathbf{H}}{\partial \mathbf{d}} = \left[\frac{\partial h_i}{\partial \delta_j}\right]_{p\times p} &= diag[\mathbf{E}_3\Phi(\delta_i T)(\mathbf{A}\mathbf{X}_i^* + \mathbf{E}_1)T] \\ &\quad + [\mathbf{E}_3\Phi(\delta_i T)\Phi((i-j)T - \delta_j T)(\mathbf{E}_1 - \mathbf{E}_2)T]_{p\times p} \end{aligned} \tag{3.45}$$

$$\begin{aligned} \frac{\partial \mathbf{H}}{\partial \mathbf{X}_1^*} &= \left[\frac{\partial h_1}{\partial \mathbf{X}_1^*}, \cdots, \frac{\partial h_p}{\partial \mathbf{X}_1^*}\right]^{\mathrm{T}} \\ &= [\mathbf{E}_3\Phi(\delta_1 T), \cdots, \mathbf{E}_3\Phi((p-1)T + \delta_p T)]^{\mathrm{T}} \end{aligned} \tag{3.46}$$

Notice that $\Phi((i-j)T - \delta_j T)$ in (3.45) becomes a zero matrix when $j \geq i$.

3.1.3 Simulation

To illustrate the derived Poincaré map, computer simulations are carried out. The simulation parameters are based on the values of a practical DC chopper-fed PMDC drive system, namely $T = 10$ ms, $g_i = 1.1$ V/A, $g_\omega = 0.54$ V/rad/s, $V_{in} = 50$ V, $R = 2.9\ \Omega$, $L = 53.7$ mH, $K_E = 0.1356$ V/rad/s, $K_T = 0.1324$ Nm/A,

$B = 0.000275$ Nm/rad/s, $J = 0.000557$ Nm/rad/s^2, $T_l = 0.39$ Nm, and $\omega_{ref} = 105$ rad/s. The resulting eigenvalues of matrix **A** are $\lambda_1 = -17.1$ and $\lambda_2 = -37.4$. This indicates that although this open-loop system shows no oscillating dynamics, the corresponding closed-loop system may exhibit not only oscillating dynamics but also subharmonics and even chaos.

3.1.3.1 Bifurcation Diagrams using Numerical Computation

A natural numerical tool to obtain the steady-state solution of the generalized Poincaré map is known as the brute-force method (Parker and Chua, 1987) – repeating the iteration of the map until the transient has died out or the steady state has been reached. This method has the advantage of simplicity, but may suffer from tedious simulation due to long-lived transients. If K iterations are spent for transient operation, and N points are needed to describe an orbit while M steps are used to depict a bifurcation diagram, then the total number of iterations for constructing a bifurcation diagram will be $(K+N)M$.

By employing a brute-force algorithm to compute the generalized Poincaré map given by (3.28)–(3.30), bifurcation diagrams of ω and i versus V_{in} and g can be produced as shown in Figures 3.3–3.6, respectively. As can be seen in these figures, the system exhibits a typical period-doubling route to chaos, which is valid for both chaotic speed and current.

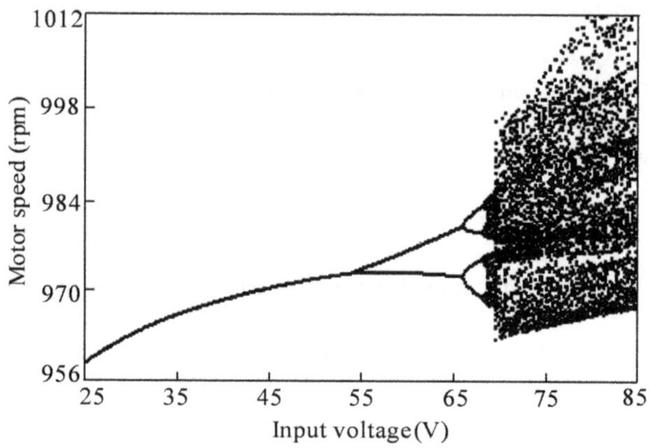

Figure 3.3 Bifurcation diagram of motor speed versus input voltage

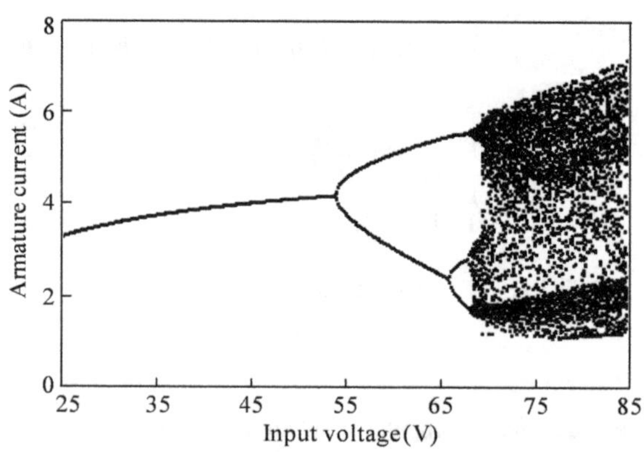

Figure 3.4 Bifurcation diagram of armature current versus input voltage

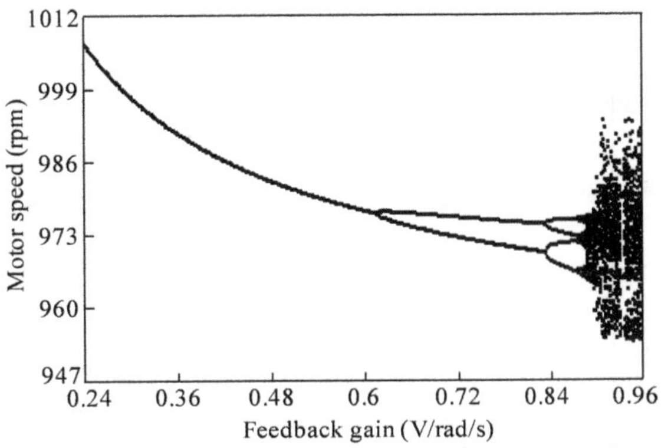

Figure 3.5 Bifurcation diagram of motor speed versus feedback gain

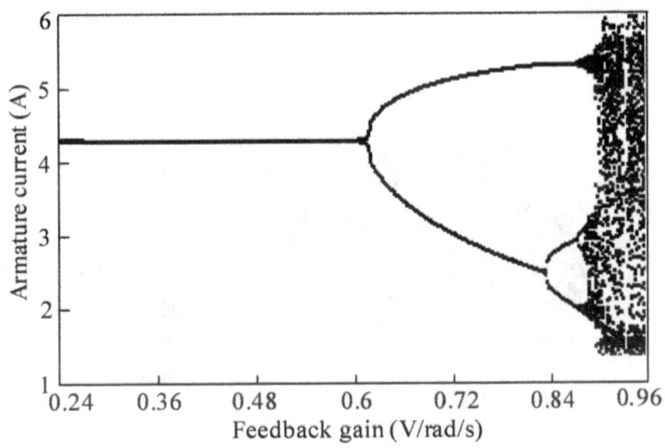

Figure 3.6 Bifurcation diagram of armature current versus feedback gain

The trajectory of i versus ω for a period-1 orbit with $V_{in} = 60$ V and $g = 0.25$ V/rad/s is shown in Figure 3.7. Moreover, the chaotic trajectory and its Poincaré section of i versus ω with $V_{in} = 60$ V and $g = 0.9$ V/rad/s are shown in Figures 3.8 and 3.9, respectively. It should be noted that the chaotic trajectory represents the phase portrait of the system solution in a continuous-time domain during chaos, whereas its Poincaré section is a set of sampled points describing the chaotic solution of the Poincaré map in the phase plane. In a Poincaré section, a period-1 orbit is denoted by a single point, a period-p orbit by p points, and a chaotic orbit by an intricate pattern of points.

Moreover, if g is large enough, a drive system with a higher switching frequency and a lower armature inductance can still exhibit chaos. For example, if the switching frequency, armature inductance, and armature resistance are changed to 2.5 kHz, 3 mH, and 1 Ω respectively, the corresponding bifurcation diagram of ω versus g shown in Figure 3.10 indicates that the drive system will operate in chaos when g reaches 24.3 V/rad/s. Its chaotic trajectory with $g = 25$ V/rad/s is shown in Figure 3.11. However, because high switching frequency operation will complicate the system behavior due to parasitic reactances, and the associated high gain value will amplify the noise caused by device switching and motor commutation,

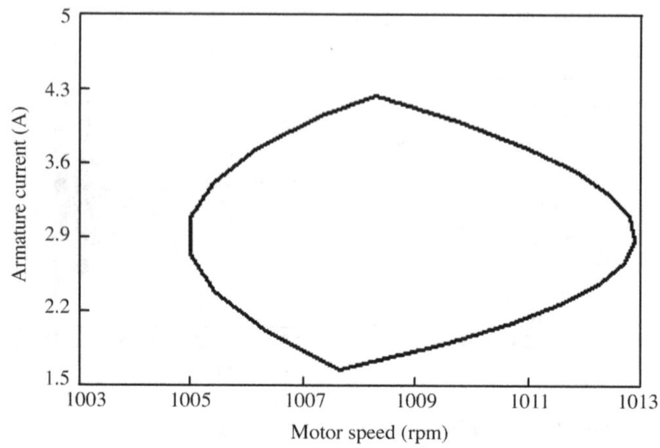

Figure 3.7 Period-1 trajectory of armature current and motor speed

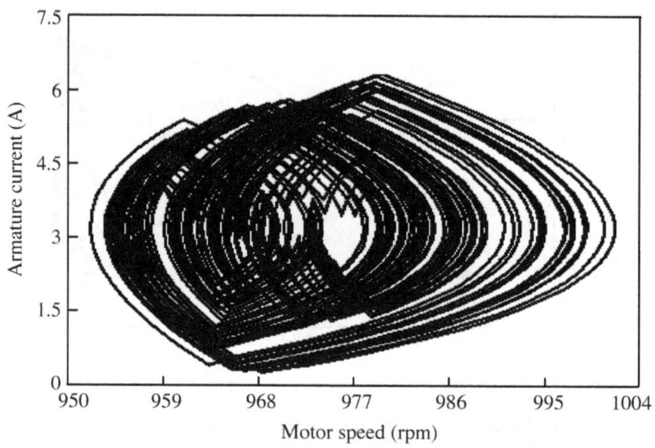

Figure 3.8 Chaotic trajectory of armature current and motor speed

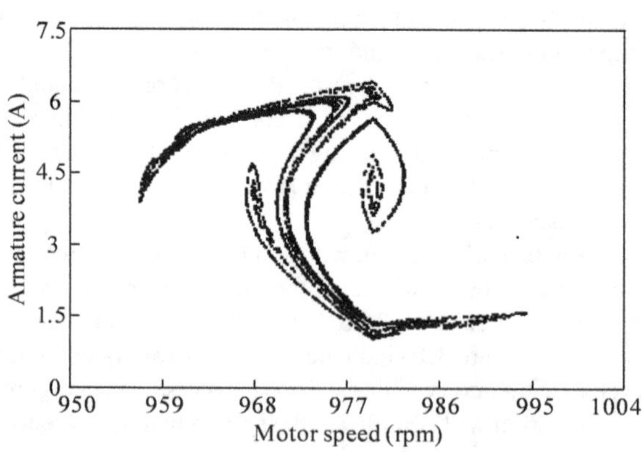

Figure 3.9 Chaotic Poincaré section of armature current and motor speed

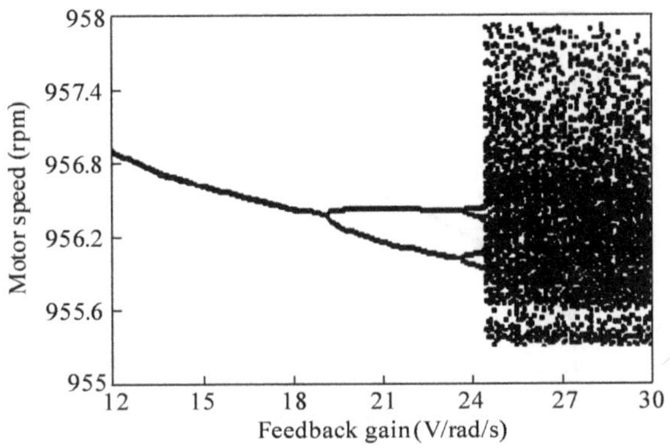

Figure 3.10 Bifurcation diagram of motor speed versus feedback gain at high switching frequency

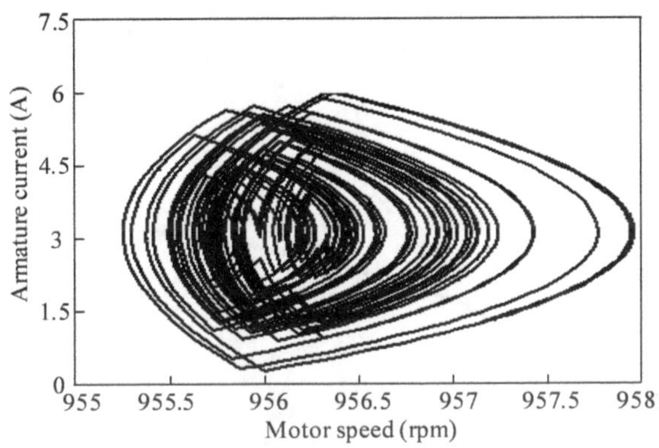

Figure 3.11 Chaotic trajectory of armature current and motor speed at high switching frequency

a lower switching frequency operation is therefore adopted to facilitate the illustration of theoretical and experimental results.

3.1.3.2 Bifurcation Diagrams using Analytical Approach

Based on the derived relationship in (3.34) and (3.35), the evolution of δ_1 and the corresponding ω of a period-1 orbit with respect to V_{in} is shown in Figures 3.12 and 3.13, respectively. Hence, by using (3.40) and (3.41), the corresponding characteristic multipliers are shown in Figure 3.14, in which one of the magnitudes is greater than unity when $V_{in} > 54.6$ V. This indicates that the period-1 orbit is unstable when $V_{in} > 54.6$ V.

Similarly, based on (3.36)–(3.39) with $p = 4$, the evolution of $\mathbf{d} = (\delta_1, \delta_2, \delta_3, \delta_4)$ and the corresponding ω of the period-4 orbit with respect to V_{in}, are shown in Figures 3.15 and 3.16, respectively. It can be found that a period-2 orbit occurs when $V_{in} > 54.6$ V and then bifurcates to the period-4 orbit when $V_{in} > 66.5$ V. Since one of the duty cycles is greater than unity after $V_{in} > 69.5$ V, the period-4 orbit only exists between 66.5 V and 69.5 V. According to the characteristic multipliers of the period-4 orbit shown in Figure 3.17, it can be found that a period-2 orbit with

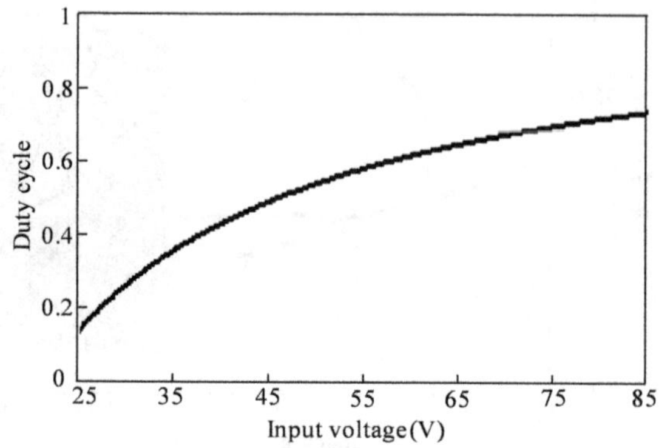

Figure 3.12 Period-1 duty cycle versus input voltage

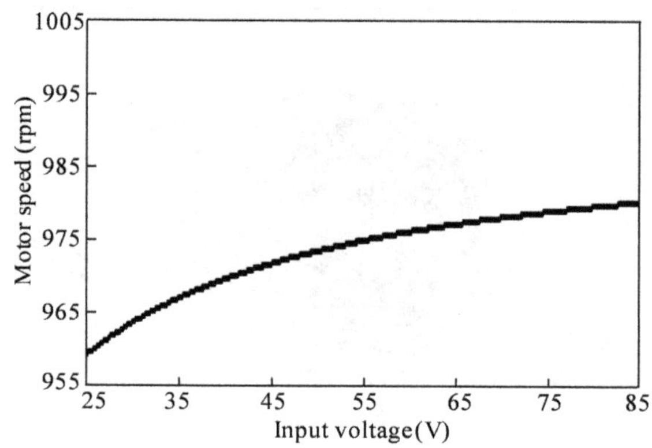

Figure 3.13 Period-1 motor speed versus input voltage

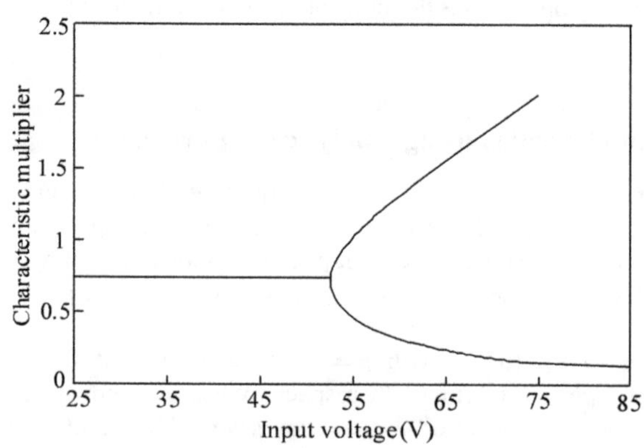

Figure 3.14 Period-1 characteristic multipliers versus input voltage

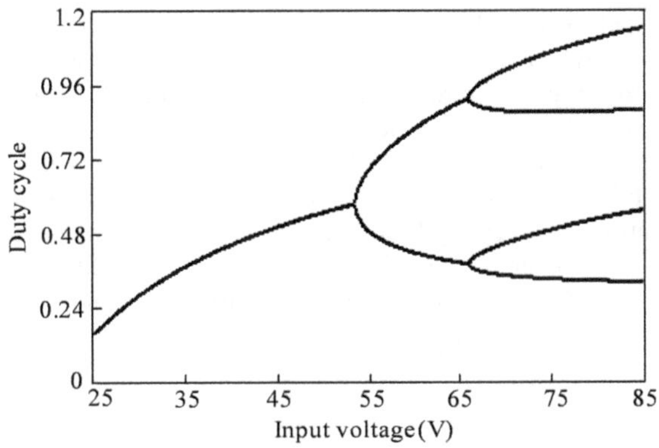

Figure 3.15 Period-4 duty cycles versus input voltage

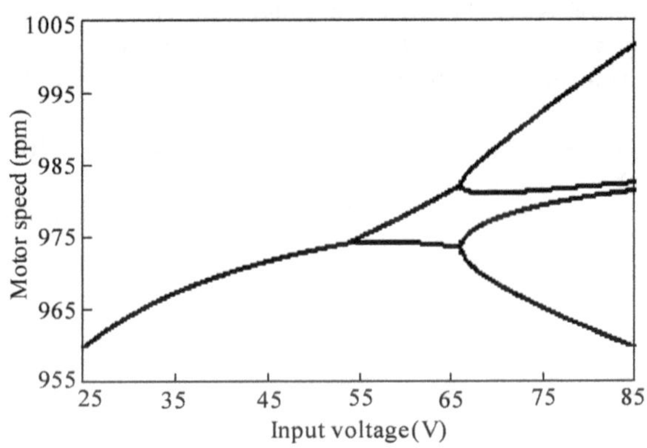

Figure 3.16 Period-4 motor speeds versus input voltage

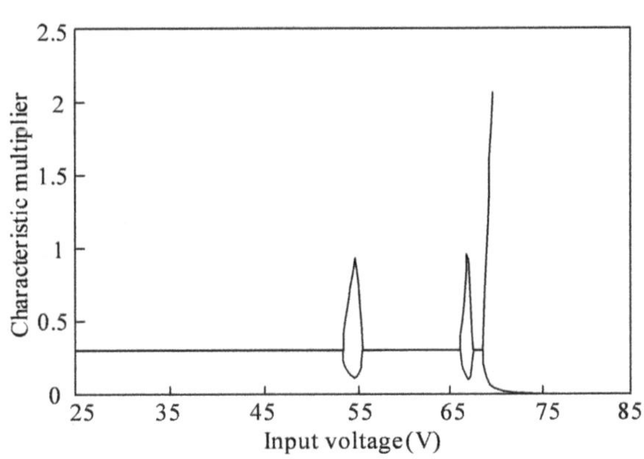

Figure 3.17 Period-4 characteristic multipliers versus input voltage

$54.6\,\text{V} < V_{in} \leq 69.5\,\text{V}$ is always stable, whereas a period-4 orbit with $66.5\,\text{V} < V_{in} \leq 69.5\,\text{V}$ is stable only when $V_{in} \leq 68.4\,\text{V}$. When $V_{in} \leq 54.6\,\text{V}$, the corresponding characteristics in Figures 3.15 and 3.16 are identical to those in Figures 3.12 and 3.13, respectively. This indicates that the stable period-1 orbit is a subset of the period-p orbit ($p > 1$).

It should be noted that the above analytical results closely agree with the bifurcation diagram (Figure 3.3) resulting from numerical computation. The required computational time based on the derived analytical solution is extremely less than that required for computation using a numerical brute-force algorithm. Moreover, the analytical solution can facilitate the identification of the desired stable operating ranges for different system parameters and conditions.

3.1.4 Experimentation

In order to further verify the theoretical analysis, an experimental DC drive system is prototyped as shown in Figure 3.18. The prototype consists of two identical PMDC motors, namely M_1 and M_2, which are directly coupled together. M_1 acts as the motor that is fed by a voltage-mode DC chopper, while M_2 performs as the generator load that is controlled by an electronic load operating as a constant current sink I_0. The built-in tachogenerator in M_1 is used to provide the speed feedback. The mechanical load torque can be represented by $T_l = K_T I_0 + T_s$, where T_s is the friction torque.

The power stage consists of a power MOSFET IRF740 driven by a driver chip DS0026, and a fast-recovery diode BYW29E-200. The electronic controller mainly consists of two operational amplifiers LM833 and a comparator LM311. Between them, an optical coupler 6N137 is adopted

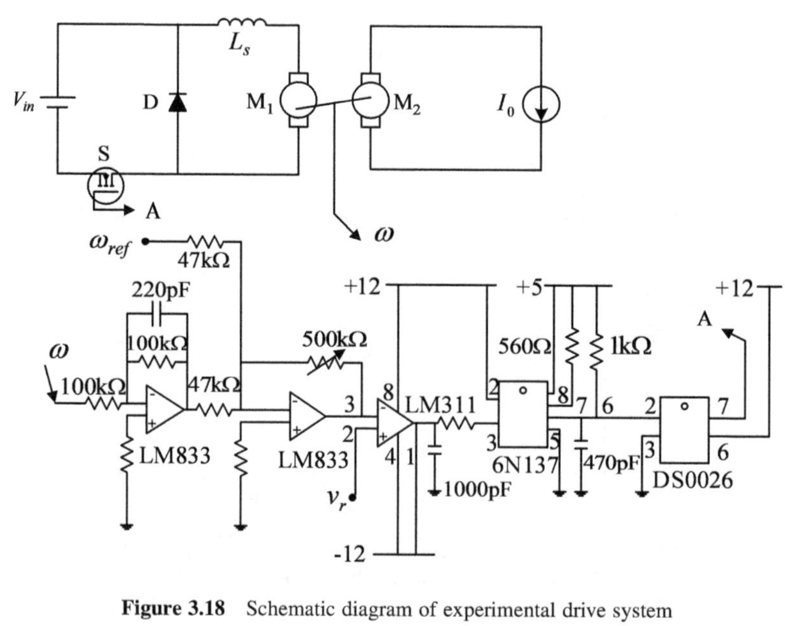

Figure 3.18 Schematic diagram of experimental drive system

in order to avoid the possible coupling of switching noise. Moreover, a switching frequency as low as 100 Hz is selected to ensure that all parasitic reactances can be neglected. Since the motor has low armature inductance, an additional inductor L_s is connected in series with the armature of M_1 to ensure that the system operates in a continuous conduction mode. All parameters of this drive system are as follows: $v_l = 0$ V, $v_u = 2.2$ V, $T = 100$ ms, $g = 0.7$ V/rad/s, $V_{in} = 60$ V, $R = 2.8\ \Omega$, $L = 537$ mH, $K_E = 0.1356$ V/rad/s, $K_T = 0.1324$ Nm/A, $B = 0.000275$ Nm/rad/s, $J = 0.000557$ Nm/rad/s^2, $T_l = 0.38$ Nm, and $\omega_{ref} = 100$ rad/s.

For the sake of simplicity and clarity, the speed feedback control signal v_c is measured instead of the actual motor speed. In fact, they have the same shape and obey a linear relationship, as given by (3.1). The measured trajectory and waveforms of v_c and i with $V_{in} = 60$ V and $g = 0.25$ V/rad/s are shown in Figure 3.19. This illustrates that the system operates in a period-1 orbit. Also, it can be found that i roughly lies between 2 A and 4.4 A while v_c is between 1.3 V and 1.6 V (equivalent to ω between 1005 rpm and 1016 rpm). Since the sampling is always at the transition from Stage 1 to 2, the sampled magnitude of i is the maximum value of 4.4 A and the corresponding v_c is roughly of 1.45 V (equivalent to ω of 1010 rpm). By comparing these experimental results with the bifurcation diagrams and trajectory shown in Figures 3.5–3.7, the theoretical analysis is verified. The superimposed high-frequency components in v_c should be due to the torsional oscillation of the shaft coupling.

Moreover, by selecting $g = 0.9$, the measured trajectory and waveforms of i and v_c shown in Figure 3.20 illustrate that the drive system operates in chaos. It can be found that the boundaries of i and ω, namely between 0.7 A and 6.5 A and between 960 rpm and 993 rpm, respectively, agree with those in the theoretical chaotic trajectory shown in Figure 3.8.

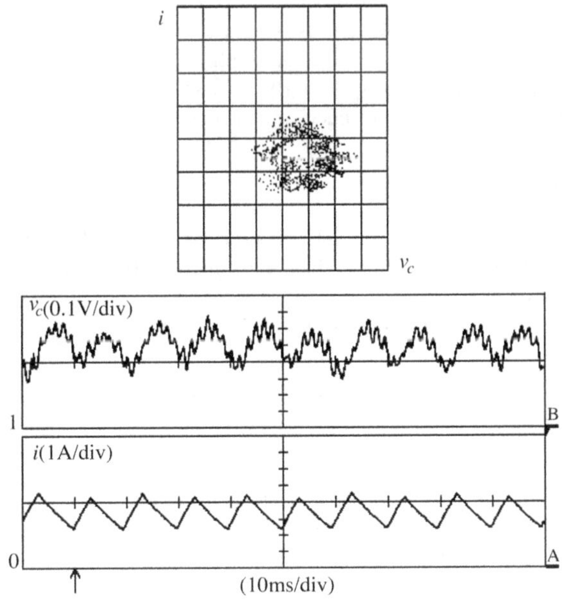

Figure 3.19 Measured period-1 trajectory and waveforms of armature current and motor speed

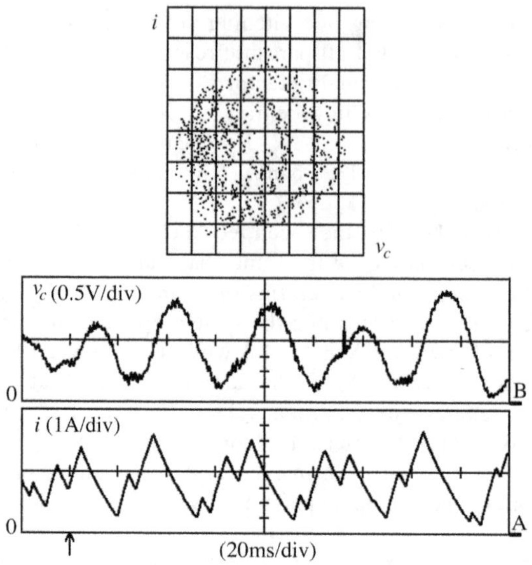

Figure 3.20 Measured chaotic trajectory and waveforms of armature current and motor speed

3.2 Current-Controlled DC Drive System

Similar to a current-controlled switching DC–DC converter (Deane, 1992; Tse, Fung, and Kwan, 1996), a current-controlled DC drive system is more prone to chaos and instability. Unlike the switching DC–DC converter, the DC drive system preferably adopts current control since it can offer direct torque control. Thus, it is highly appropriate for the investigation, both numerically and analytically, of the chaotic behavior of a current-controlled DC drive system. By this method the stable and chaotic ranges of system parameters can be obtained, and the occurrence of chaos can be avoided.

3.2.1 Modeling

Similar to the voltage control scheme, the DC buck-chopper-fed PMDC drive system is adopted to exemplify the current control scheme (Chen, Chau, and Chan, 2000b). The corresponding schematic and equivalent circuit is shown in Figure 3.21.

Considering that the operational amplifier A_1 and A_2 have gains g_ω and g_i, the speed and current control signals v_ω and v_i can be expressed as:

$$v_\omega(t) = g_\omega(\omega_{ref} - \omega(t)) \tag{3.47}$$

$$v_i(t) = g_i i(t) \tag{3.48}$$

where $i(t)$, $\omega(t)$, and ω_{ref} are armature current, rotor speed, and reference speed of the DC motor, respectively. Then, both v_ω and v_i are fed into the comparator A_3 which outputs the pulse to the reset of an R-S latch. The power switch S is controlled by this R-S latch which is set by clock pulses of period T. Once the latch is set by the clock pulse, S is turned on and diode D is off. S then remains closed until v_i exceeds v_ω, where the latch begins to reset. When the latch is reset, S is turned off and D is on. S then remains open until the arrival of the next clock pulse, where it closes again. If both set and reset signals occur simultaneously, the reset will dominate the set and S will remain open until the occurrence of another clock pulse.

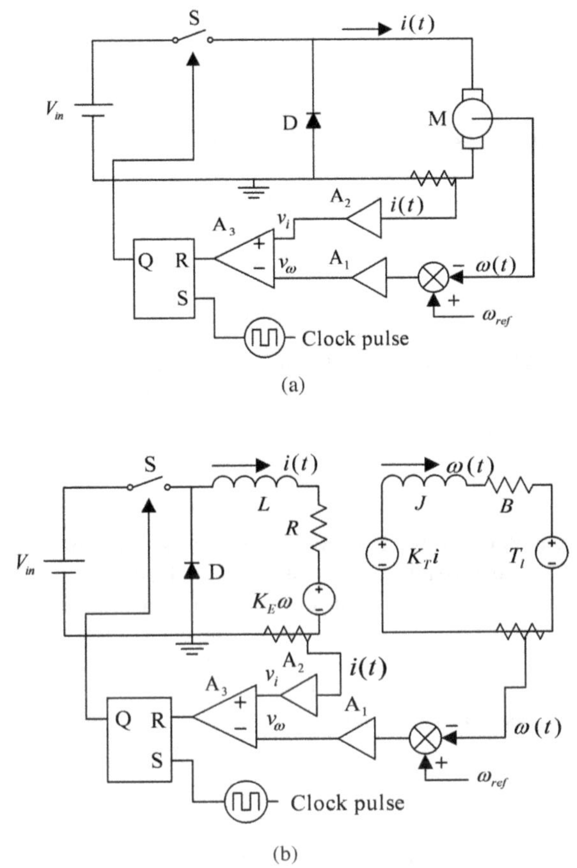

Figure 3.21 Current-controlled DC drive system. (a) Schematic. (b) Equivalent circuit

Therefore, the system equation can be divided into two stages, as given by:

Stage 1 (setting the latch):

$$\frac{\mathrm{d}}{\mathrm{d}t}\begin{pmatrix}\omega(t)\\ i(t)\end{pmatrix}=\begin{pmatrix}-B/J & K_T/J\\ -K_E/L & -R/L\end{pmatrix}\begin{pmatrix}\omega(t)\\ i(t)\end{pmatrix}+\begin{pmatrix}-T_l/J\\ V_{in}/L\end{pmatrix} \tag{3.49}$$

Stage 2 (resetting the latch at $v_i = v_\omega$)

$$\frac{\mathrm{d}}{\mathrm{d}t}\begin{pmatrix}\omega(t)\\ i(t)\end{pmatrix}=\begin{pmatrix}-B/J & K_T/J\\ -K_E/L & -R/L\end{pmatrix}\begin{pmatrix}\omega(t)\\ i(t)\end{pmatrix}+\begin{pmatrix}-T_l/J\\ 0\end{pmatrix} \tag{3.50}$$

where R is the armature resistance, L is the armature inductance, V_{in} is the DC supply voltage, K_E is the back-EMF constant, K_T is the torque constant, B is the viscous damping, J is the load inertia, and T_l is the load torque.

By defining the state vector $\mathbf{X}(t)$ and the matrices $\mathbf{A}$, $\mathbf{E}_1$, $\mathbf{E}_2$, $\mathbf{E}_3$, $\mathbf{E}_4$ as:

$$\mathbf{X}(t)=\begin{pmatrix}\omega(t)\\ i(t)\end{pmatrix},\ \mathbf{A}=\begin{pmatrix}-B/J & K_T/J\\ -K_E/L & -R/L\end{pmatrix} \tag{3.51}$$

$$\mathbf{E}_1=\begin{pmatrix} -T_l/J \\ V_{in}/L \end{pmatrix},\ \mathbf{E}_2=\begin{pmatrix} -T_l/J \\ 0 \end{pmatrix},\ \mathbf{E}_3=(\, g_\omega \quad g_i\,),\ \mathbf{E}_4=\begin{pmatrix} \omega_{ref} \\ 0 \end{pmatrix} \tag{3.52}$$

the system equation given by (3.49) and (3.50) can be rewritten as:

$$\dot{\mathbf{X}}(t)=\mathbf{A}\,\mathbf{X}(t)+\mathbf{E}_k \quad (k=1,2) \tag{3.53}$$

By using (3.47) and (3.48), the switching condition $v_i(t)-v_\omega(t)=0$ can be expressed as:

$$\mathbf{E}_3(\mathbf{X}(t)-\mathbf{E}_4)=0 \tag{3.54}$$

It should be noted that (3.53) is a time-varying linear state equation switching between two stages. As its switching condition, given by (3.54), depends on the external speed reference and the internal state vector, the whole system described by both (3.53) and (3.54) exhibits nonlinear dynamics.

3.2.2 *Analysis*

The analysis of a system's chaotic behavior begins with the solution of (3.53) in a continuous-time domain, namely $\mathbf{X}(t)$. Then, the solution of the system is described by a sequence of samples $\{\mathbf{X}_n\}$ $n=0,\ 1,\ 2,\ \cdots$. Differing from the conventional discretization of a continuous-time state equation, the successive sample $\mathbf{X}_{n+1}$ may not be taken at $(n+1)T$. The sample $\mathbf{X}_{n+1}$ occurs at $(n+m)T$ when there is a change of switching state after m $(m\geq 1)$ clock pulses, because there is no intersection between v_ω and v_i within $(m-1)$ clock pulses (Hamill, Deane, and Jefferies, 1992). The corresponding mapping from $\mathbf{X}_n$ to its successive sample $\mathbf{X}_{n+1}$ is known as a Poincaré map. Hence, a Poincaré map that maps a sample $\mathbf{X}_n$ of $\mathbf{X}(t)$ at $t=nT$ to its successive one $\mathbf{X}_{n+1}$ at $t=(n+m)T$ is defined as P: $\Re^2\rightarrow\Re^2$:

$$\mathbf{X}_{n+1}=P(\mathbf{X}_n) \tag{3.55}$$

It should be noted that the Poincaré map given by (3.55) is a generalized case with $m\geq 1$, which can fully describe the system behavior. When $m>1$, the solution of (3.55) can only be solved by using numerical simulation. On the other hand, focusing on a specific case with $m=1$, the solution can be analytically solved to allow the system behavior to be described in terms of periodic orbits and stability.

3.2.2.1 Solution of System Equation

Given an initial value $\mathbf{X}(t_0)$, the continuous-time solution of the system equation given by (3.53) can be expressed as:

$$\mathbf{X}(t)=\Phi(t-t_0)\mathbf{X}(t_0)+\int_{t_0}^{t}\Phi(t-\tau)\mathbf{E}_k d\tau \quad (k=1,2) \tag{3.56}$$

For a practical DC drive system, the matrix $\mathbf{A}$ given by (3.51) always yields a positive $\det\mathbf{A}=(BR+K_EK_T)/(LJ)$ and hence is always invertible. Moreover, even ignoring the positive BR term, $(K_EK_T)/(LJ)$ can be written as $1/(\tau_m\tau_a)$ where $\tau_m=(JR)/(K_EK_T)$ is the mechanical time constant and $\tau_a=L/R$ is the electrical time constant. As τ_m is from tens of milliseconds to several seconds and τ_a is from tens of microseconds to tens of milliseconds, $1/(\tau_m\tau_a)$ and hence $\det\mathbf{A}$ are seldom close to zero. Thus, the continuous-time solution given by (3.56) can be rewritten as:

$$\mathbf{X}(t)=-\mathbf{A}^{-1}\mathbf{E}_k+\Phi(t-t_0)(\mathbf{X}(t_0)+\mathbf{A}^{-1}\mathbf{E}_k) \quad (k=1,2) \tag{3.57}$$

where $\Phi(t) = e^{\mathbf{A}t}$ is the state transition matrix. By defining the parameters α and Δ as:

$$\alpha = \frac{1}{2}\left(\frac{R}{L}+\frac{B}{J}\right),\ \Delta = \frac{1}{4}\left(\frac{R}{L}-\frac{B}{J}\right)^2 - \frac{K_E K_T}{LJ} \tag{3.58}$$

the eigenvalues λ_1, λ_2 of $\mathbf{A}$ can be expressed as:

$$\Delta = 0,\ \lambda_1 = \lambda_2 = \lambda = -\alpha \tag{3.59}$$

$$\Delta > 0,\ \lambda_1 = -\alpha + \sqrt{\Delta},\ \lambda_2 = -\alpha - \sqrt{\Delta} \tag{3.60}$$

$$\Delta < 0,\ \lambda_1 = -\alpha + j\sqrt{-\Delta},\ \lambda_2 = -\alpha - j\sqrt{-\Delta} \tag{3.61}$$

Hence, the corresponding $\Phi(t)$ can be obtained as:

$$\Delta = 0,\ \Phi(t) = \exp(-\alpha t)[\mathbf{1} - t(2\alpha\mathbf{1} - \mathbf{A})] \tag{3.62}$$

$$\Delta > 0,\ \Phi(t) = \frac{1}{\lambda_2 - \lambda_1}[\exp(\lambda_1 t)(\lambda_2\mathbf{1} - \mathbf{A}) - \exp(\lambda_2 t)(\lambda_1\mathbf{1} - \mathbf{A})] \tag{3.63}$$

$$\Delta < 0,\ \Phi(t) = \exp(-\alpha t)\left[\mathbf{1}\cos(\beta t) + \frac{1}{\beta}(\alpha\,\mathbf{1} + \mathbf{A})\sin(\beta t)\right] \tag{3.64}$$

where $\mathbf{1}$ is the identity matrix and $\beta = \sqrt{-\Delta}$.

3.2.2.2 Derivation of Generalized Poincaré Map

Since $\mathbf{X}_n$ is always sampled at the beginning of the clock pulse that changes S from off to on, the drive system always operates in Stage 1 first and then in Stage 2 for each sampling interval. By defining the intervals of Stage 1 and Stage 2 as δT and $\delta' T$, respectively, the interval of the Poincaré map becomes $mT = (\delta + \delta')T$. Thus, $\mathbf{X}(\delta T)$ and $\mathbf{X}_{n+1}$ can be directly deduced from (3.57) as given by:

$$\mathbf{X}(\delta T) = -\mathbf{A}^{-1}\mathbf{E}_1 + \Phi(\delta T)(\mathbf{X}_n + \mathbf{A}^{-1}\mathbf{E}_1) \tag{3.65}$$

$$\mathbf{X}_{n+1} = -\mathbf{A}^{-1}\mathbf{E}_2 + \Phi(mT - \delta T)(\mathbf{X}(\delta T) + \mathbf{A}^{-1}\mathbf{E}_2) \tag{3.66}$$

By substituting (3.65) into (3.54), δ can be determined based on the solution of the following transcendental equation:

$$h(\delta) = \mathbf{E}_3(-\mathbf{A}^{-1}\mathbf{E}_1 + \Phi(\delta T)(\mathbf{X}_n + \mathbf{A}^{-1}\mathbf{E}_1) - \mathbf{E}_4) = 0 \tag{3.67}$$

On the other hand, m can be deduced as the minimum integer that is larger than δ while fulfilling $h(\delta) < 0$. Hence, the Poincaré map can be written as:

$$\mathbf{X}_{n+1} = -\mathbf{A}^{-1}\mathbf{E}_1 + \Phi(mT)(\mathbf{X}_n + \mathbf{A}^{-1}\mathbf{E}_1) + (\mathbf{1} - \Phi(mT - \delta T))\mathbf{A}^{-1}(\mathbf{E}_1 - \mathbf{E}_2) \tag{3.68}$$

It should be note that the derived Poincaré map is a generalized mapping which covers all possible solutions such as real and complex roots due to different system parameters and operating conditions. Thus, this generalized Poincaré map can be considered as the mapping for any second-order dynamical systems using similar mathematical models, such as other DC drive systems. Moreover, the derivation can readily be extended to those higher-order dynamical systems involving power switches.

3.2.2.3 Analysis of Periodic Orbits

Based on the above generalized Poincaré map, the dynamic bifurcation of the drive system can readily be investigated by employing a brute-force method. Thus, different bifurcation diagrams with respect to different system parameters can be obtained. However, as each iterative computation of the generalized Poincaré map needs to solve the transcendental equation, the corresponding numerical simulation is usually very tedious. In order to avoid the lengthy computation and to attain an insight into the periodic solution, the analysis can be focused on the case where the interval of mapping is the same as the clock cycle, mathematically $m = 1$. The corresponding mapping is known as the specific Poincaré map.

The steady-state periodic solution of the drive can be a period-1 orbit $\mathbf{X}^*$, or a period-p orbit $\{\mathbf{X}_1^*, \cdots, \mathbf{X}_p^*\}$ with $p > 1$. The corresponding specific Poincaré maps are described as:

$$\mathbf{X}^* = P(\mathbf{X}^*) \tag{3.69}$$

$$\mathbf{X}_{k+1}^* = P(\mathbf{X}_k^*) \quad (k = 1, \cdots, p-1),\ \mathbf{X}_1^* = P(\mathbf{X}_p^*) \tag{3.70}$$

Firstly, the period-1 orbit is analyzed. Since $m = 1$, the corresponding Poincaré map can be obtained from (3.68) as given by:

$$\mathbf{X}_{n+1} = -\mathbf{A}^{-1}\mathbf{E}_1 + \Phi(T)(\mathbf{X}_n + \mathbf{A}^{-1}\mathbf{E}_1) + (\mathbf{1} - \Phi(T - \delta T))\mathbf{A}^{-1}(\mathbf{E}_1 - \mathbf{E}_2) \tag{3.71}$$

Substituting (3.71) into the mapping given by (3.69) and taking $\mathbf{X}_{n+1} = \mathbf{X}_n$, the period-1 orbit can be deduced as:

$$\mathbf{X}^* = -\mathbf{A}^{-1}\mathbf{E}_1 + (\mathbf{1} - \Phi(T))^{-1}(\mathbf{1} - \Phi(T - \delta T))\mathbf{A}^{-1}(\mathbf{E}_1 - \mathbf{E}_2) \tag{3.72}$$

After substitution of (3.72) into (3.67), δ can be determined from the corresponding transcendental equation:

$$h(\delta) = \mathbf{E}_3[-\mathbf{A}^{-1}\mathbf{E}_1 + \Phi(\delta T)(\mathbf{1} - \Phi(T))^{-1}(\mathbf{1} - \Phi(T - \delta T))\mathbf{A}^{-1}(\mathbf{E}_1 - \mathbf{E}_2) - \mathbf{E}_4] = 0 \tag{3.73}$$

Hence, provided that $\delta \in (0,1)$, $\mathbf{X}^*$ can be obtained from (3.72).

Secondly, the period-p orbit $\{\mathbf{X}_1^*, \cdots, \mathbf{X}_p^*\}$ is analyzed. By defining $\delta_1, \cdots, \delta_p$ as the p duty cycles within p periods of clock pulses with $p > 1$, the p-fold iterative mapping can be formulated from (3.71) as:

$$\begin{aligned}\mathbf{X}_{n+p} &= P^{(p)}(\mathbf{X}_n)\\ &= -\mathbf{A}^{-1}\mathbf{E}_1 + \Phi(pT)(\mathbf{X}_n + \mathbf{A}^{-1}\mathbf{E}_1)\\ &\quad + \sum_{j=1}^{p}[\Phi((p-j)T)(\mathbf{1} - \Phi(T - \delta_j T))]\mathbf{A}^{-1}(\mathbf{E}_1 - \mathbf{E}_2)\end{aligned} \tag{3.74}$$

By using the definition in (3.70) and (3.71), the period-p orbit can be obtained as:

$$\mathbf{X}_1^* = -\mathbf{A}^{-1}\mathbf{E}_1 + (\mathbf{1} - \Phi(pT))^{-1}\sum_{j=1}^{p}[\Phi((p-j)T)\,(\mathbf{1} - \Phi(T - \delta_j T))]\mathbf{A}^{-1}(\mathbf{E}_1 - \mathbf{E}_2) \tag{3.75}$$

$$\begin{aligned}\mathbf{X}_i^* &= -\mathbf{A}^{-1}\mathbf{E}_1 + \Phi((i-1)T)(\mathbf{X}_1^* + \mathbf{A}^{-1}\mathbf{E}_1)\\ &\quad + \sum_{j=1}^{i-1}[\Phi((i-j-1)T)(\mathbf{1} - \Phi(T - \delta_j T))]\mathbf{A}^{-1}(\mathbf{E}_1 - \mathbf{E}_2) \quad (i = 2, \cdots, p)\end{aligned} \tag{3.76}$$

$$h_i(\mathbf{d}) = \mathbf{E}_3[-\mathbf{A}^{-1}\mathbf{E}_1 + \Phi(\delta_i T)(\mathbf{X}_i^* + \mathbf{A}^{-1}\mathbf{E}_1) - \mathbf{E}_4] = 0 \quad (i = 1, \cdots, p) \tag{3.77}$$

where $\Phi(0) = \mathbf{1}$ and $\mathbf{d} = (\delta_1, \cdots, \delta_p)$. By substituting (3.75) and (3.76) into (3.77), $\mathbf{d}$ can be determined. Hence, provided that $\delta_i \in (0, 1)$ for $i = 1, \cdots, p$, $\mathbf{X}_1^*, \cdots, \mathbf{X}_p^*$ can then be obtained from (3.75) and (3.76).

Due to the cyclic property of the period-p orbit, $\{\mathbf{X}_2^*, \cdots, \mathbf{X}_p^*, \mathbf{X}_1^*\}, \cdots, \{\mathbf{X}_p^*, \mathbf{X}_1^*, \cdots, \mathbf{X}_{p-1}^*\}$ are other period-p orbits which correspond to the same subharmonic frequency. If $\mathbf{X}_1^* = \cdots = \mathbf{X}_p^*$, the period-$p$ orbit becomes a period-1 orbit, indicating that the period-1 orbit is a particular case of the period-p orbit.

3.2.2.4 Stability and Characteristic Multipliers

Both period-1 and period-p orbits may be stable or unstable. Hence, the corresponding characteristic multipliers must be further calculated in order to test the stability of the orbits.

For a period-1 orbit that is a fixed point of the specific Poincaré map given by (3.71), its characteristic multipliers are eigenvalues of the Jacobian matrix of that mapping, which is given by:

$$\begin{aligned} DP(\mathbf{X}^*) &= \Phi(T) - \frac{\partial \Phi(T - \delta T)}{\partial \delta} \mathbf{A}^{-1}(\mathbf{E}_1 - \mathbf{E}_2) \frac{\partial \delta}{\partial \mathbf{X}^*} \\ &= \Phi(T) + \Phi(T - \delta T)(\mathbf{E}_1 - \mathbf{E}_2) T \frac{\partial \delta}{\partial \mathbf{X}^*} \end{aligned} \tag{3.78}$$

where $\frac{\partial \delta}{\partial \mathbf{X}^*} = \left(\frac{\partial \delta}{\partial \omega}, \frac{\partial \delta}{\partial i}\right)$. According to the implicit-function theorems, $\frac{\partial \delta}{\partial \mathbf{X}^*}$ can be deduced from (3.67) as:

$$\begin{aligned} \frac{\partial \delta}{\partial \mathbf{X}^*} &= -\left(\frac{\partial h}{\partial \delta}\right)^{-1} \frac{\partial h}{\partial \mathbf{X}^*} \\ &= -[\mathbf{E}_3 \Phi(\delta T)(\mathbf{A}\mathbf{X}^* + \mathbf{E}_1) T]^{-1} \mathbf{E}_3 \Phi(\delta T) \end{aligned} \tag{3.79}$$

For the period-p orbit $\{\mathbf{X}_1^*, \cdots, \mathbf{X}_p^*\}$, $\mathbf{X}_1^*$ is a fixed point of the p-fold iterative specific Poincaré map. Therefore, its characteristic multipliers are the eigenvalues of the Jacobian matrix of that mapping, which is given by:

$$DP^{(p)}(\mathbf{X}_1^*) = \Phi(pT) + \sum_{j=1}^{p} \Phi((p-j)T)\Phi(T - \delta_j T)(\mathbf{E}_1 - \mathbf{E}_2) T \frac{\partial \delta_j}{\partial \mathbf{X}_1^*} \tag{3.80}$$

By substituting (3.76) into (3.77):

$$\mathbf{H}(\mathbf{d}, \mathbf{X}_1^*) = [h_1(\mathbf{d}, \mathbf{X}_1^*), \cdots, h_p(\mathbf{d}, \mathbf{X}_1^*)]^{\mathrm{T}} = \mathbf{0} \tag{3.81}$$

Hence, according to the implicit-function theorems, the partial derivative in (3.80) can be expressed as:

$$\frac{\partial \mathbf{d}}{\partial \mathbf{X}_1^*} = \left(\frac{\partial \delta_1}{\partial \mathbf{X}_1^*}, \cdots, \frac{\partial \delta_p}{\partial \mathbf{X}_1^*}\right)^{\mathrm{T}} = -\left[\frac{\partial \mathbf{H}}{\partial \mathbf{d}}\right]^{-1} \left[\frac{\partial \mathbf{H}}{\partial \mathbf{X}_1^*}\right] \tag{3.81}$$

where

$$\begin{aligned} \frac{\partial \mathbf{H}}{\partial \mathbf{d}} = \left[\frac{\partial h_i}{\partial \delta_j}\right]_{p \times p} &= diag[\mathbf{E}_3 \Phi(\delta_i T)(\mathbf{A}\mathbf{X}_i^* + \mathbf{E}_1) T] \\ &\quad + [\mathbf{E}_3 \Phi(\delta_i T)\Phi((i-j)T - \delta_j T)(\mathbf{E}_1 - \mathbf{E}_2) T]_{p \times p} \end{aligned} \tag{3.82}$$

$$\frac{\partial \mathbf{H}}{\partial \mathbf{X}_1^*} = \left[\frac{\partial h_1}{\partial \mathbf{X}_1^*}, \cdots, \frac{\partial h_p}{\partial \mathbf{X}_1^*}\right]^{\mathrm{T}} = [\mathbf{E}_3\Phi(\delta_1 T), \cdots, \mathbf{E}_3\Phi((p-1)T+\delta_p T)]^{\mathrm{T}} \tag{3.83}$$

Notice that $\Phi((i-j)T-\delta_j T)$ in (3.82) becomes a zero matrix when $j \geq i$.

3.2.3 Simulation

To illustrate the derived Poincaré map, computer simulations are carried out. The simulation parameters are based on the same PMDC drive system that was adopted for the voltage-controlled simulation. Similar to Section 3.1.3, the resulting eigenvalues of matrix **A** are $\lambda_1 = -17.1$ and $\lambda_2 = -37.4$, indicating that although this open-loop system shows no oscillating dynamics, the corresponding closed-loop system may exhibit not only oscillating dynamics but also subharmonics, and even chaos.

3.2.3.1 Bifurcation Diagrams using Numerical Computation

By employing the brute-force method to compute the generalized Poincaré map (Parker and Chua, 1989), the bifurcation diagrams of ω and i versus V_{in} and g_ω can be calculated, as shown in Figure 3.22. The

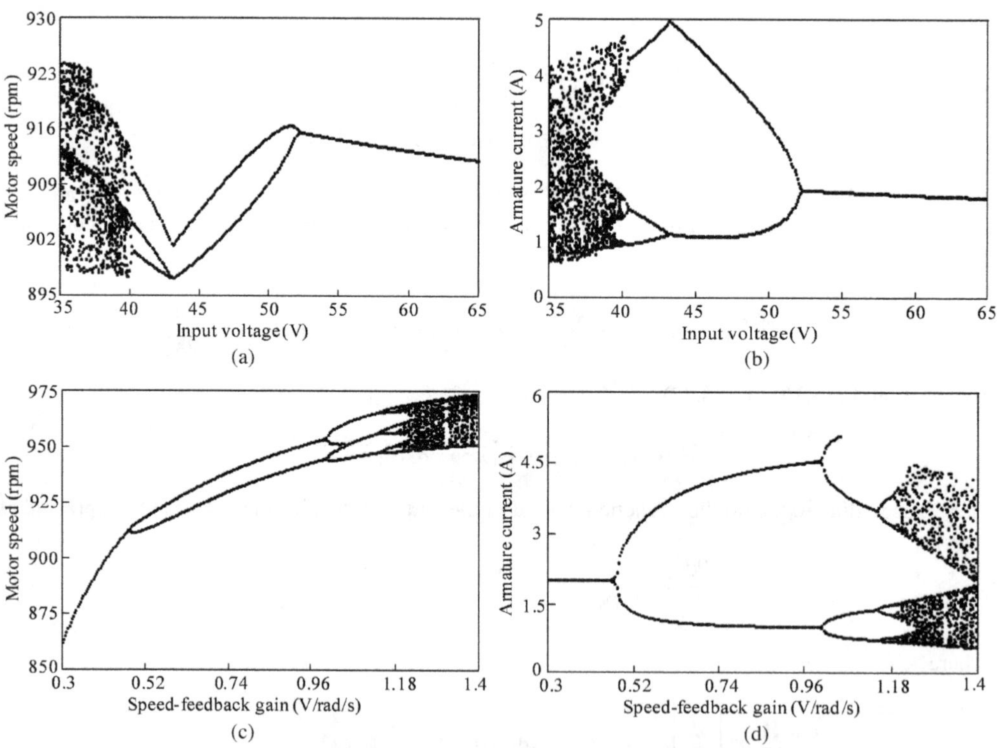

Figure 3.22 Bifurcation diagrams. (a) Speed versus input voltage. (b) Current versus input voltage. (c) Speed versus speed-feedback gain. (d) Current versus speed-feedback gain

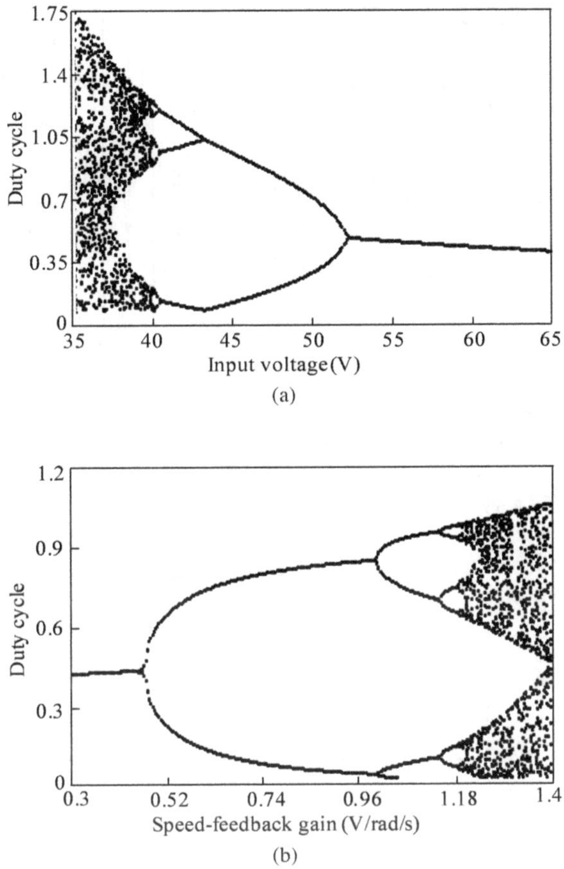

Figure 3.23 Distribution diagrams. (a) Duty cycle versus input voltage. (b) Duty cycle versus speed-feedback gain

corresponding δ with respect to V_{in} and g_ω are also shown in Figure 3.23. As can be seen in these figures, the system exhibits a chaotic behavior, which is valid for both chaotic speed and current. It is interesting to note that the period-2 orbit of both ω and i versus V_{in} bifurcates to a period-3 orbit when V_{in} is reduced to 43.4 V, whereas one branch of the period-4 orbit of both ω and i versus g_ω terminates when g_ω is increased to 1.05 V/rad/s. The reason is due to the fact that δ has to be positive, resulting in the discontinuities at 43.4 V and 1.05 V/rad/s, as shown in Figure 3.23.

The bifurcation diagrams function to illustrate the occurrence of subharmonics and chaos with respect to the variation of system parameters. In order to attain an insight into the subharmonic and chaotic behaviors, both time-domain waveforms and trajectories are investigated. For the sake of clarity, the speed and current control signals, v_ω and v_i, are used to represent the speed ω and current i, respectively. In fact, they simply obey linear relationships as given by (3.47) and (3.48). The simulation waveforms of v_ω, v_i and clock pulses, as well as the trajectory of v_ω versus v_i for the period-1 orbit with $V_{in} = 60$ V and $g_\omega = 0.54$ V/rad/s, are shown in Figure 3.24. This illustrates that the period-1 trajectory has boundaries of v_i from 2.2 V to 5 V and v_ω from 4.7 V to 5.2 V. When $V_{in} = 51$ V, the system operates in a period-2 orbit, as shown in Figure 3.25, in which v_i lies between 1.8 V and 5.2 V while v_ω is between 4.5 V and 5.5 V. Moreover, when $V_{in} = 35$ V, the system is in chaos. The corresponding chaotic waveforms and trajectory

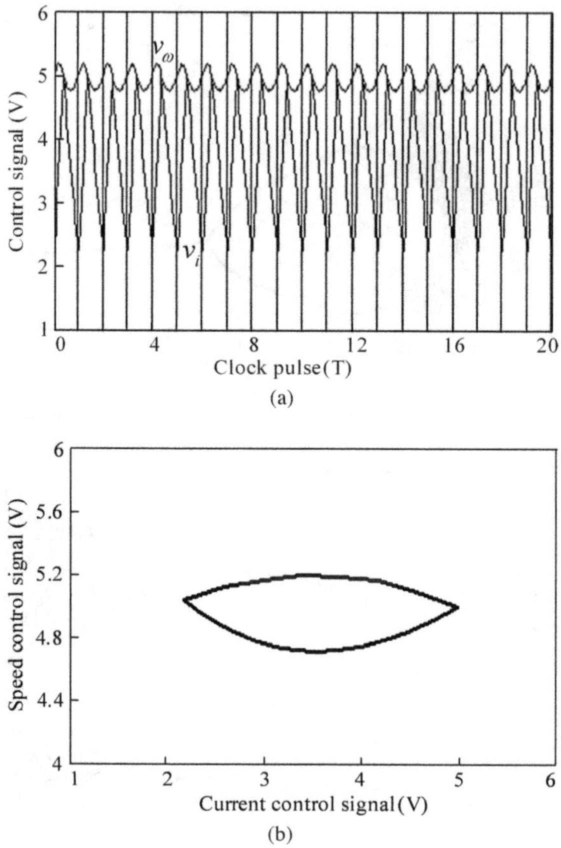

Figure 3.24 Control signals of speed and current during period-1 operation. (a) Waveforms. (b) Trajectory

are shown in Figure 3.26. To further illustrate its chaotic behavior, a system Poincaré section consisting of 4000 sampling points is also shown in Figure 3.26.

3.2.3.2 Bifurcation Diagrams using Analytical Approach

Based on the derived relationship in (3.75)–(3.77) and (3.80)–(3.83), the bifurcation diagram of i with respect to g_ω, as well as the corresponding duty cycle δ and characteristic multipliers (λ_1, λ_2), are shown in Figure 3.27 in which period-1 and period-2 orbits are both involved. It can be found that the system operates in period-1 when $g_\omega < 0.48$ V/rad/s. While g_ω gradually increases to 0.48 V/rad/s, one of the magnitudes of (λ_1, λ_2) approaches unity, and the system begins to bifurcate to a period-2 orbit. Since one of the duty cycles of the period-2 orbit is equal to unity when $g_\omega = 1.1$ V/rad/s, the orbit only exists between 0.48 V/rad/s and 1.1 V/rad/s. According to the characteristic multipliers of the period-2 orbit, it can be found that the period-2 orbit lying 0.48 V/rad/s $< g_\omega \leq 1.1$ V/rad/s is stable only when $g_\omega \leq 1$ V/rad/s. This bifurcation diagram, resulting from analytical modeling, closely agrees with that shown in Figure 3.22(d) obtained by numerical computation. In fact, without duplicating the figures, other bifurcation diagrams, namely i versus V_{in} as well as ω versus V_{in} and g_ω, can be obtained via analytical modeling and have the same patterns as shown in Figure 3.22.

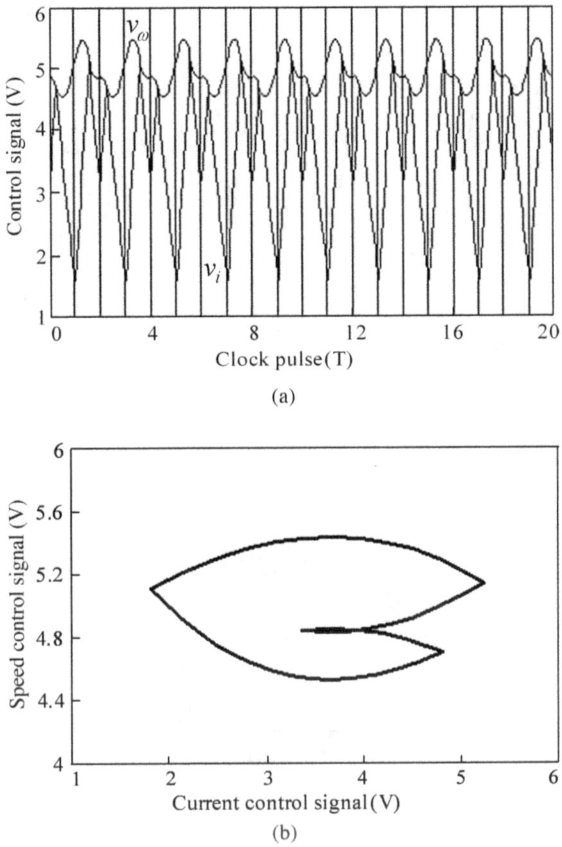

Figure 3.25 Control signals of speed and current during period-2 operation. (a) Waveforms. (b) Trajectory

It should be noted that the required computational time based on the derived analytical solution is much less than that required for computation using the numerical algorithm. Increasingly, the analytical solution can facilitate the identification of the desired stable operating ranges for different system parameters and conditions.

3.2.3.3 Identification of Stable Operating Ranges

For a practical DC drive system, the operating point should be designed to locate on the stable period-1 orbit. The corresponding stability is governed by the period-1 orbit of the specific Poincaré map described by (3.72) and (3.73) as well as the characteristic multipliers given by (3.78) and (3.79). By substituting $\Phi(\delta T) = \mathbf{1} + \delta T\mathbf{A}$ into (3.73), δ is simply expressed as:

$$\delta = \frac{\mathbf{E}_3(\mathbf{A}^{-1}\mathbf{E}_2 + \mathbf{E}_4)}{T\,\mathbf{E}_3(\mathbf{1} - \Phi(T))^{-1}(\mathbf{E}_1 - \mathbf{E}_2)} \tag{3.84}$$

Due to this explicit expression given by (3.84), the period-1 orbit and its characteristic multipliers can be easily calculated by using (3.72), (3.78), and (3.79). For a given set of system parameters, the

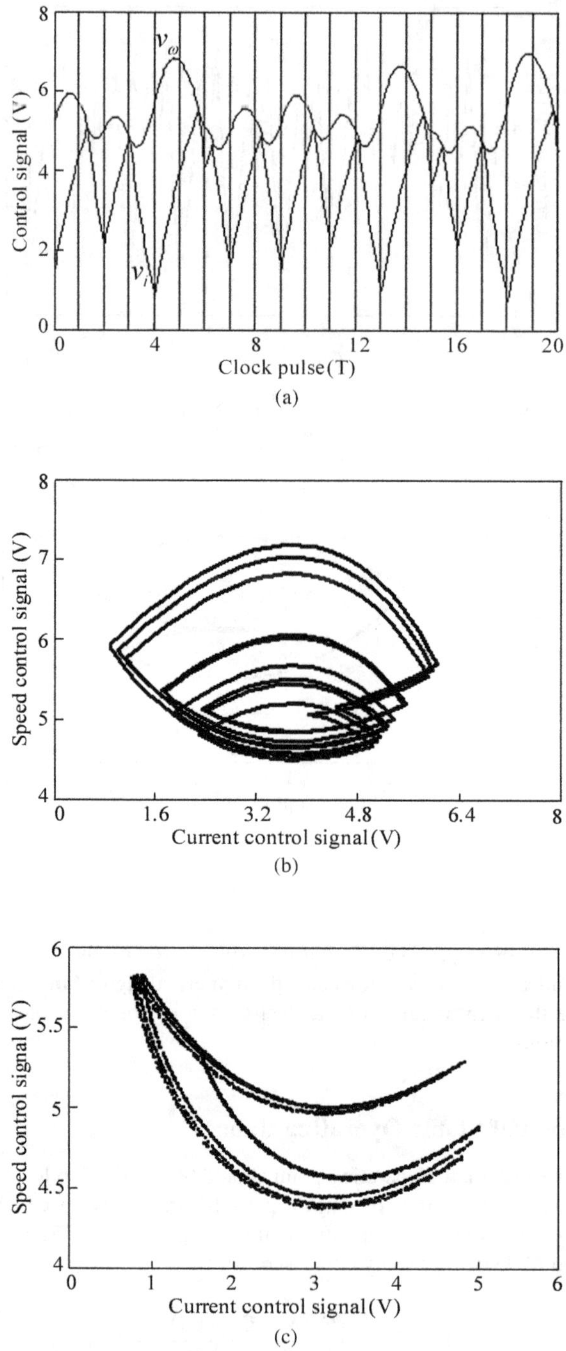

Figure 3.26 Control signals of speed and current during chaotic operation. (a) Waveforms. (b) Trajectory. (c) Poincaré section

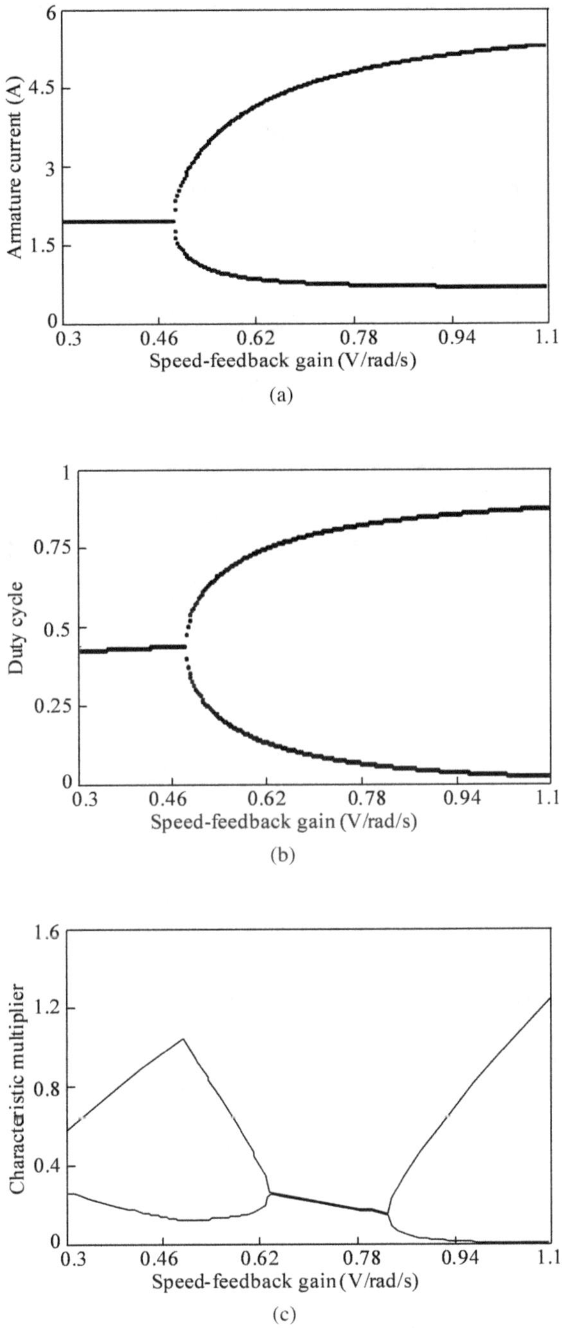

Figure 3.27 Analytical modeling results. (a) Armature current versus speed-feedback gain. (b) Duty cycle versus speed-feedback gain. (c) Characteristic multipliers versus speed-feedback gain

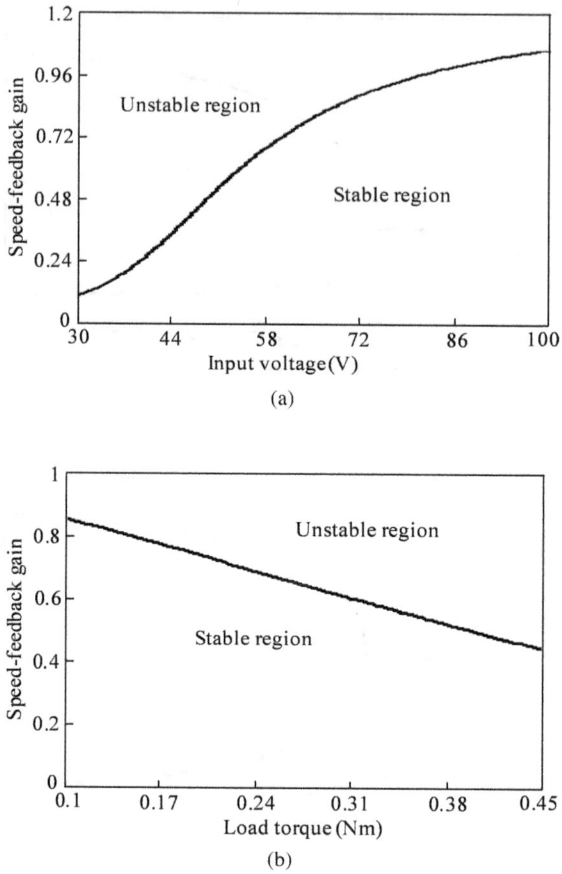

Figure 3.28 Stable regions of typical system parameters. (a) Speed-feedback gain versus input voltage. (b) Speed-feedback gain versus load torque

period-1 orbit is stable if and only if the magnitudes of all characteristic multipliers resulting from (3.78) are less than unity.

Based on the above procedure, the stable operating regions for typical system parameters, namely g_ω, V_{in}, and T_l, are determined. As shown in Figure 3.28, it indicates that there are different relationships between g_ω and V_{in} as well as g_ω and T_l governing the system's stability. Namely, the stable range of the speed-feedback gain not only depends on the load torque but also on the input voltage. Stable ranges of other system parameters can similarly be determined by using (3.72), (3.78), (3.79), and (3.84).

3.2.4 Experimentation

The experimental set-up is based on the current-controlled DC drive system shown in Figure 3.21(a). Similar to Section 3.1.4, the mechanical load T_l is realized by another DC machine which operates in the generator mode. The armature circuit of this machine is then connected to an electronic load which serves as a controllable current sink. Thus, T_l can be electronically controlled to keep at the desired value. As the DC motor and the mechanical load are directly coupled together by a shaft coupling unit, this two-mass mechanical system inevitably exhibits mechanical vibration, also called torsional oscillation. Based on

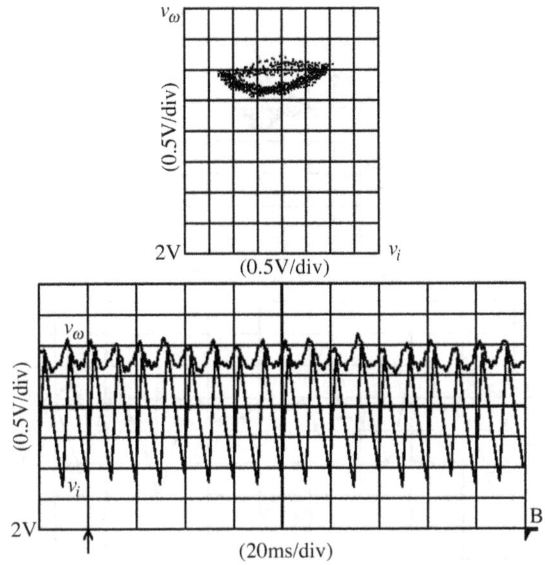

Figure 3.29 Measured trajectory and waveforms of speed and current control signals during period-1 operation

the system parameters, the corresponding mechanical resonant frequency is found to be about 1.1 kHz (Sugiura and Hori, 1996). Since the system switching frequency is selected as 100 Hz, which is much less than the mechanical resonant frequency, the effect of torsional oscillation on the system dynamics is ignored in both theoretical analysis and numerical simulation. As shown in Figures 3.29–3.31, there are high-frequency ripples (about 1.1 kHz), due to torsional oscillation, superimposing on the measured

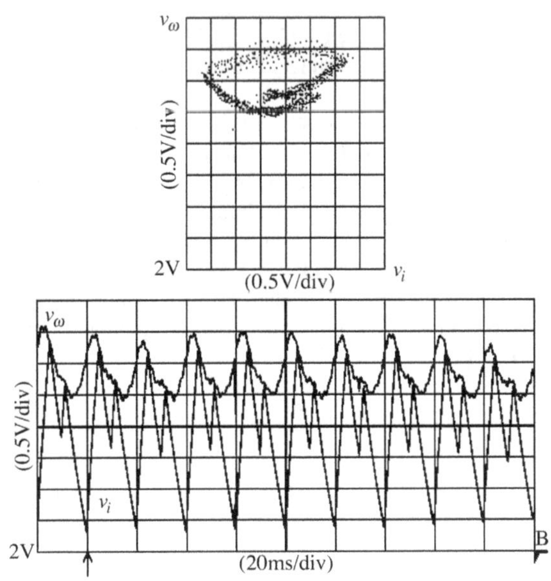

Figure 3.30 Measured trajectory and waveforms of speed and current control signals during period-2 operation

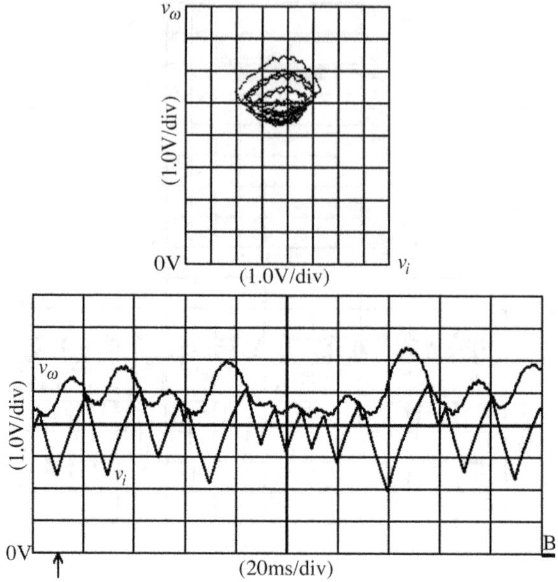

Figure 3.31 Measured trajectory and waveforms of speed and current control signals during chaotic operation

waveforms of v_ω. The magnitude of these ripples is insignificant compared with that of the steady-state periodic solutions.

The measured trajectory and waveforms of v_ω and v_i with $V_{in} = 60$ V and $g_\omega = 0.54$ V/rad/s are shown in Figure 3.29. This illustrates that the system operates in a period-1 orbit in which v_i lies between 2.7 V and 5 V while v_ω lies between 4.7 V and 5.2 V. When $V_{in} = 51$ V, the system operates in a period-2 orbit, as shown in Figure 3.30, in which v_i lies between 2.3 V and 5.2 V while v_ω lies between 4.5 V and 5.5 V. In a comparison with those shown in Figures 3.24 and 3.25, the measured results and the theoretical prediction have a good agreement.

Moreover, by selecting $V_{in} = 35$ V, the measured trajectory and waveforms of v_ω and v_i shown in Figure 3.31 illustrate that the drive system is in chaotic operation. For the period-1 and period-2 orbits, the measured trajectory and waveforms were directly compared with the theoretical predictions, but in the chaotic case the measured chaotic trajectory and waveforms could not be compared with the theoretical predictions because the chaotic behavior was not periodic and, therefore, the experimental measuring period could not be the same as the theoretical analyzing period. Also, its characteristics are extremely sensitive to the system initial conditions. Nevertheless, it can be found that the measured boundaries of the chaotic trajectory shown in Figure 3.31 resemble the theoretical prediction in Figure 3.26.

References

Chau, K.T., Chen, J.H., Chan, C.C. *et al.* (1997a) Chaotic behavior in a simple DC drive. Proceedings of IEEE Power Electronics and Drive Systems Conference, pp. 473–479.

Chau, K.T., Chen, J.H., and Chan, C.C. (1997b) Dynamic bifurcation in DC drives. Proceedings of IEEE Power Electronics Specialists Conference, pp. 1330–1336.

Chau, K.T., Chen, J.H., Chan, C.C., and Chan, D.T.W. (1997c) Modeling of subharmonics and chaos in DC motor drives. Proceedings of IEEE International Industrial Electronics Conference, pp. 523–528.

Chen, J.H., Chau, K.T., and Chan, C.C. (2000a) Chaos in voltage-mode controlled DC drive systems. *International Journal of Electronics*, **86**, 857–874.

Chen, J.H., Chau, K.T., and Chan, C.C. (2000b) Analysis of chaos in current-mode controlled dc drive systems. *IEEE Transactions on Industrial Electronics*, **47**, 67–76.

Deane, J.H.B. (1992) Chaos in a current-mode controlled boost DC-DC converter. *IEEE Transactions on Circuits and Systems - I: Fundamental Theory and Applications*, **39**, 680–683.

Hamill, D.C. and Jefferies, D.J. (1988) Subharmonics and chaos in a controlled switched-mode power converter. *IEEE Transactions on Circuits and Systems*, **35**, 1059–1061.

Hamill, D.C., Deane, J.H.B., and Jefferies, D.J. (1992) Modeling of chaotic DC-DC converters by iterated nonlinear mappings. *IEEE Transactions on Power Electronics*, **7**, 25–36.

Parker, T.S. and Chua, L.O. (1987) Chaos: a tutorial for engineers. *Proceedings of IEEE*, **75**, 982–1008.

Parker, T.S. and Chua, L.O. (1989) *Practical Numerical Algorithm for Chaotic Systems*, Springer-Verlag, New York.

Sugiura, K. and Hori, Y. (1996) Vibration suppression in 2- and 3-mass system based on the feedback of imperfect derivative of the estimated torsional torque. *IEEE Transactions on Industrial Electronics*, **43**, 56–64.

Tse, C.K., Fung, S.C., and Kwan, M.W. (1996) Experimental confirmation of chaos in a current-programmed Ćuk converter. *IEEE Transactions on Circuits and Systems - I: Fundamental Theory and Applications*, **43**, 605–608.

4

Chaos in AC Drive Systems

AC drive systems have been widely accepted for industrial applications. In general, they take the advantages of a higher power density and a higher efficiency than DC drive systems. AC drive systems are composed of two major groups, namely the induction drive systems and synchronous drive systems. Among the induction drive systems, the cage-rotor is almost exclusively used for industrial applications. Among the synchronous drive systems, the permanent magnet (PM) brushless AC drive system (usually termed PM synchronous drive system) is becoming popular, whereas the synchronous reluctance drive system is receiving attention.

In this chapter, chaos is investigated in three representative AC drive systems, namely a cage-rotor induction drive system, a PM synchronous drive system, and a synchronous reluctance (SynR) drive system.

4.1 Induction Drive Systems

It is well known that the motor parameter variations, especially the increase in rotor resistance due to prolonged operation, in field-oriented control (FOC) based induction drive systems may violate the necessary decoupling condition, causing an unexpected speed fluctuation. In recent years, it has been identified that the compensation of this rotor resistance variation may cause chaotic phenomena (Gao and Chau, 2003a). Therefore, in this section a practical induction drive system using FOC is first modeled, then Poincaré mapping and bifurcation analysis are conducted. Consequently, computer simulation and experimental verification are given to verify the chaotic phenomena.

4.1.1 Modeling

The general *d–q* model of a cage-rotor induction motor is well known (Leonhard, 1996), and can be expressed as:

$$\frac{\mathrm{d}i_{ds}}{\mathrm{d}t} = -\gamma i_{ds} + \omega i_{qs} + \eta\beta\psi_{dr} + \beta\omega_r\psi_{qr} + \frac{v_{ds}}{\sigma L_s} \tag{4.1}$$

$$\frac{\mathrm{d}i_{qs}}{\mathrm{d}t} = -\omega i_{ds} - \gamma i_{qs} - \beta\omega_r\psi_{dr} + \eta\beta\psi_{qr} + \frac{v_{qs}}{\sigma L_s} \tag{4.2}$$

$$\frac{\mathrm{d}\psi_{dr}}{\mathrm{d}t} = \eta L_m i_{ds} - \eta\psi_{dr} + (\omega - \omega_r)\psi_{qr} \tag{4.3}$$

Chaos in Electric Drive Systems: Analysis, Control and Application, First Edition. K.T. Chau and Zheng Wang.
© 2011 John Wiley & Sons (Asia) Pte Ltd. Published 2011 by John Wiley & Sons (Asia) Pte Ltd.

$$\frac{\mathrm{d}\psi_{qr}}{\mathrm{d}t} = \eta L_m i_{qs} - (\omega - \omega_r)\psi_{dr} - \eta\psi_{qr} \tag{4.4}$$

$$\frac{\mathrm{d}\omega_r}{\mathrm{d}t} = \frac{3}{2}\frac{P}{2}\frac{L_m}{L_r}\frac{1}{J}(\psi_{dr} i_{qs} - \psi_{qr} i_{ds}) - \frac{B\omega_r}{J} \tag{4.5}$$

where i_{ds}, i_{qs}, and v_{ds}, v_{qs} are the stator currents and voltages, respectively; ψ_{dr} and ψ_{qr} are the rotor fluxes; R_s and R_r are the stator and rotor resistances, respectively; L_s, L_r, and L_m are the stator inductance, rotor inductance, and mutual inductance, respectively; η is defined as $1/T_r$ in which $T_r = L_r/R_r$ is the rotor time constant; γ is defined as $R_s/\sigma L_s + (1-\sigma)/\sigma T_r$; and β is defined as $L_m/\sigma L_s L_r$ in which $\sigma = 1 - L_m^2/L_s L_r$ is the coupling factor, ω_r is the rotor speed, ω is the speed of the reference frame, J is the rotor inertia, B is the coefficient of viscous friction, and P is the number of poles.

The control diagram of the FOC-based induction drive system for speed tracking is shown in Figure 4.1. The compensator, which functions to compensate the increase of rotor resistance due to prolonged operation, is represented by:

$$\omega - \omega_r = \delta\eta L_m g_\omega(\omega_r^* - \omega_r) \tag{4.6}$$

where ω_r^* is the rotor speed command, g_ω is the gain of the speed controller, and δ is the gain of the compensator. By rearranging (4.3) and (4.4), i_{ds} and i_{qs} can be expressed in terms of ψ_{dr} and ψ_{qr}. Hence, the stator current commands are given by:

$$i_{ds}^* = \frac{T_r\dot{\psi}_{dr}^* + \psi_{dr}^*}{L_m} - \frac{\psi_{qr}^*(\omega - \omega_r)}{\eta L_m} \tag{4.7}$$

$$i_{qs}^* = \frac{T_r\dot{\psi}_{qr}^* + \psi_{qr}^*}{L_m} + \frac{\psi_{dr}^*(\omega - \omega_r)}{\eta L_m} \tag{4.8}$$

Since the rotor fluxes cannot be directly measured, they are generally estimated on the basis of (4.3) and (4.4), as given by:

$$\frac{\mathrm{d}\hat{\psi}_{dr}}{\mathrm{d}t} = \eta L_m i_{ds} - \eta\hat{\psi}_{dr} + (\omega - \omega_r)\hat{\psi}_{qr} \tag{4.9}$$

$$\frac{\mathrm{d}\hat{\psi}_{qr}}{\mathrm{d}t} = \eta L_m i_{qs} - (\omega - \omega_r)\hat{\psi}_{dr} - \eta\hat{\psi}_{qr} \tag{4.10}$$

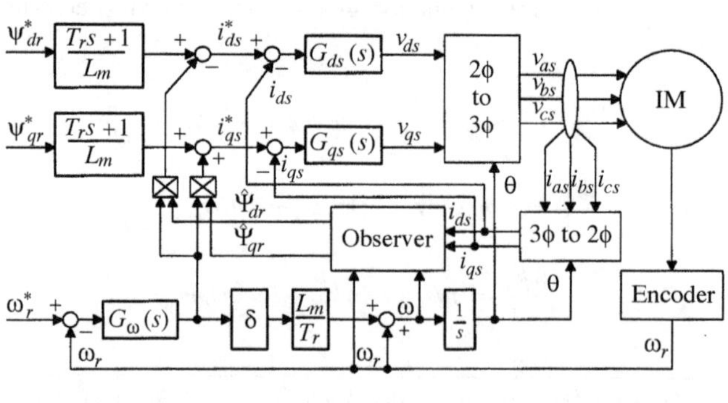

Figure 4.1 Control diagram of FOC-based induction drive system

By using these estimated rotor fluxes together with (4.6) under the normal operation of $\delta = 1$, the stator current commands given by (4.7) and (4.8) can be rewritten as:

$$i_{ds}^* = \frac{T_r \dot{\psi}_{dr}^* + \psi_{dr}^*}{L_m} - \hat{\psi}_{qr} g_\omega (\omega_r^* - \omega_r) \tag{4.11}$$

$$i_{qs}^* = \frac{T_r \dot{\psi}_{qr}^* + \psi_{qr}^*}{L_m} + \hat{\psi}_{dr} g_\omega (\omega_r^* - \omega_r) \tag{4.12}$$

It should be noted that δ is a control variable. Namely, when δ deviates from the condition of $\delta = 1$, the rotor resistance compensation may cause chaos to occur. As depicted in Figure 4.1, the stator voltages can then be expressed as:

$$v_{ds} = g_{ds}(i_{ds}^* - i_{ds}) \tag{4.13}$$

$$v_{qs} = g_{qs}(i_{qs}^* - i_{qs}) \tag{4.14}$$

where g_{ds} and g_{qs} are the gains of the current controllers. Together with the speed controller, they are governed by proportional-integral (PI) control, which has the following form in the frequency domain:

$$G(s) = k_p + \frac{k_i}{s} \tag{4.15}$$

where k_p is the proportional constant and k_i is the integral constant.

By defining ten state variables, namely $x_1 = i_{ds}$, $x_2 = i_{qs}$, $x_3 = \psi_{dr}$, $x_4 = \psi_{qr}$, $x_5 = \omega_r$, $x_6 = \int (i_{ds}^* - i_{ds})\, \mathrm{d}t$, $x_7 = \int (i_{qs}^* - i_{qs})\, \mathrm{d}t$, $x_8 = \int (\omega_r^* - \omega_r)\, \mathrm{d}t$, $x_9 = \hat{\psi}_{dr}$, and $x_{10} = \hat{\psi}_{qr}$, the closed-loop system depicted in Figure 4.1 can be represented by the following first-order differential equations:

$$\begin{aligned} \frac{\mathrm{d}x_1}{\mathrm{d}t} = & -\gamma x_1 + x_2 x_5 + \delta \eta L_m x_2 [k_{\omega p}(\omega_r^* - x_5) + k_{\omega i} x_8] \\ & + \eta \beta x_3 + \beta x_4 x_5 + \frac{1}{\sigma L_s} k_{di} x_6 \\ & + \frac{1}{\sigma L_s} k_{dp} \left[\frac{T_r \dot{\psi}_{dr}^* + \psi_{dr}^*}{L_m} - k_{\omega p}(\omega_r^* - x_5) x_{10} - k_{\omega i} x_8 x_{10} - x_1 \right] \end{aligned} \tag{4.16}$$

$$\begin{aligned} \frac{\mathrm{d}x_2}{\mathrm{d}t} = & -\gamma x_2 - x_1 x_5 - \delta \eta L_m x_1 [k_{\omega p}(\omega_r^* - x_5) + k_{\omega i} x_8] \\ & + \eta \beta x_4 - \beta x_3 x_5 + \frac{1}{\sigma L_s} k_{qi} x_7 \\ & + \frac{1}{\sigma L_s} k_{qp} \left[\frac{T_r \dot{\psi}_{qr}^* + \psi_{qr}^*}{L_m} + k_{\omega p}(\omega_r^* - x_5) x_9 + k_{\omega i} x_8 x_9 - x_2 \right] \end{aligned} \tag{4.17}$$

$$\frac{\mathrm{d}x_3}{\mathrm{d}t} = \eta L_m x_1 - \eta x_3 + \delta \eta L_m [k_{\omega p}(\omega_r^* - x_5) + k_{\omega i} x_8] x_4 \tag{4.18}$$

$$\frac{\mathrm{d}x_4}{\mathrm{d}t} = \eta L_m x_2 - \eta x_4 - \delta \eta L_m [k_{\omega p}(\omega_r^* - x_5) + k_{\omega i} x_8] x_3 \tag{4.19}$$

$$\frac{dx_5}{dt}=\frac{3}{2}\frac{P}{2}\frac{L_m}{L_r}\frac{1}{J}(x_2x_3-x_1x_4)-\frac{B_m}{J}x_5 \tag{4.20}$$

$$\frac{dx_6}{dt}=\frac{T_r\dot{\psi}_{dr}^*+\psi_{dr}^*}{L_m}-k_{\omega p}(\omega_r^*-x_5)x_{10}-k_{\omega i}x_8x_{10}-x_1 \tag{4.21}$$

$$\frac{dx_7}{dt}=\frac{T_r\dot{\psi}_{qr}^*+\psi_{qr}^*}{L_m}+k_{\omega p}(\omega_r^*-x_5)x_9+k_{\omega i}x_8x_9-x_2 \tag{4.22}$$

$$\frac{dx_8}{dt}=\omega_r^*-x_5 \tag{4.23}$$

$$\frac{dx_9}{dt}=\eta L_m x_1-\eta x_9+\delta\eta L_m\left[k_{\omega p}(\omega_r^*-x_5)+k_{\omega i}x_8\right]x_{10} \tag{4.24}$$

$$\frac{dx_{10}}{dt}=\eta L_m x_2-\eta x_{10}-\delta\eta L_m\left[k_{\omega p}(\omega_r^*-x_5)+k_{\omega i}x_8\right]x_9 \tag{4.25}$$

Based on the above differential equations, the dynamical behavior of the FOC-based induction drive system can readily be analyzed. Obviously, because of its high order of dimensions, numerical analysis is employed.

4.1.2 Analysis

In order to analyze the chaotic behavior of FOC-based induction drive systems, Poincaré mapping and bifurcation analysis are employed. The fundamental concept of this mapping was due to Poincaré. It replaces the solution of a continuous-time dynamical system by an iterative map, the so-called Poincaré map. This map acts like a stroboscope, and produces a sequence of samples of the continuous-time solution. Thus, the steady-state behavior of the Poincaré map, termed the orbit, corresponds to the steady-state waveform of the continuous-time dynamical system. On the other hand, it is essential to know about the formation of chaos when there is a variation in the system parameters. As a parameter is varied, a bifurcation indicates an abrupt change in the steady-state behavior of the system. A plot of the steady-state orbit against a bifurcation parameter is known as a bifurcation diagram. Thus, a bifurcation analysis facilitates the appraisal of the steady-state system behavior at a glance.

For the FOC-based induction drive system, the speed command may be fixed or time-varying. When this speed command is time-varying, it can be expressed as:

$$\omega_r^*=\bar{\omega}+\hat{\omega}\sin\left(\frac{2\pi}{T}t\right) \tag{4.26}$$

where $\bar{\omega}$ and $\hat{\omega}$ denote the DC component and AC amplitude, respectively, and T is the period of the oscillation. With this periodic speed command, the drive is a nonautonomous system. To derive the Poincaré map of the dynamical system described by (4.16)–(4.26), a ten-dimensional surface $\sum\in\Re^{10}\times S^1$ is defined as:

$$\sum:=\{(\mathbf{x},t)\in\Re^{10}\times S^1 : t=t_0\} \tag{4.27}$$

where $\mathbf{x}=[x_1,x_2,\ldots,x_{10}]$ is the solution of the state vector. The trajectory of $\mathbf{x}(t)$ repeatedly passes the surface $\sum$ for every period T (namely the period of the periodic command). The sequence of surface crossing (known as the orbit) defines the Poincaré map as given by:

$$\mathrm{P}:\Re^{10}\to\Re^{10},\ \mathbf{x}_{n+1}=\mathrm{P}(\mathbf{x}_n) \tag{4.28}$$

where $\mathbf{x}_n$ and $\mathbf{x}_{n+1}$ are the nth and $(n+1)$th samples of $\mathbf{x}(t)$, respectively.

By selecting δ as a bifurcation parameter, the steady-state orbits resulting from the Poincaré map (4.28) of the drive system described by (4.16)–(4.26) can be numerically determined, hence generating the desired bifurcation diagram.

4.1.3 *Simulation*

A practical three-phase FOC-based induction drive system is employed for exemplification. The corresponding motor parameters are $\psi^*_{dr} = 0.4\,\text{Wb}$, $\psi^*_{qr} = 0\,\text{Wb}$, $P = 4$, $R_s = 0.760\,\Omega$, $R_r = 0.675\,\Omega$, $L_s = 0.2248\,\text{H}$, $L_r = 0.2235\,\text{H}$, $L_m = 0.2176\,\text{H}$, $J = 0.0111\,\text{kgm}^2$, and $B = 7.355 \times 10^{-4}\,\text{Nm/rad/s}$. The control parameters of the PI current controllers are $k_{dp} = k_{qp} = 100$ and $k_{di} = k_{qi} = 50$, whereas those of the PI speed controller are $k_{\omega p} = 20$ and $k_{\omega i} = 5$. Based on these parameters, computer simulation can be performed to illustrate the occurrence of chaos in a three-phase FOC-based induction drive system.

With different values of periodic speed command, as given by (4.26), the speed bifurcation diagrams can readily be obtained from (4.16)–(4.28). Figure 4.2 shows two bifurcation diagrams of the motor speed with respect to the variation of δ under $\omega^*_r = 20\sin(\pi t)\text{rad/s}$ and $\omega^*_r = 20 + 20\sin(\pi t)\text{rad/s}$. It can be

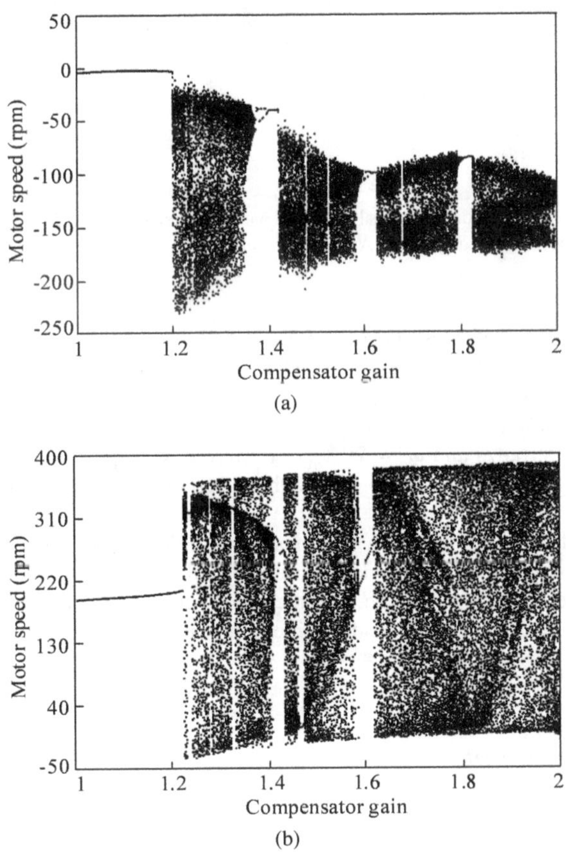

Figure 4.2 Speed bifurcation diagrams. (a) Without speed offset. (b) With speed offset

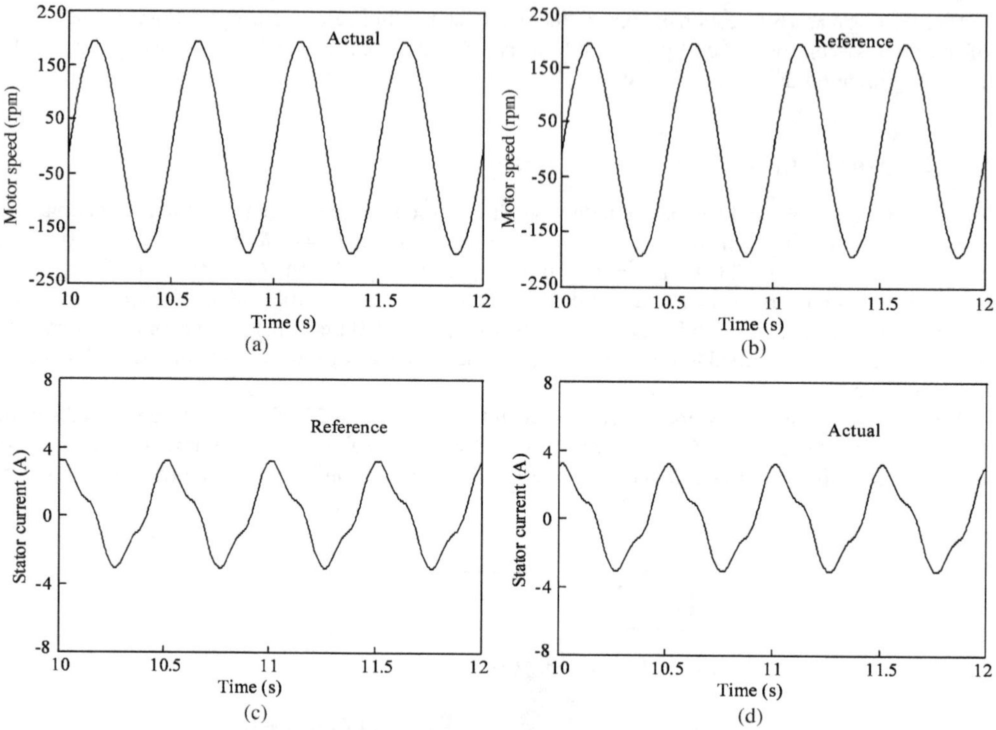

Figure 4.3 Simulated waveforms under normal operation without speed offset. (a) Actual motor speed. (b) Reference motor speed. (c) Reference stator current. (d) Actual stator current

seen that the drive system becomes chaotic from its normal operation when δ is increased beyond a threshold (about $\delta = 1.2$) for both cases, with and without DC speed offset.

When $\delta = 1$, as can be seen in Figure 4.2, the drive system performs its normal operation. Figures 4.3 and 4.4 show the corresponding motor speed and stator current waveforms under the cases of $\omega_r^* = 20\sin(\pi t)\text{rad/s}$ and $\omega_r^* = 20 + 20\sin(\pi t)\text{rad/s}$, respectively. It can be found that the motor speeds (marked "actual") closely follow the speed commands (marked "reference") during normal operation.

With an increase of δ, the drive system becomes chaotic. When $\delta = 1.3$, as reflected by Figure 4.2, the system performs a chaotic operation. Figures 4.5 and 4.6 show the chaotic waveforms of the motor speed and stator current under the aforementioned two speed commands. It can be observed that both speed and current exhibit a random-like, but bounded, oscillation that is actually the key nature of chaos.

4.1.4 Experimentation

For experimentation, the whole FOC-based drive system is implemented. Under the same operation conditions as for simulation, the measured waveforms under normal operations are shown in Figures 4.7 and 4.8. It can be observed that they are in good agreement with the simulated waveforms in Figures 4.3 and 4.4, respectively. Also, the measured waveforms under chaotic

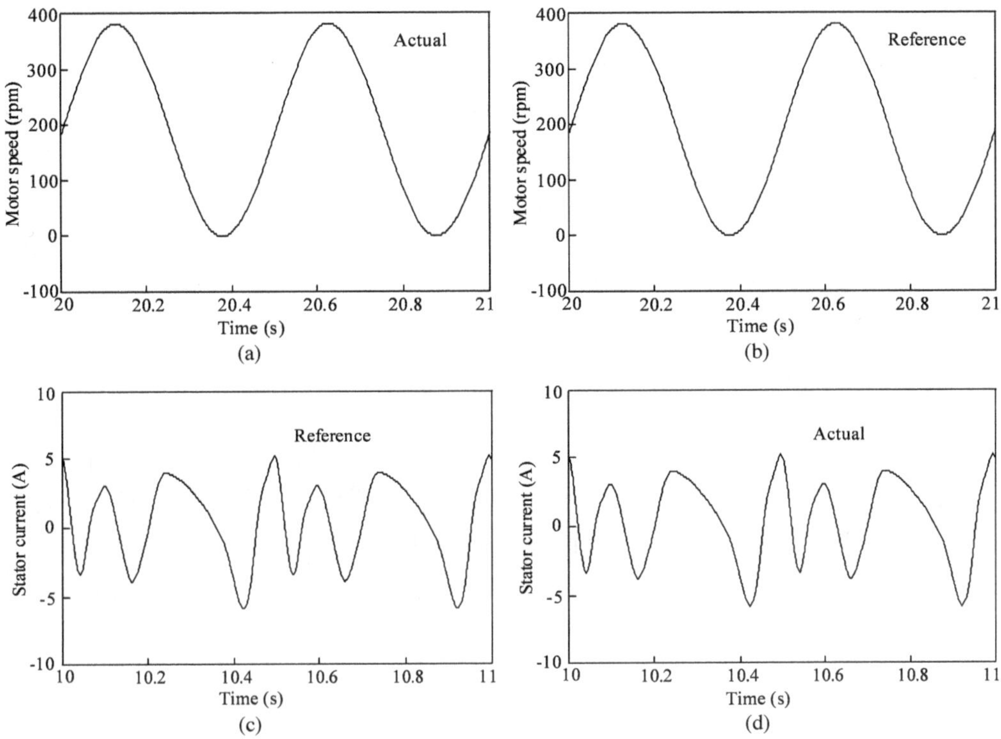

Figure 4.4 Simulated waveforms under normal operation with speed offset. (a) Actual motor speed. (b) Reference motor speed. (c) Reference stator current. (d) Actual stator current

operation are shown in Figures 4.9 and 4.10. They both offer the random-like but bounded nature. It should be noted that the simulated and measured chaotic patterns cannot be compared because the chaotic pattern is aperiodic and very sensitive to initial conditions.

From the above results, it can be confirmed that the FOC-based induction drive system exhibits chaotic behavior when δ deviates sufficiently from the condition of $\delta = 1$, namely the deviation from normal rotor resistance compensation.

4.2 Permanent Magnet Synchronous Drive Systems

With the advent of high-energy PM materials, PM synchronous motors can offer higher power density and higher efficiency than induction motors (Gieras and Wing, 2002), and PM synchronous drive systems are therefore becoming more attractive for modern industrial applications. However, due to the nonlinear dynamics of PM synchronous motors, some kinds of their abnormal operation may exhibit a strange behavior. Recently, chaotic behavior in PM synchronous machines has been reported by Gao and Chau (2003b) and Ye and Chau (2005).

This section analyzes the effect of PMs on an abnormal operation of PM synchronous drive systems. After modeling the system dynamics as a periodically forced nonautonomous equation, the corresponding Poincaré map and bifurcation diagram show that the design of the PMs significantly

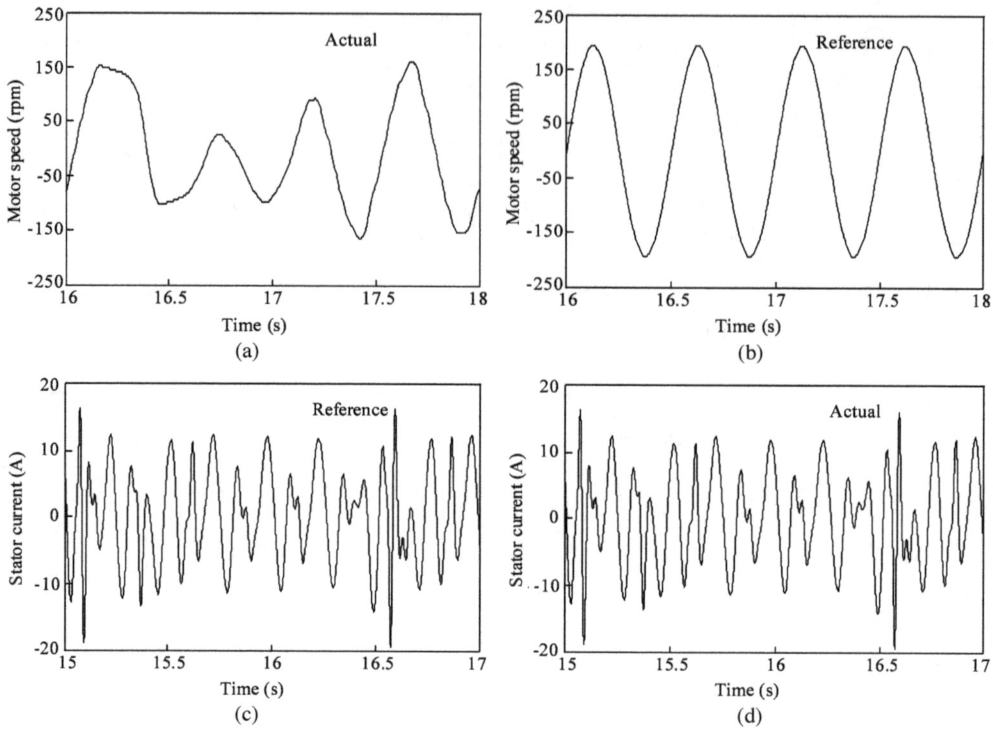

Figure 4.5 Simulated waveforms under chaotic operation without speed offset. (a) Actual motor speed. (b) Reference motor speed. (c) Reference stator current. (d) Actual stator current

affects the behavior of operation. Namely, chaos may occur if the PMs are not properly designed. Both computer simulation and experimental results are given to illustrate the design criterion.

4.2.1 Modeling

A three-phase PM synchronous drive system can be modeled in the d–q frame, as given by Fitzgerald, Kingsley and Umans (1991):

$$\begin{cases} L_{ds}\dfrac{\mathrm{d}i_{ds}}{\mathrm{d}t} = u_{ds} - R_s i_{ds} + \dfrac{P}{2}\omega_r L_{qs} i_{qs} \\ L_{qs}\dfrac{\mathrm{d}i_{qs}}{\mathrm{d}t} = u_{qs} - R_s i_{qs} + \dfrac{P}{2}\omega_r L_{ds} i_{ds} - \dfrac{P}{2}\omega_r \psi_{PM} \\ J\dfrac{\mathrm{d}\omega_r}{\mathrm{d}t} = \dfrac{3}{2}\dfrac{P}{2}\psi_{PM} i_{qs} + \dfrac{3}{2}\dfrac{P}{2}(L_{ds} - L_{qs}) i_{ds} i_{qs} - T_l - B\omega_r \end{cases} \tag{4.29}$$

where i_{ds} and i_{qs} are the stator currents, u_{ds} and u_{qs} are the stator voltages, L_{ds} and L_{qs} are the stator inductances, R_s is the stator resistance, P is the number of poles, ψ_{PM} is the PM flux, ω_r is the motor speed, T_l is the load torque, J is the rotor inertia, and B is the viscous friction coefficient.

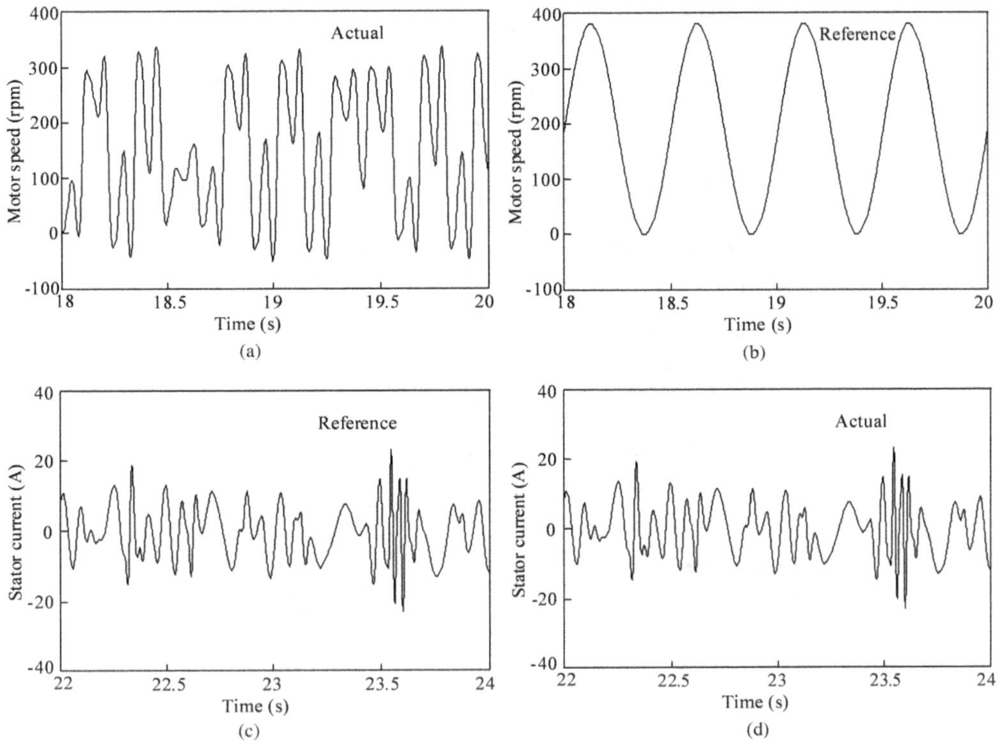

Figure 4.6 Simulated waveforms under chaotic operation with speed offset. (a) Actual motor speed. (b) Reference motor speed. (c) Reference stator current. (d) Actual stator current

When the PM synchronous motor is normally fed by a three-phase balanced sinusoidal supply, u_{ds} and u_{qs} are DC quantities. Under abnormal operation, these stator voltages can be time-varying and oscillating, as given by:

$$\begin{cases} u_{ds} = V_m \sin(2\pi ft - \theta) \\ u_{qs} = V_m \cos(2\pi ft - \theta) \end{cases} \tag{4.30}$$

where V_m is the voltage amplitude, f is the oscillation frequency, and θ is the spatial angle. By substituting (4.30) into (4.29), the system dynamical equation can be written in a periodically driven nonautonomous form:

$$\begin{cases} L_{ds}\dfrac{\mathrm{d}i_{ds}}{\mathrm{d}t} = V_m\sin(2\pi ft - \theta) - R_s i_{ds} + \dfrac{P}{2}\omega_r L_{qs} i_{qs} \\ L_{qs}\dfrac{\mathrm{d}i_{qs}}{\mathrm{d}t} = V_m\cos(2\pi ft - \theta) - R_s i_{qs} + \dfrac{P}{2}\omega_r L_{ds} i_{ds} - \dfrac{P}{2}\omega_r \psi_{PM} \\ J\dfrac{\mathrm{d}\omega_r}{\mathrm{d}t} = \dfrac{3}{2}\dfrac{P}{2}\psi_{PM} i_{qs} + \dfrac{3}{2}\dfrac{P}{2}(L_{ds} - L_{qs}) i_{ds} i_{qs} - T_l - B\omega_r \\ \dfrac{\mathrm{d}\theta}{\mathrm{d}t} = \dfrac{P}{2}\omega_r \end{cases} \tag{4.31}$$

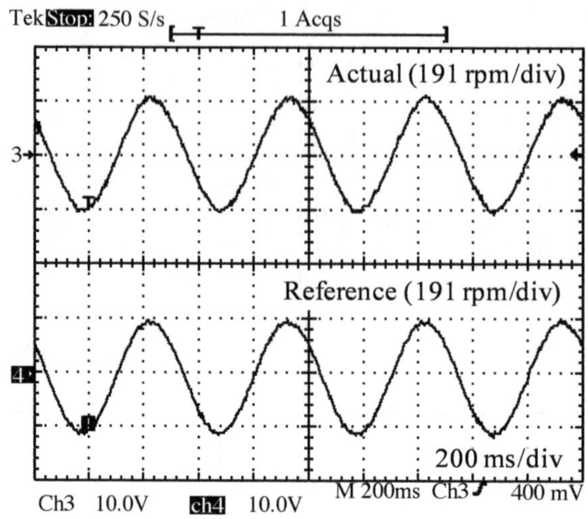

(a)

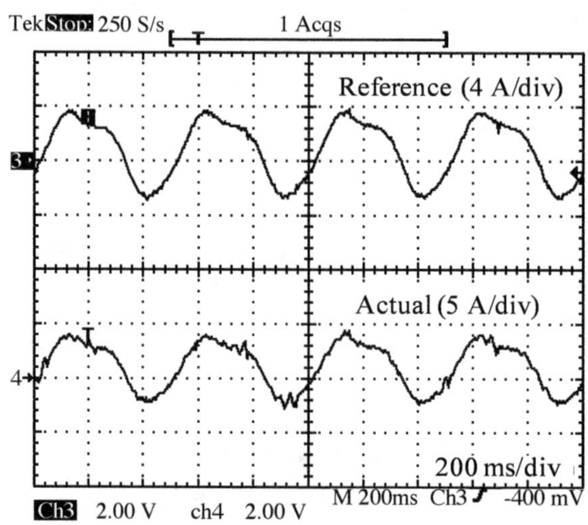

(b)

Figure 4.7 Experimental waveforms under normal operation without speed offset. (a) Motor speeds. (b) Stator currents

4.2.2 Analysis

It is well known that the periodically forced nonautonomous system is prone to subharmonics resonance and chaos (Thompson and Stewart, 2002). Closed-form analytical solutions are usually not available, whereas numerical means such as the Poincaré map and bifurcation analysis are generally employed.

Poincaré mapping functions to replace the solution of a continuous-time dynamical system by an iterative map. The steady-state behavior of the Poincaré map, termed the orbit, corresponds to the steady-state waveform of the continuous-time dynamical system. For the nonautonomous system

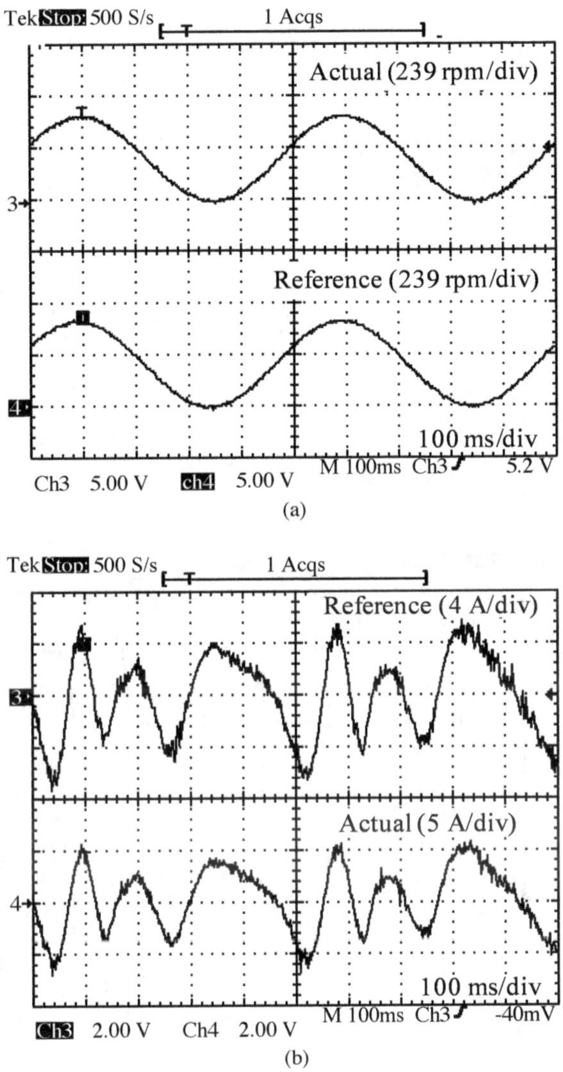

Figure 4.8 Experimental waveforms under normal operation with speed offset. (a) Motor speeds. (b) Stator currents

described by (4.3.1), a natural way to construct the Poincaré map is to sample the trajectory with a frequency f. Thus, the Poincaré surface $\sum \in \Re^4 \times S^1$ is defined as:

$$\sum := \{(x, t) \in \Re^4 \times S^1 : t = t_0\} \tag{4.32}$$

where $\mathbf{x} = [i_{ds}, i_{qs}, \omega_r, \theta]$ is the solution of the state vector. The trajectory of $\mathbf{x}(t)$ repeatedly passes the surface $\sum$ for every period T (namely, the period of the stator voltage oscillation). The sequence of

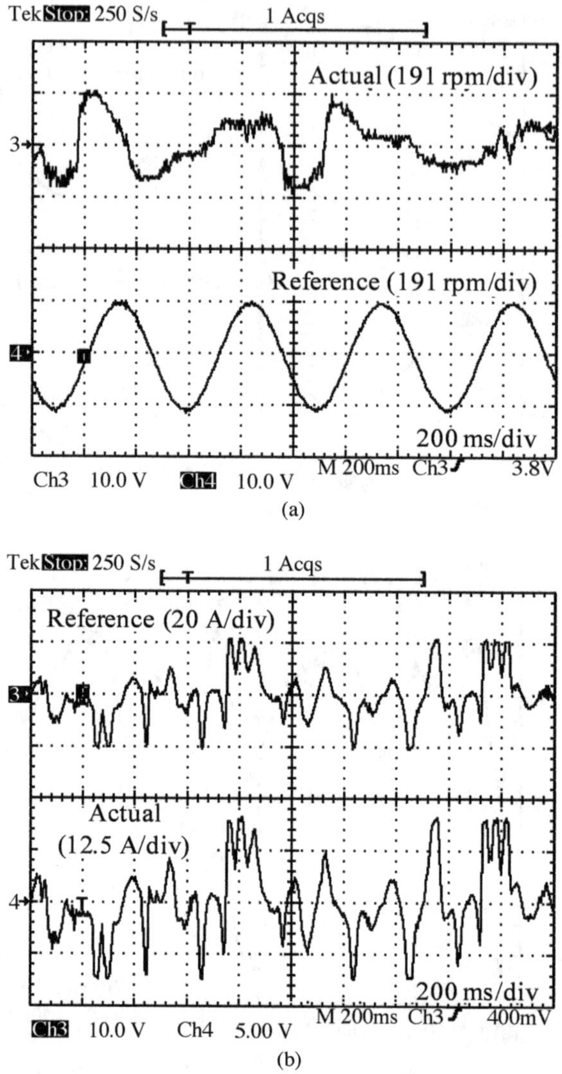

Figure 4.9 Experimental waveforms under chaotic operation without speed offset. (a) Motor speeds. (b) Stator currents

surface crossings defines the Poincaré map, as given by:

$$P : \Re^4 \rightarrow \Re^4, \ \mathbf{x}_{n+1} = P(\mathbf{x}_n) \tag{4.33}$$

where $\mathbf{x}_n$ and $\mathbf{x}_{n+1}$ are the nth and $(n + 1)$th samples of $\mathbf{x}(t)$, respectively.

A bifurcation analysis is used to measure the changes in the system's steady-state behavior with each variation in the system's parameters. As a parameter is varied, a bifurcation is an abrupt change in the steady-state behavior of the system. A plot of the steady-state orbit against a bifurcation parameter is

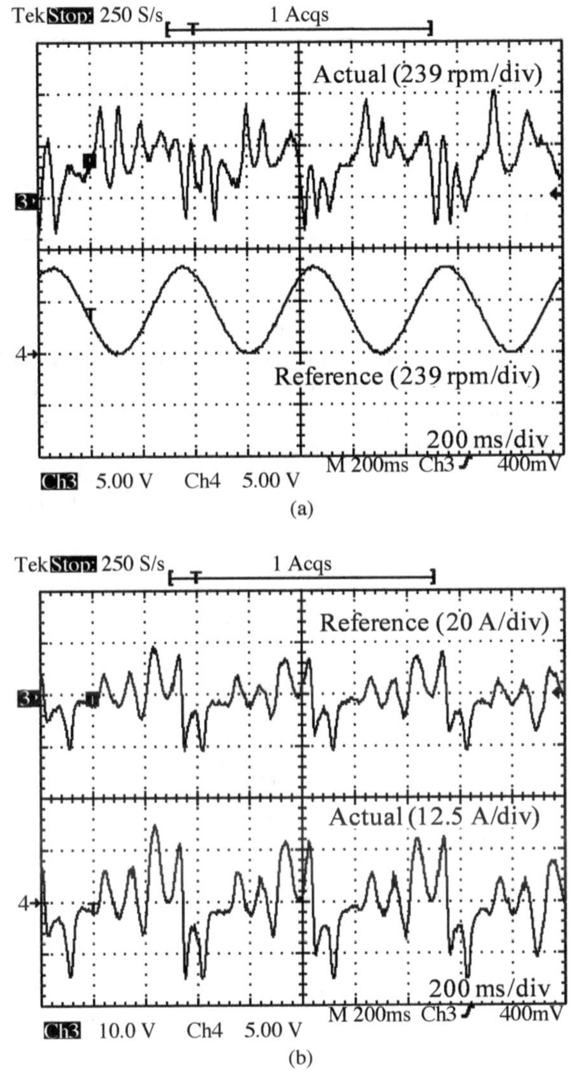

Figure 4.10 Experimental waveforms under chaotic operation with speed offset. (a) Motor speeds. (b) Stator currents

known as a bifurcation diagram, and the bifurcation analysis facilitates the appraisal of the steady-state behavior of the system at a glance.

4.2.3 Simulation

Different parameters affect the operation of a PM synchronous drive system. The PM size, and hence the PM flux, are of particular concern. A practical three-phase 6-pole interior PM synchronous motor with parameters listed in Table 4.1 is used for exemplification.

Table 4.1 Key parameters of PM synchronous motor

d-axis stator inductance L_{ds}	0.237 mH
q-axis stator inductance L_{qs}	0.152 mH
Stator resistance R_s	0.327 Ω
Number of poles P	6
PM flux ψ_{PM}	0.1472 Wb
Rotor inertia J	5.100×10^{-4} kgm^2
Viscous friction coefficient B	2.140×10^{-3} Nm/rad/s

When selecting $V_m = 30$ V and $T_l = 0$, the bifurcation diagram of the motor speed with respect to f is shown in Figure 4.11. Obviously, it reveals that synchronism can only be achieved when f is lower than 14.06 Hz. There exists a region (14.06 Hz $\leq f \leq$ 17.62 Hz) in which the motor behavior begins to either oscillate periodically or tremble chaotically, and PM sizing provides an explanation for this loss of synchronism. In order to verify this observation, the PM synchronous drive purposely selects the conditions of $V_m = 30$ V, $f = 15$ Hz, and $T_l = 0$. The corresponding bifurcation diagram of motor speed with respect to ψ_{PM} is shown in Figure 4.12. It illustrates that

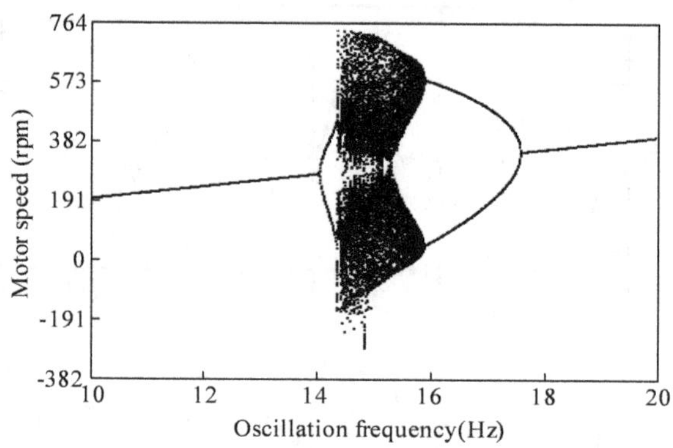

Figure 4.11 Bifurcation diagram of motor speed with respect to oscillation frequency

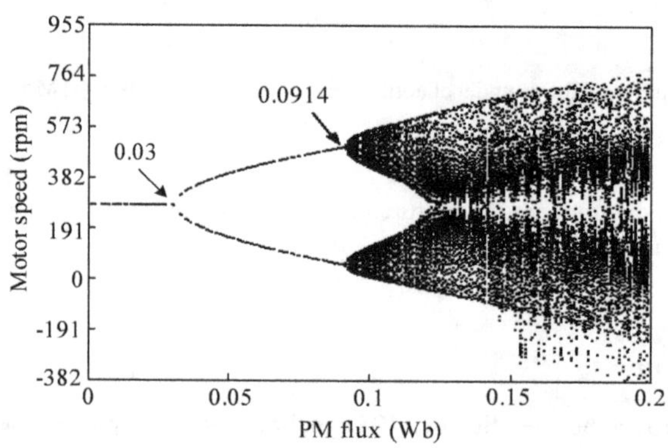

Figure 4.12 Bifurcation diagram of motor speed with respect to PM flux

there exists a critical value of ψ_{PM} for stable operation. Thus, the PM synchronous drive system operating at the above conditions exhibits a chaotic behavior because the corresponding PM flux (0.1472 Wb) is larger than the critical value of 0.0914 Wb. Synchronism can be guaranteed only if the PM flux is smaller than 0.03 Wb; the motor will oscillate periodically in the region of $0.03\,\text{Wb} \leq \psi_{PM} \leq 0.0914\,\text{Wb}$.

4.2.4 Experimentation

Based on the aforementioned PM synchronous motor and operating conditions, the resulting chaotic waveforms of i_{ds}, i_{qs}, and ω_r are measured as shown in Figures 4.13 and 4.14, respectively. Consequently,

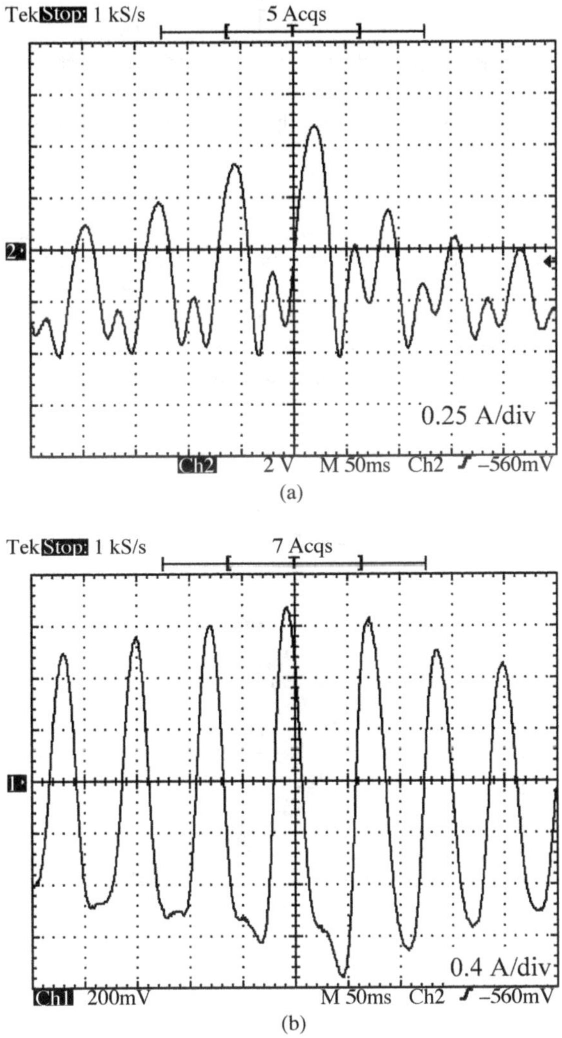

Figure 4.13 Measured chaotic current waveforms. (a) d-Axis current. (b) q-Axis current

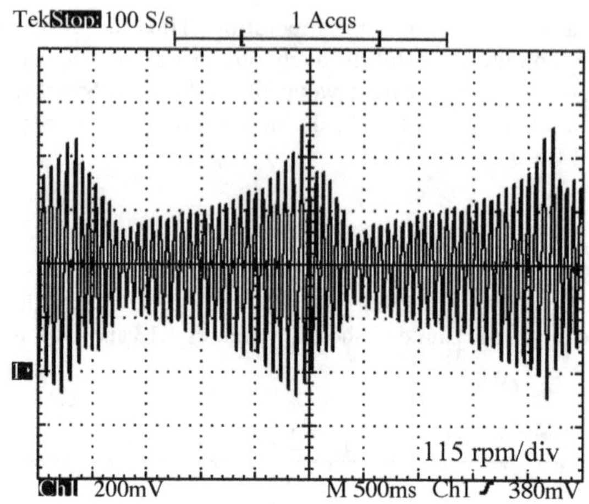

Figure 4.14 Measured chaotic motor speed waveform

the corresponding chaotic trajectories on the i_{ds}–i_{qs}, i_{ds}–ω_r and i_{qs}–ω_r planes are measured as depicted in Figure 4.15. It can be found that the waveforms exhibit a well-known property of chaos (namely, random-like but bounded), while the trajectories resemble the well-known Rössler attractor (Thompson and Stewart, 2002), especially the one on the i_{ds}–ω_r plane.

Based on the above findings, the designer for PM synchronous drive systems should not only size the PMs for the sake of maximizing the torque, but also take into account the critical value of PM flux to avoid the formation of chaos during abnormal operation.

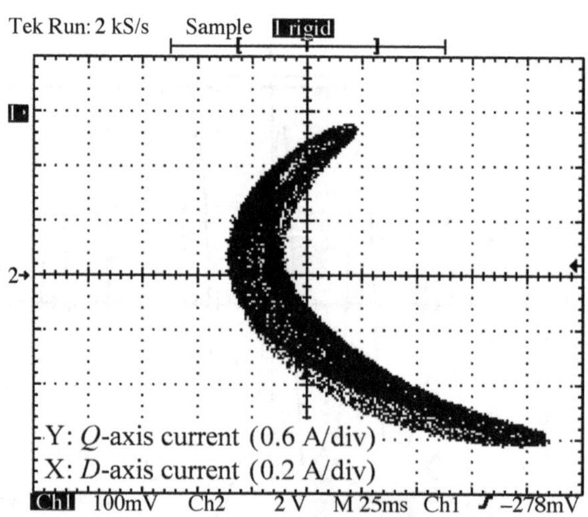

(a)

Figure 4.15 Measured chaotic trajectories. (a) Plane of d-axis current versus q-axis current. (b) Plane of d-axis current versus motor speed. (c) Plane of q-axis current versus motor speed

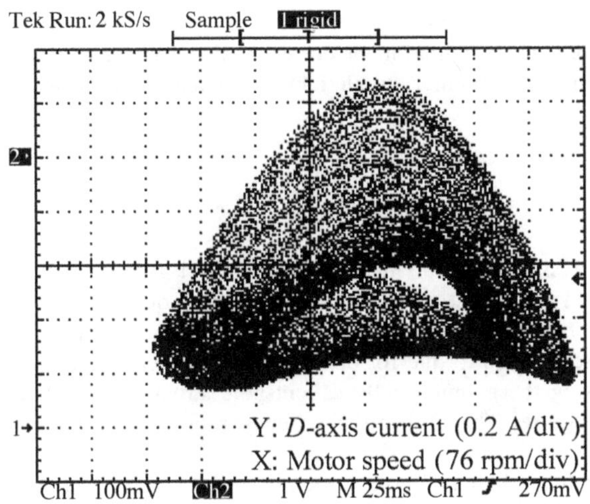

(b)

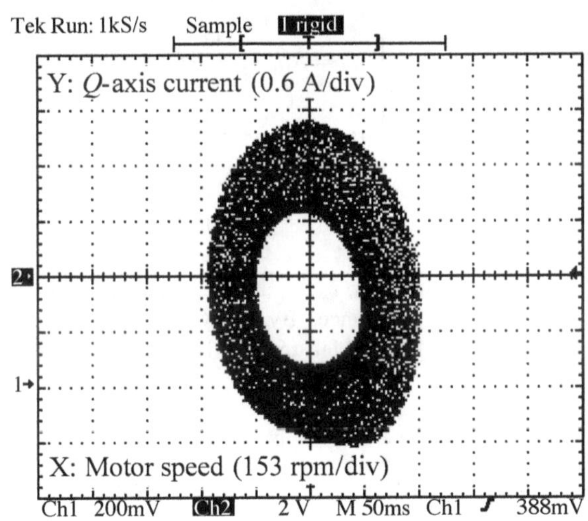

(c)

Figure 4.15 *(Continued)*

4.3 Synchronous Reluctance Drive Systems

The SynR drive systems have the advantages of high mechanical integrity, high-current operation ability, high-temperature working tolerance, and low material cost (Lipo, 1991). With the use of field-oriented control (FOC) or vector control, SynR drive systems can compete favorably with induction drive systems in high-performance applications (Betz *et al.*, 1993; Sharaf-Eldin *et al.*, 1999).

The purpose of this section is to discuss the occurrence of bifurcation and chaotic behavior in SynR drive systems. A practical SynR drive system adopting FOC is used for exemplification. Based on the derived nonlinear system equation, a bifurcation analysis shows that the system loses stability via Hopf bifurcation when the d-axis component of its three-phase motor voltages loses its control (Gao and Chau, 2004). Moreover, calculation of corresponding Lyapunov exponent further proves the existence of chaos. Finally, computer simulations and experimental results are used to support the theoretical analysis.

4.3.1 Modeling

The SynR motor is a singly salient machine. Its stator is typically equipped with three-phase sinusoidally distributed windings, which is similar to that of other AC motors such as the induction motor or the PM synchronous motor. Its rotor is purposely constructed with salient poles so as to produce the desired reluctance torque for electromechanical energy conversion. This salient rotor can be derived by using the geometrically salient-pole structure, the flux-barrier structure or the axially laminated structure. The higher the saliency ratio, the larger the reluctance torque can be produced.

Figure 4.16 shows a three-phase two-pole SynR motor with an axially laminated rotor structure. Since both the field winding and damper winding are absent in the structure, the corresponding system equations are very simple, as given by:

$$\begin{cases} v_{as} = R_s i_{as} + \dfrac{d\psi_{as}}{dt} \\ v_{bs} = R_s i_{bs} + \dfrac{d\psi_{bs}}{dt} \\ v_{cs} = R_s i_{cs} + \dfrac{d\psi_{cs}}{dt} \end{cases} \tag{4.34}$$

where v_{as}, v_{bs}, v_{cs} are the stator voltages, i_{as}, i_{bs}, i_{cs} are the stator currents, ψ_{as}, ψ_{bs}, ψ_{cs} are the stator flux linkages, and R_s is the stator resistance. By applying the well-known Park transformation (Fitzgerald, Kingsley and Umans, 1991; Matsuo and Lipo, 1993), the system equations given

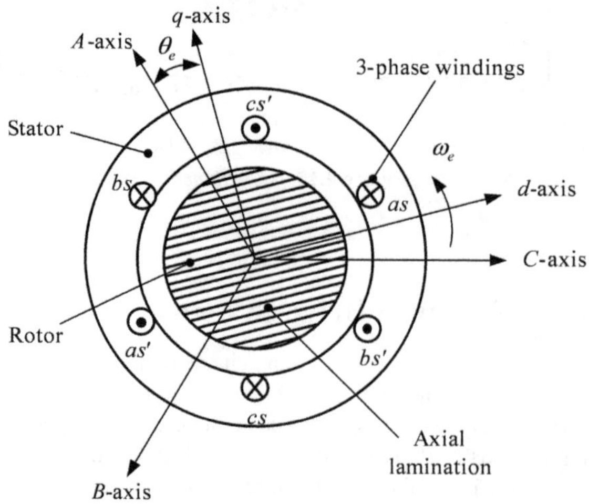

Figure 4.16 Schematic diagram of three-phase SynR motor

by (4.34) can be rewritten as:

$$\begin{cases} v_{ds} = R_s i_{ds} + \dfrac{\mathrm{d}\psi_{ds}}{\mathrm{d}t} - \omega_e L_{qs} i_{qs} \\ v_{qs} = R_s i_{qs} + \dfrac{\mathrm{d}\psi_{qs}}{\mathrm{d}t} + \omega_e L_{ds} i_{ds} \\ \psi_{ds} = L_{md} i_{ds} + L_{ls} i_{ds} = L_{ds} i_{ds} \\ \psi_{qs} = L_{mq} i_{qs} + L_{ls} i_{qs} = L_{qs} i_{qs} \end{cases} \tag{4.35}$$

where the subscripts ds and qs represent the corresponding d-axis and q-axis quantities, respectively, L_{md} and L_{mq} are the magnetizing inductances, L_{ls} is the stator leakage inductance, and ω_e is the synchronous speed. Hence, the electromagnetic torque is expressed as:

$$T_e = \frac{3}{2}\frac{P}{2}(\psi_{ds} i_{qs} - \psi_{qs} i_{ds}) \tag{4.36}$$

where P is the number of poles. In terms of L_{ds} and L_{qs}, this can be rewritten as:

$$T_e = \frac{3}{2}\frac{P}{2}(L_{ds} - L_{qs}) i_{ds} i_{qs} \tag{4.37}$$

From (4.37), it can be found that the higher the saliency (L_{ds}/L_{qs}) of the SynR motor, and the greater the difference between L_{ds} and L_{qs}, the larger becomes the value of T_e. The motion equation is then given by:

$$J\frac{\mathrm{d}\omega_r}{\mathrm{d}t} = T_e - B\omega_r - T_l \tag{4.38}$$

where J is the moment of inertia of the drive system, B is the viscous friction coefficient, T_l is the load torque and $\omega_r = 2\omega_e/P$ is the motor speed. Based on the above d–q model, the SynR drive system can employ FOC as illustrated in Figure 4.17. Hence, the dynamical model can be deduced from (4.35), (4.37),

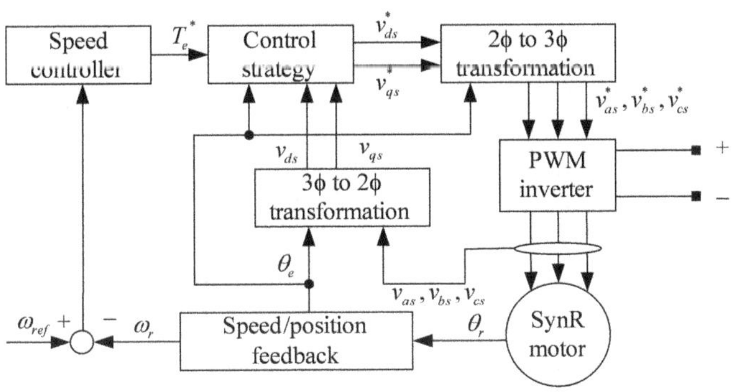

Figure 4.17 Control diagram of three-phase SynR drive system

and (4.38) as given by:

$$\begin{cases} v_{ds} = L_{ds}\dfrac{\mathrm{d}i_{ds}}{\mathrm{d}t} + R_s i_{ds} - \omega_e L_{qs} i_{qs} \\ v_{qs} = L_{qs}\dfrac{\mathrm{d}i_{qs}}{\mathrm{d}t} + R_s i_{qs} + \omega_e L_{ds} i_{ds} \\ J\dfrac{\mathrm{d}\omega_r}{\mathrm{d}t} = \dfrac{3}{2}\dfrac{P}{2}(L_{ds} - L_{qs}) i_{ds} i_{qs} - B\omega_r - T_l \end{cases} \tag{4.39}$$

The key merit of FOC is to decouple the field component (d-axis stator current i_{ds}) and the torque component (q-axis stator current i_{qs}) for high-performance operation. In general, the SynR motor is maintained at full flux by exciting it with a constant i_{ds} while the torque is regulated by controlling i_{qs}. Based on a voltage-fed inverter, i_{ds} and i_{qs} are governed by v_{ds} and v_{qs}, respectively. The constant value of v_{ds} can be easily deduced by substituting $\mathrm{d}i_{ds}/\mathrm{d}t = 0$ into (4.39), whereas v_{qs} is regulated by the speed error using proportional-integral-derivative (PID) control. The corresponding control criteria are set below:

$$\begin{cases} v_{ds} = R_s i_{ds} - \omega_e L_{qs} i_{qs} \\ v_{qs} = \left(k_p + \dfrac{k_i}{s} + s k_d\right)(\omega_r - \omega_{ref}) \end{cases} \tag{4.40}$$

where ω_{ref} is the reference mechanical rotor speed.

The above FOC works well provided that i_{ds} can be kept constant and under proper control. If v_{ds} loses its control, the SynR drive system may exhibit strange behavior. For the sake of simplicity, the analytical derivation is based on the assumption of $k_i = k_d = 0$. Thus, based on (4.39) and (4.40), the SynR drive system can be expressed as:

$$\begin{cases} v_{ds} = L_{ds}\dfrac{\mathrm{d}i_{ds}}{\mathrm{d}t} + R_s i_{ds} - \omega_e L_{qs} i_{qs} \\ k_p(\omega_r - \omega_{ref}) = L_{qs}\dfrac{\mathrm{d}i_{qs}}{\mathrm{d}t} + R_s i_{qs} + \omega_e L_{ds} i_{ds} \\ J\dfrac{\mathrm{d}\omega_r}{\mathrm{d}t} = \dfrac{3}{2}\dfrac{P}{2}(L_{ds} - L_{qs}) i_{ds} i_{qs} - B\omega_r - T_l \end{cases} \tag{4.41}$$

After rearranging i_{ds}, i_{qs}, and ω_r as the state variables, (4.41) can be rewritten as:

$$\begin{cases} \dfrac{\mathrm{d}i_{ds}}{\mathrm{d}t} = (v_{ds} - R_s i_{ds} + \omega_e L_{qs} i_{qs})/L_{ds} \\ \dfrac{\mathrm{d}i_{qs}}{\mathrm{d}t} = \left[-R i_{qs} - \omega_e L_{ds} i_{ds} + k_p(\omega_r - \omega_{ref})\right]/L_{qs} \\ \dfrac{\mathrm{d}\omega_r}{\mathrm{d}t} = \left[\dfrac{3}{2}\dfrac{P}{2}(L_{ds} - L_{qs}) i_{ds} i_{qs} - B\omega_r - T_l\right]/J \end{cases} \tag{4.42}$$

The system equation can be further simplified by transforming t to $\tau t'$ and $(i_{ds}, i_{qs}, \omega_e)$ to (x, y, z) as defined by:

$$\begin{bmatrix} i_{ds} \\ i_{qs} \\ \omega_e \end{bmatrix} = \begin{bmatrix} bk & 0 & 0 \\ 0 & k & 0 \\ 0 & 0 & 1/\tau \end{bmatrix} \begin{bmatrix} x \\ y \\ z \end{bmatrix} \tag{4.43}$$

where $b = L_{qs}/L_{ds}$, $k = \sqrt{8J/3P^2(L_{ds}-L_{qs})b\tau^2}$ and $\tau = L_{qs}/R_s$. Hence, the SynR drive system can be expressed as:

$$\begin{cases} \dfrac{dx}{dt'} = v'_{ds} - bx + yz \\ \dfrac{dy}{dt'} = -y - xz + c(z - z_{ref}) \\ \dfrac{dz}{dt'} = xy - az + T'_l \end{cases} \tag{4.44}$$

where $v'_{ds} = v_{ds}/kR_s$, $c = 2k_p/kPL_{qs}$, $z_{ref} = \tau\omega_{ref}$, $a = BL_{qs}/JR_s$ and $T'_l = P\tau^2 T_l/2J$.

4.3.2 *Analysis*

It is essential to analyze the stability of the equilibrium point and the trajectory around the equilibrium point. The key is to derive the eigenvalues of the system at the equilibrium point.

For the special case when $v_{ds} = 0$, $\omega_{ref} = 0$, and $T_l = 0$, the system equation (4.44) becomes:

$$\begin{cases} \dfrac{dx}{dt'} = -bx + yz \\ \dfrac{dy}{dt'} = -y - xz + cz \\ \dfrac{dz}{dt'} = xy - az \end{cases} \tag{4.45}$$

The equilibrium point can be deduced by setting the derivatives in (4.45) equal to zero. Obviously, the origin is a trivial equilibrium point. The nonzero equilibria can be solved by the following equations:

$$\begin{cases} x^2 - cx + a = 0 \\ y^2 - ab = 0 \\ az - xy = 0 \end{cases} \tag{4.46}$$

Table 4.2 Routh–Hurwitz stability criterion

s^3	1	$b(a+x_{eq}{}^2)/a$
s^2	$a+b+1$	$2b(x_{eq}^2-a)$
s^1	$\dfrac{b(x_{eq}{}^2+a)}{a}-\dfrac{2b(x_{eq}{}^2-a)}{a+b+1}$	0
s^0	$2b(x_{eq}{}^2-a)$	0

As i_{ds}, i_{qs}, ω_r and their transformed quantities x, y, z are all realistic parameters, there are three possible cases:

- If $\Delta = c^2-4a < 0$, there is one equilibrium point (0, 0, 0).
- If $\Delta = c^2-4a = 0$, there are three equilibrium points (0, 0, 0) and $(c/2, \pm\sqrt{ab}, \pm\sqrt{b/a})$.
- If $\Delta = c^2-4a > 0$, there are five equilibrium points (0, 0, 0) and $((c\pm\sqrt{c^2-4a})/2, \pm\sqrt{ab}, x_{eq}y_{eq}/a)$.

The local stability of the equilibrium point is described by the eigenvalues of the system characteristic equation:

$$\det(\lambda\mathbf{I}-\mathbf{J}) = 0 \tag{4.47}$$

where λ denotes the eigenvalues, $\mathbf{I}$ is the identity matrix, and $\mathbf{J}$ is the Jacobian matrix of the transformed system evaluated at the equilibrium point (x_{eq}, y_{eq}, z_{eq}). Hence, the eigenvalues can be deduced from an explicit cubic equation, as given by:

$$\lambda^3+r_2\lambda^2+r_1\lambda+r_0 = 0 \tag{4.48}$$

It is easy to check that the origin is a locally stable equilibrium point since the corresponding eigenvalues are all negative, that is $-a$, $-b$. and -1. For the nonzero equilibrium case, an explicit form of (4.48) can be obtained using (4.46):

$$\lambda^3+(a+b+1)\lambda^2+b(a+x_{eq}^2)\lambda/a+2b(x_{eq}^2-a) = 0 \tag{4.49}$$

Applying the Routh–Hurwitz stability criterion, as shown in Table 4.2, the local stability is guaranteed by the following condition:

$$\begin{cases} 2b(x_{eq}{}^2-a) > 0 \\ \dfrac{ab(3a+b+1)-b(a-b-1)x_{eq}{}^2}{a(a+b+1)} > 0 \end{cases} \tag{4.50}$$

Accordingly, the system exhibits Hopf bifurcation (Alligood, Sauer and Yorke, 1996) if there exist a pair of complex eigenvalues satisfying the following criteria:

$$\begin{cases} \mathrm{Re}(\lambda)|_{c=c_0} = 0 \\ \mathrm{Im}(\lambda)|_{c=c_0} \neq 0 \\ \dfrac{\mathrm{d}}{\mathrm{d}c}\mathrm{Re}(\lambda)|_{c=c_0} \neq 0 \end{cases} \tag{4.51}$$

where c_0 is the critical value of c at which the Hopf bifurcation occurs.

Considering that there exists a purely imaginary nonzero characteristic root $\lambda = nj$ ($n \neq 0$) in (4.49), it yields:

$$-n^3 j - (a+b+1)n^2 + \frac{b}{a}(a+x_{eq}^2)nj + 2b(x_{eq}^2 - a) = 0 \tag{4.52}$$

Separating the real part and the imaginary part results in:

$$\begin{cases} n^2 = b(a+x_{eq}{}^2)/a \\ n^2 = 2b(x_{eq}{}^2 - a)/(a+b+1) \end{cases} \tag{4.53}$$

For positive a, b, and c, the existence of n is guaranteed by the following condition:

$$\begin{cases} a-b-1 > 0 \\ x_{eq}{}^2 - a > 0 \\ x_{eq}{}^2 = a(3a+b+1)/(a-b-1) \end{cases} \tag{4.54}$$

This condition shows that the system parameters L_{ds}, L_{qs}, J, and B of the SynR drive system have a critical effect on the existence of bifurcation. If their values are not properly set, Hopf bifurcation and chaos may occur.

It is straightforward to check that $((c-\sqrt{c^2-4a})/2)^2 < a$ for positive a and c. For this reason, the equilibrium point pair $((c-\sqrt{c^2-4a})/2, \pm\sqrt{ab}, x_{eq}y_{eq}/a)$, if they exist, do not meet the bifurcation criteria in (4.51). From (4.50), they are always locally unstable.

By substituting $x_{eq} = (c+\sqrt{c^2-4a})/2$ into (4.54), the critical value of c at which Hopf bifurcation occurs can be obtained:

$$c_0 = 4a\sqrt{a/[(a-b-1)(3a+b+1)]} \tag{4.55}$$

The corresponding critical eigenvalues are given by:

$$\lambda_1 = -(a+b+1), \; \lambda_{2,3} = \pm j\sqrt{bc_0(c_0+\sqrt{c_0{}^2-4a})/2a} \tag{4.56}$$

Based on the stability criterion in (4.50), when c is large enough, but still smaller than the critical value c_0, the equilibrium point pair $((c+\sqrt{c^2-4a})/2, \pm\sqrt{ab}, x_{eq}y_{eq}/a)$ are stable. Once the value of c exceeds the critical value, they become unstable saddle points. Furthermore, they may even become chaotic attractors.

In general, the solution of system (4.44) leads to a fifth order polynomial. Consequently, the equilibrium point (x_{eq}, y_{eq}, z_{eq}) has five possible roots: five real roots; three real roots and a pair of complex conjugate

Table 4.3 Relationship between signs of Lyapunov exponents and type of attractors

Signs	Type of attractors
(−, −, −)	Fixed point
(0, −, −)	Limit cycle
(0, 0, −)	Quasiperiodic torus
(+, 0, −)	Chaotic

roots; or one real root and two pairs of complex conjugate roots. Analytical results for this case are almost impossible, nevertheless a numerical analysis can always be performed.

Under the occurrence of Hopf bifurcation, the dynamical system may demonstrate a complicated behavior – that is, chaos. To further identify the chaotic behavior, the calculation of Lyapunov exponents plays an important role. Namely, a system exhibits chaotic behavior if at least one of its Lyapunov exponents is positive (Hilborn, 1994). For a three-dimensional state-space system described by a set of three first-order differential equations, the corresponding type of attractors (fixed point, limit cycle, quasiperiod torus, and chaotic) can be directly determined by the signs of Lyapunov exponents (Parker and Chua, 1989) as listed in Table 4.3. Obviously, the SynR drive system described by (4.44) belongs to the aforementioned three-dimensional state-space system. Thus, this SynR drive system demonstrates chaotic behavior if its Lyapunov exponents are one zero, one positive, and one negative.

The notion of a Lyapunov exponent is a generalization of the concept of an eigenvalue as a measure of the stability of a fixed point or a characteristic exponent as a measure of the stability of a periodic orbit. For a chaotic trajectory, it is not sensible to examine the instantaneous eigenvalue of a trajectory. The next best quantity, therefore, is an eigenvalue averaged over the whole trajectory. The Lyapunov exponent is best defined by measuring the evolution (under a flow) of the tangent manifold:

$$\lambda_i = \lim_{t\to\infty} \frac{1}{t} \ln|m_i(t)|, \quad (i = 1, \ldots, l) \tag{4.57}$$

where $m_1(t), \ldots, m_l(t)$ is the solution of the system characteristic equation (4.47), and l is the system dimension. The Gram–Schmidt orthonormalization algorithm can readily be used to calculate the Lyapunov exponents.

4.3.3 Simulation

To illustrate the strange behavior of an SynR drive system, computer simulations of both waveforms and trajectories are carried out. A three-phase 4-pole SynR drive system is used for exemplification, which has the parameters $P=4$, $L_{ds}=133.3$ mH, $L_{qs}=25.1$ mH, $R_s=0.029\ \Omega$, $T_l=0$ Nm, $J=1.988\times10^{-3}$ kgm^2, $B=3.513\times10^{-3}$ Nm/rads^{-1}, $v_{ds}=0$ V, and $\omega_{ref}=0$ rad/s. Notice that the values of v_{ds}, T_l, and ω_{ref} are simply chosen for illustration purposes, and can readily be altered without affecting the final conclusion. Most importantly, the parameters of this practical drive system meet the requirements of Hopf bifurcation, as described by (4.54).

The locus of eigenvalues obtained by the equilibrium point pair $((c+\sqrt{c^2-4a})/2, \pm\sqrt{ab}, x_{eq}y_{eq}/a)$ is depicted in Figure 4.18, in which the parameter c starts from $\Delta=0$, namely $c_i = 2\sqrt{a} = 2.473$.

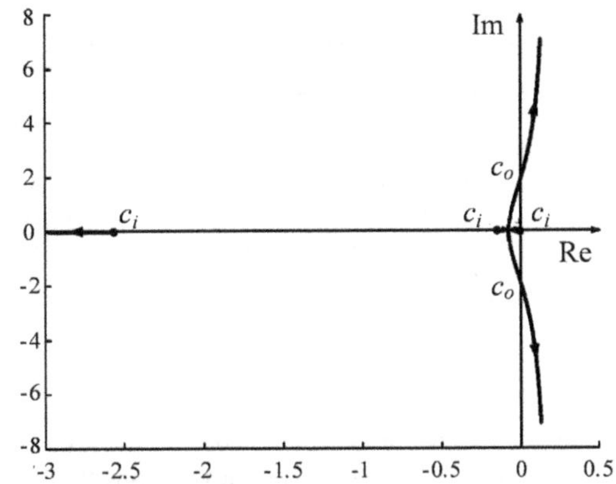

Figure 4.18 Locus of eigenvalues via parameter c

As discussed above, this pair of equilibrium points lose their stability once the parameter c exceeds its critical value. Clearly, at this critical value of $c_0 = 5.421$, the locus crosses the imaginary axis and Hopf bifurcation occurs.

Figure 4.19 shows the Lyapunov exponents of the SynR drive system. From these three curves, the underlying system behavior can be easily distinguished by using Table 4.3. With a small value of c, the SynR drive system has only one stable equilibrium point (fixed point). The corresponding Lyapunov exponents are all negative. Thus, five equilibria occur with an increase in c. Once c exceeds its critical value ($c_0 = 5.421$), a pair of stable equilibria lose their stability and become a pair of saddle points. This pair of saddle points cause the SynR drive system to demonstrate a complex behavior. There are chaotic attractors in the regions of $5.421 \leq c < 15.385$ and

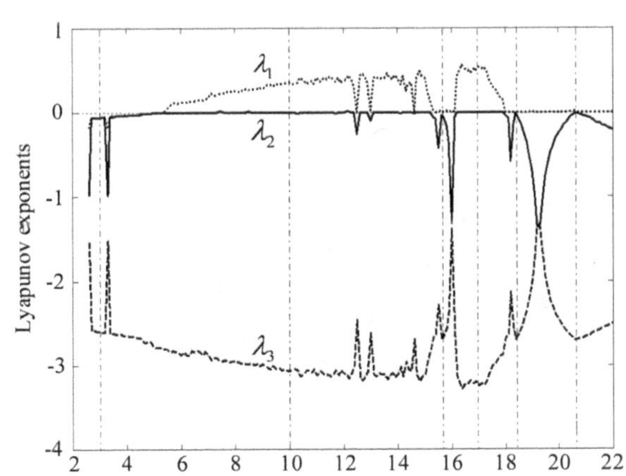

Figure 4.19 Lyapunov exponents via parameter c

Table 4.4 Lyapunov exponents at typical values of c

c	Lyapunov exponents	Attractor type
3.00	−0.056, −0.056, −2.598	Fixed point
10.00	0.336, 0.000, −3.045	Chaotic
15.52	0.000, −0.014, −2.678	Limit cycle
17.00	0.514, 0.000, −3.224	Chaotic
18.40	0.000, −0.012, −2.699	Limit cycle
20.74	0.000, −0.010, −2.700	Limit cycle

$16.125 \leq c \leq 17.995$. On the other hand, in the regions of $15.385 \leq c < 16.125$ and $c \geq 18.005$, the system exhibits limit cycles. Table 4.4 lists the Lyapunov exponents under different typical values of c.

Figure 4.20 shows the speed bifurcation diagram via the parameter c. This diagram is obtained by plotting the successive 200 crossing points of the steady-state trajectory with a fixed Poincaré section via the parameter c. Therefore, the underlying chaotic attractor (CA) or limit cycle (LC) can be easily identified, as labeled in the figure. It can be found that this diagram matches the phenomena obtained by Figure 4.19 and Table 4.4. Namely, when $15.385 \leq c < 16.125$ or $c \geq 18.005$, the system exhibits limit-cycle operation; when $5.421 \leq c < 15.385$ or $16.125 \leq c \leq 17.995$, the system offers chaotic operation.

Figure 4.21 shows the simulated chaotic waveforms of i_{ds}, i_{qs}, and ω_r of the SynR drive system when $c = 10$ and the corresponding trajectories on the i_{ds}–i_{qs}, i_{ds}–ω_r and i_{qs}–ω_r planes. It can be found that the waveforms offer the well-known chaotic properties, namely random-like but bounded, while the trajectories resemble a butterfly (like the well-known Lorenz attractor). On the other hand, the SynR drive system exhibits limit-cycle operation when $c = 22$. Figure 4.22 depicts the corresponding time-domain waveforms of i_{ds}, i_{qs}, and ω_r, and its trajectories on the i_{ds}–i_{qs}, i_{ds}–ω_r and i_{qs}–ω_r planes. It can easily be observed that both the waveforms and trajectories offer a periodic property.

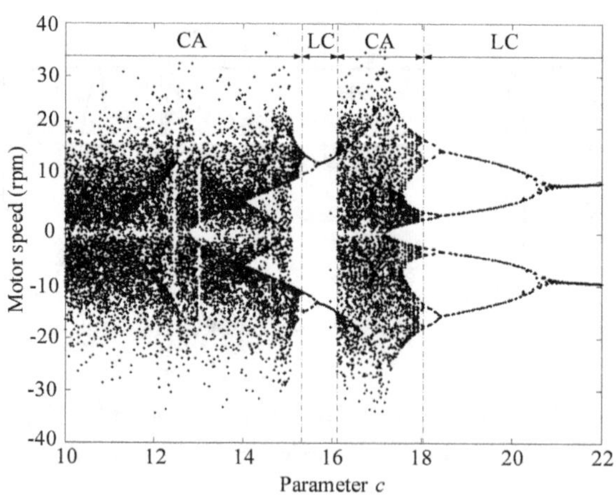

Figure 4.20 Speed bifurcation diagram via parameter c

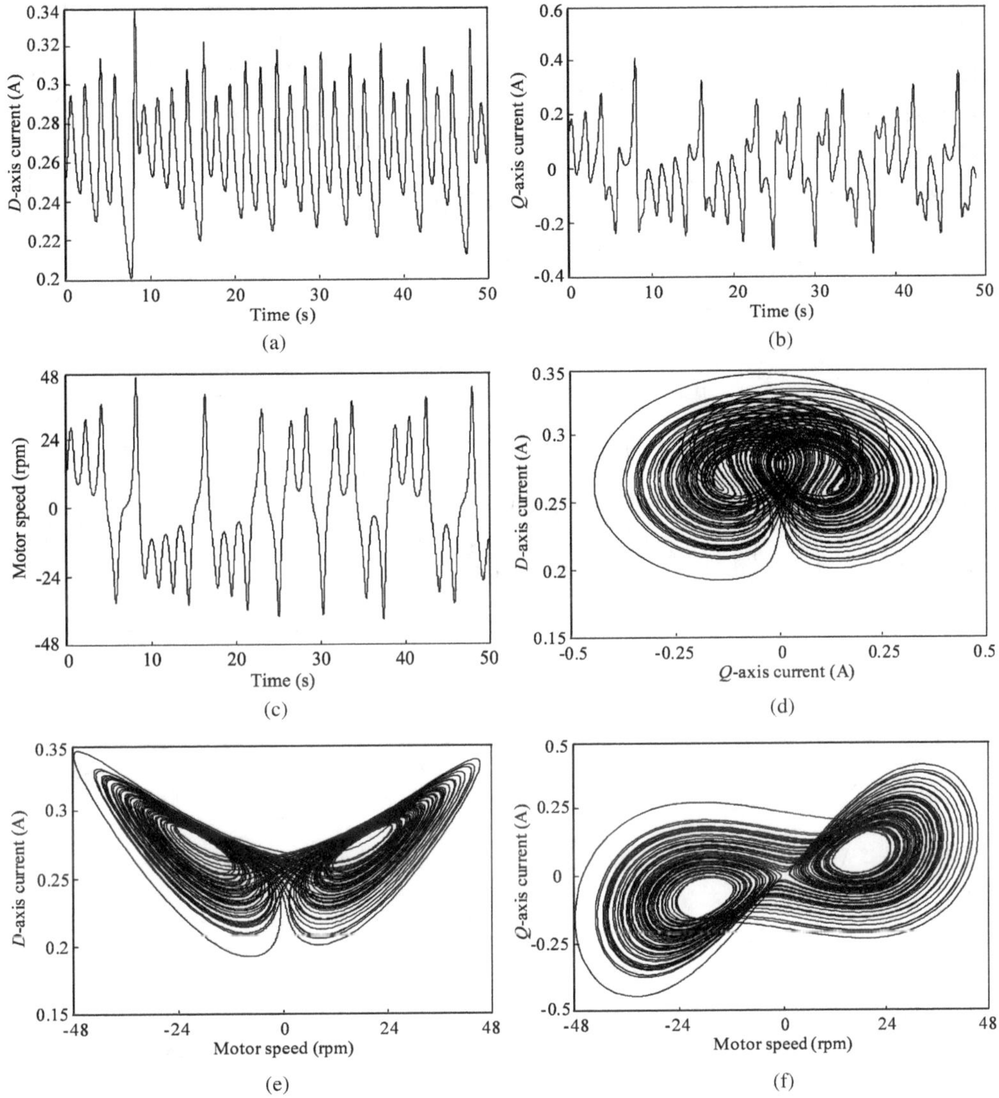

Figure 4.21 Simulated chaotic waveforms and trajectories when $c = 10$. (a) d-Axis current. (b) q-Axis current. (c) Motor speed. (d) Plane of d-axis current versus q-axis current. (e) Plane of d-axis current versus motor speed. (f) Plane of q-axis current versus motor speed

4.3.4 Experimentation

Based on the aforementioned SynR drive system and operating conditions, the resulting waveforms and trajectories are measured, as shown in Figures 4.23 and 4.24. Comparing the periodic waveforms and trajectories shown in Figure 4.22 with those in Figure 4.24, the measured results are in agreement with the simulation results. On the other hand, comparing the chaotic waveforms shown in Figures 4.21 and 4.23, the simulated and measured patterns cannot be matched because the chaotic pattern is aperiodic and very sensitive to the initial conditions. Nevertheless, the random-like but

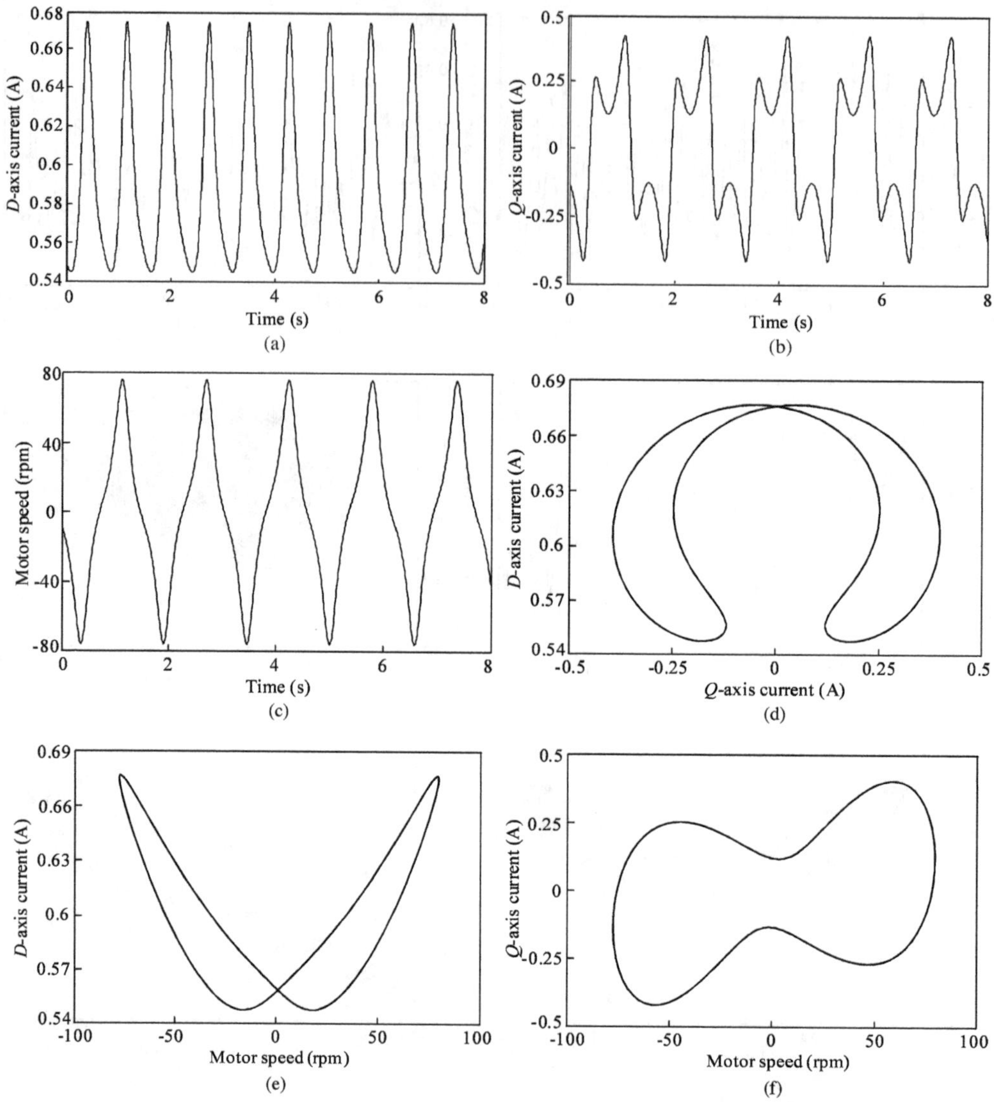

Figure 4.22 Simulated periodic waveforms and trajectories when $c = 22$. (a) d-Axis current. (b) q-Axis current. (c) Motor speed. (d) Plane of d-axis current versus q-axis current. (e) Plane of d-axis current versus motor speed. (f) Plane of q-axis current versus motor speed

bounded nature can be easily observed from their trajectories. The measured boundary values are also in good agreement with the simulated values.

It should be noted that the above analysis regulates the classical concept of SynR drive system operation. In addition to fixed-point operation, the SynR drive system can offer chaotic operation and limit-cycle operation with variations of controllable parameters. Although the chaotic operation has not yet been widely used for industrial applications, it is anticipated that the resulting chaotic motion of this SynR drive system is beneficial to some niche areas, such as industrial mixing or domestic washing.

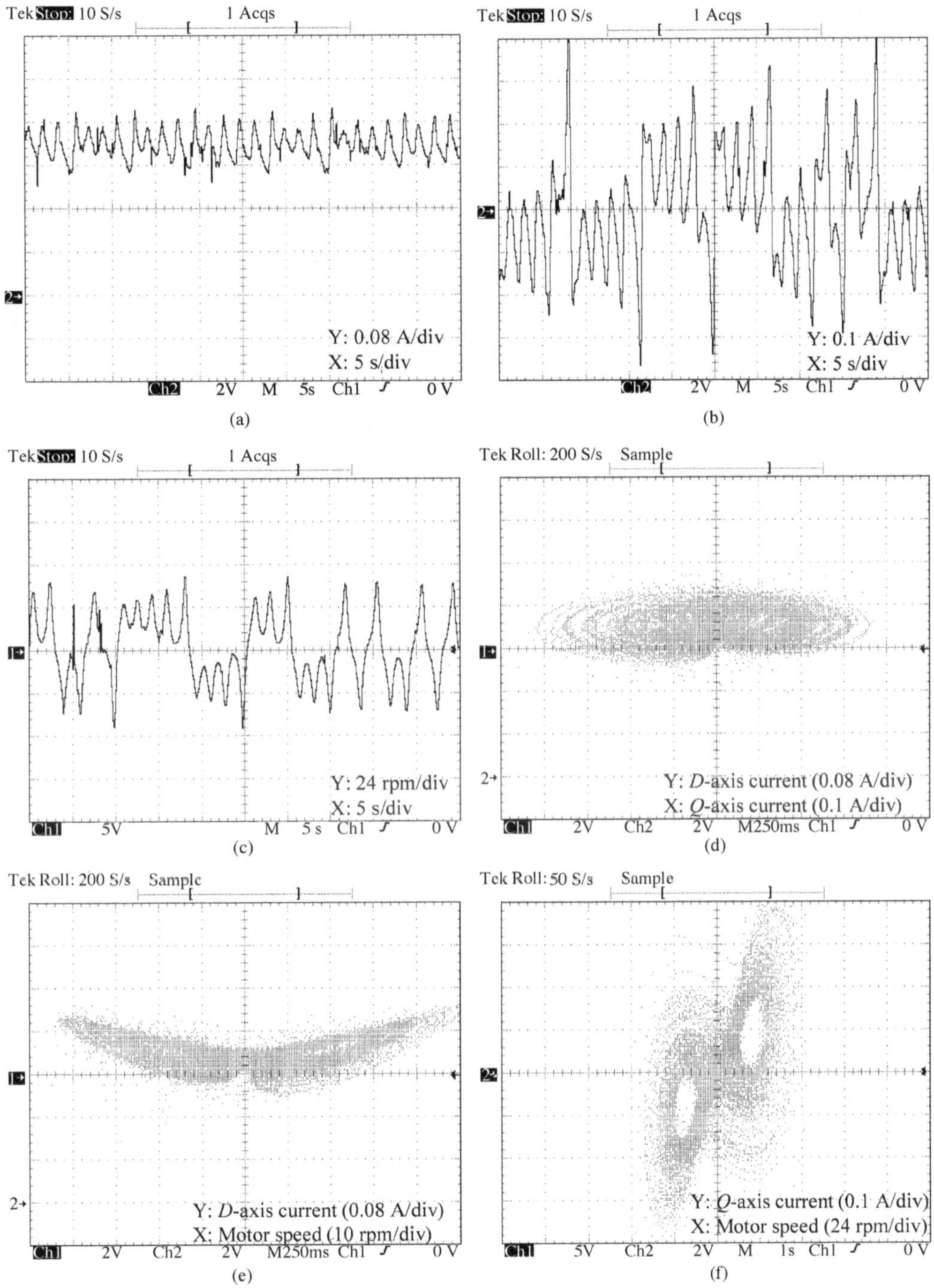

Figure 4.23 Measured chaotic waveforms and trajectories when $c = 10$. (a) d-Axis current. (b) q-Axis current. (c) Motor speed. (d) Plane of d-axis current versus q-axis current. (e) Plane of d-axis current versus motor speed. (f) Plane of q-axis current versus motor speed

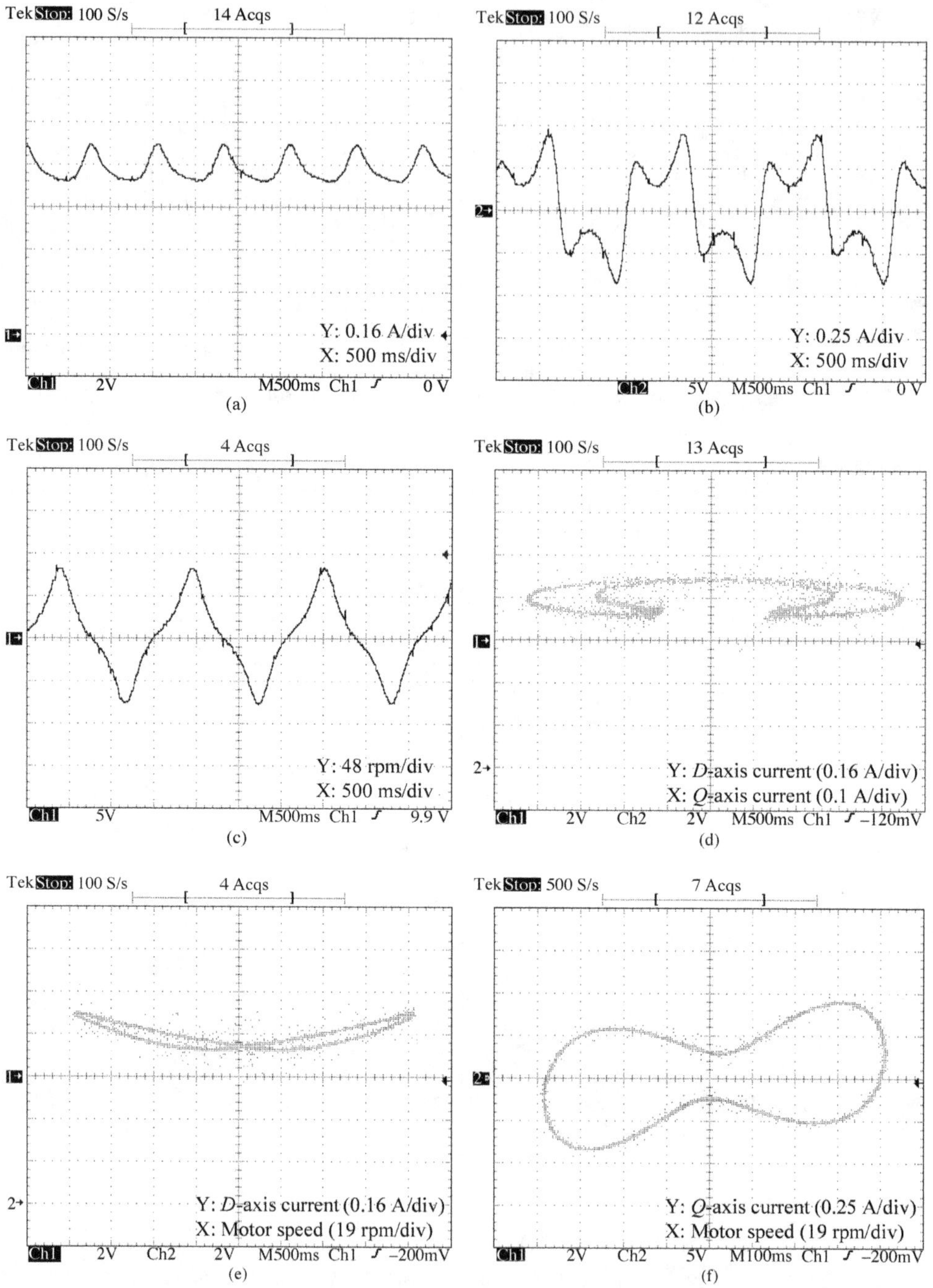

Figure 4.24 Measured periodic waveforms and trajectories when $c = 22$. (a) d-Axis current. (b) q-Axis current. (c) Motor speed. (d) Plane of d-axis current versus q-axis current. (e) Plane of d-axis current versus motor speed. (f) Plane of q-axis current versus motor speed

References

Alligood, K.T., Sauer, T.D., and Yorke, J.A. (1996) *Chaos: An Introduction to Dynamical Systems*, Springer-Verlag, New York.

Betz, R.E., Lagerquist, R., Jovanovic, M. *et al.* (1993) Control of synchronous reluctance machines. *IEEE Transactions on Industry Applications*, **29**, 1110–1122.

Fitzgerald, A.E., Kingsley, C. Jr., and Umans, S.D. (1991) *Electric Machinery*, McGraw-Hill, Singapore.

Gao, Y. and Chau, K.T. (2003a) Chaotification of induction motor drives under periodic speed command. *Electric Power Components and Systems*, **31**, 1083–1099.

Gao, Y. and Chau, K.T. (2003b) Design of permanent magnets to avoid chaos in PM synchronous machines. *IEEE Transactions on Magnetics*, **39**, 2995–2997.

Gao, Y. and Chau, K.T. (2004) Hopf bifurcation and chaos in synchronous reluctance motor drives. *IEEE Transactions on Energy Conversion*, **19**, 296–302.

Gieras, J.F. and Wing, M. (2002) *Permanent Magnet Motor Technology: Design and Applications*, Marcel Dekker, New York.

Hilborn, R.C. (1994) *Chaos and Nonlinear Dynamics: An Introduction for Scientists and Engineers*, Oxford University Press, Oxford.

Leonhard, W. (1996) *Control of Electrical Drives*, Springer-Verlag, New York.

Lipo, T.A. (1991) Synchronous reluctance machines – a viable alternative for AC drives? *Electric Machines and Power Systems*, **19**, 659–671.

Matsuo, T. and Lipo, T.A. (1993) Field oriented control of synchronous reluctance machine. Proceedings of IEEE Power Electronics Specialists Conference, pp. 425–431.

Parker, T.S. and Chua, L.O. (1989) *Practical Numerical Algorithms for Chaotic Systems*, Springer-Verlag, New York.

Sharaf-Eldin, T., Dunnigan, M.W., Fletcher, J.E., and Williams, B.W. (1999) Nonlinear robust control of a vector-controlled synchronous reluctance machine. *IEEE Transactions on Power Electronics*, **14**, 1111–1121.

Thompson, J.M.T. and Stewart, H.B. (2002) *Nonlinear Dynamics and Chaos*, Wiley, New York.

Ye, S. and Chau, K.T. (2005) Design of permanent magnets to chaoize PM synchronous motors for industrial mixer. Proceedings of IEEE International Magnetics Conference, pp. 723–724.

References

5

Chaos in Switched Reluctance Drive Systems

The switched reluctance (SR) drive is a specific kind of doubly-salient drive. The SR drive system has the advantages of robust structure, high reliability, simple control, wide range of constant power operation, and low manufacturing cost (Lawrenson, 1992). On the other hand, it suffers from the drawbacks of large noise, large torque ripple, special converter configuration, and high nonlinearities (Miller, 1993; Zhan, Chan and Chau, 1999). Thus, it is anticipated that the SR drive system is more prone to chaos due to the high nonlinearities.

In the SR drive, there are two driving periods, namely the commutation period and the pulse width modulation (PWM) period (Miller, 1993). The commutation period is the stroke angle determined by the phase commutation, in which the phase current always has an initial value of zero. Hence, for each phase winding, the drive system operates in a discontinuous conduction mode for the term of the commutation period. The PWM period is the period of the carrier signal for PWM regulation, which is also the switching period of power devices. This PWM period is usually short enough to force the phase current to be continuous within the commutation period. The phase current oscillation, and hence the torque oscillation, are governed by the distribution of those PWM pulses within the commutation period. At normal operation, the PWM pulses are uniformly distributed. Hence, the corresponding oscillation is regular and its oscillating frequency follows the PWM frequency. However, if some pulse widths are zero, which are caused by the skipping cycles, both the current and torque oscillations will be very severe. The commutation period will become nonperiodic and generate a low-frequency oscillation. Apart from the system nonlinearities, this low-frequency oscillation is another factor contributing to the chaotic operation of the SR drive. Furthermore, the synchronization between the commutation and PWM periods is an additional factor contributing to chaotic operation. When these two periods are synchronous, asynchronous, or incommensurate – namely the ratio of the commutation period to the PWM period is respectively an integer, a rational number, or an irrational number – the corresponding operation is respectively a fundamental, subharmonic, or quasiperiodic solution.

Instead of using voltage PWM regulation, the SR drive system can be controlled by the current hysteresis regulation. Actually, this current hysteresis regulation generally prefers a voltage PWM regulation for the speed control of the SR drive. Contrary to the voltage PWM regulation, the current hysteresis regulation can be easily implemented by comparing the reference current and actual current with a hysteresis loop. Therefore, the switching frequency of the current hysteresis regulation is variable and determined by the operating conditions of the SR drive. Since chaos has been well verified to exist in a

Chaos in Electric Drive Systems: Analysis, Control and Application, First Edition. K.T. Chau and Zheng Wang.
© 2011 John Wiley & Sons (Asia) Pte Ltd. Published 2011 by John Wiley & Sons (Asia) Pte Ltd.

dynamical system with a hysteresis loop (Zhusubaliyev and Mosekilde, 2003), an SR drive system using current hysteresis regulation can also exhibit chaotic behavior under certain operating conditions.

The subharmonic and chaotic behavior was first identified in a voltage-controlled closed-loop rotational SR drive system (Chen *et al.*, 2002). By tuning the speed gain, subharmonic and chaotic behaviors occur in the drive system. Consequently, the chaotic behavior in a voltage-controlled open-loop linear SR drive system has also been investigated (De Castro, Robert and Goeldel, 2008). By tuning the step frequency, which is defined to be three times the switching frequency, bifurcation and chaos phenomenon can be observed in the drive system.

In this chapter, chaos in two major classes of SR drive systems, namely the voltage-control mode and the current-control mode, are modeled, analyzed, and validated by computer simulation and practical experimentation.

5.1 Voltage-Controlled Switched Reluctance Drive System

5.1.1 Modeling

Figure 5.1 shows a typical 3-phase voltage-controlled SR drive system. The corresponding speed control is achieved by applying PWM chopping to its motor voltages. The commutation strategy uses rotor position feedback to select the turn-on angle θ_o and turn-off angle θ_c of those lower-leg power switches (A_2, B_2, and C_2). When the phase windings are conducted in turn, the dwell interval $\theta_d = \theta_c - \theta_o$ of each phase winding is selected to be equal to the commutation angle $\theta_s = 2\pi/(mN_r)$, where m is the number of phases and N_r is the number of rotor poles. As shown in Figure 5.2, the stator phase-A winding starts to conduct at $\theta = \theta_o$ and ends at $\theta = \theta_o + \theta_s$. Subsequently, the phase-B winding conducts from $\theta = \theta_o + \theta_s$ to $\theta = \theta_o + 2\theta_s$, and the phase-C winding conducts from $\theta = \theta_o + 2\theta_s$ to $\theta = \theta_o + 3\theta_s$. For each cycle of conduction of all phase windings, the rotational angle of the rotor is a total of $2\pi/N_r$. Figure 5.2(a) also indicates that each phase winding conducts at the instant of decreasing magnetic reluctance between the stator and the rotor, hence producing a positive torque to drive the rotor. For synchronizing the voltage PWM regulation with the phase commutation, the ramp voltage v_r for each phase winding is a function of the instantaneous rotor displacement θ, as given by:

$$v_r = v_l + (v_u - v_l)[(\theta - \theta_o) \bmod \theta_T]/\theta_T \tag{5.1}$$

where v_l and v_u are the lower and upper bounds of the ramp voltage, $\theta_T = \theta_d/n_\theta$ is its period, n_θ is an integer, and $(\theta - \theta_o) \bmod \theta_T$ is defined as the remainder of $(\theta - \theta_o)$ divided by θ_T. As shown in Figure 5.1, since v_ω and v_ω^* are linear functions of the instantaneous speed ω and reference speed ω^*, respectively, the speed control signal v_c can be expressed as:

$$v_c = g(\omega - \omega^*) \tag{5.2}$$

where g is the overall feedback gain incorporating both the frequency-to-voltage (F/V) converter and the operational amplifier (OA). Then, both v_r and v_c are fed into the comparator (CM) which outputs the signal to turn on or off those upper-leg power switches (A_1, B_1, and C_1), depending on the phase commutation. When v_c exceeds v_r, the upper-leg switch, being the same phase of the on-state lower-leg switch, is turned off; otherwise, it is turned on. The other phase switches remain off. The corresponding waveforms of v_c and v_r, as well as switching signals, are shown in Figure 5.2(b).

Instead of the phase current i_k, the phase flux linkage ψ_k is chosen as the state variable so that the system differential equation does not involve the calculation of $\partial\psi_k(\theta, i_k)/\partial\theta$ and $\partial\psi_k(\theta, i_k)/\partial i_k$. Since m phase windings of the SR motor are conducted in turn, only two adjacent phase windings have currents at the same time when $m > 2$. For the sake of clarity and simplicity, m-phase windings conducted in turn are

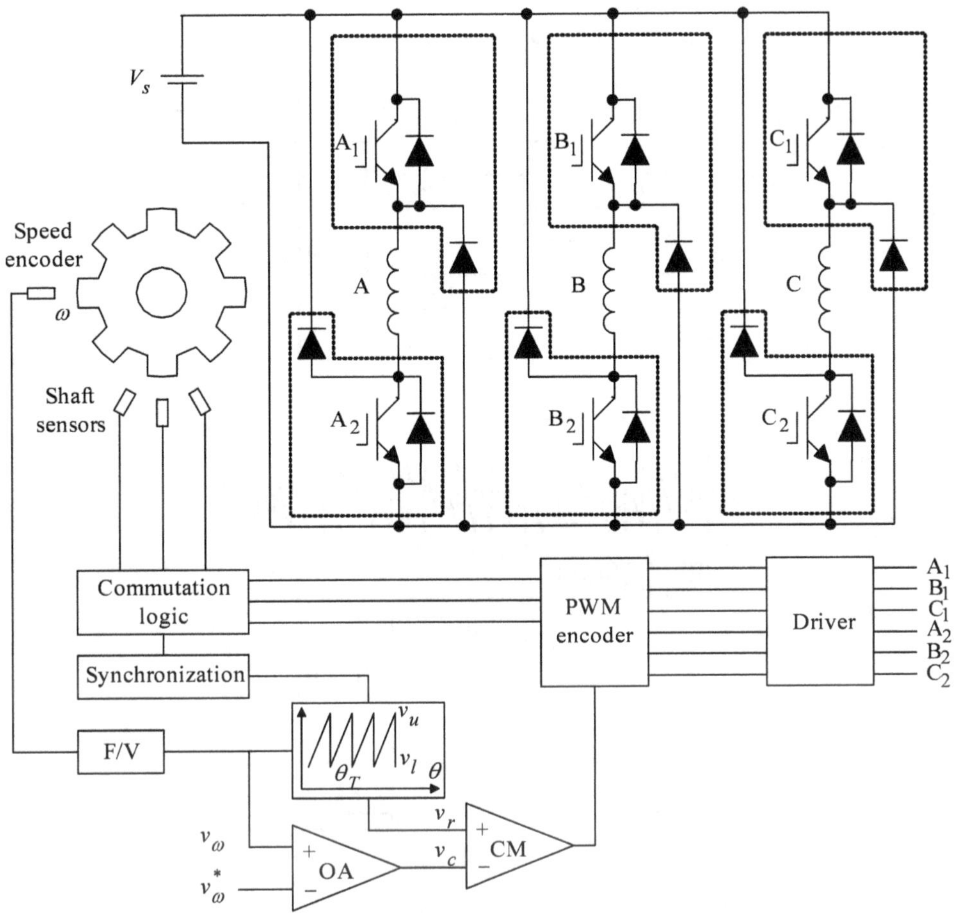

Figure 5.1 Voltage-controlled SR drive system

represented by only two-phase windings (namely 1 and 2) activated alternately. When the phase winding is controlled by PWM regulation, it is called the activated winding; otherwise, it is called the inactivated winding. If we consider that winding 2 lags behind winding 1 by θ_s, the system equation of the SR drive can be expressed as:

$$\begin{cases} \dfrac{\mathrm{d}\theta}{\mathrm{d}t} = \omega \\ \dfrac{\mathrm{d}\omega}{\mathrm{d}t} = (-B\omega + T_e(\theta, \psi_1, \psi_2, \theta_s) - T_l)/J \\ \dfrac{\mathrm{d}\psi_1}{\mathrm{d}t} = -Ri_1(\theta, \psi_1) + u_1(\theta) \\ \dfrac{\mathrm{d}\psi_2}{\mathrm{d}t} = -Ri_2(\theta - \theta_s, \psi_2) + u_2(\theta - \theta_s) \end{cases} \tag{5.3}$$

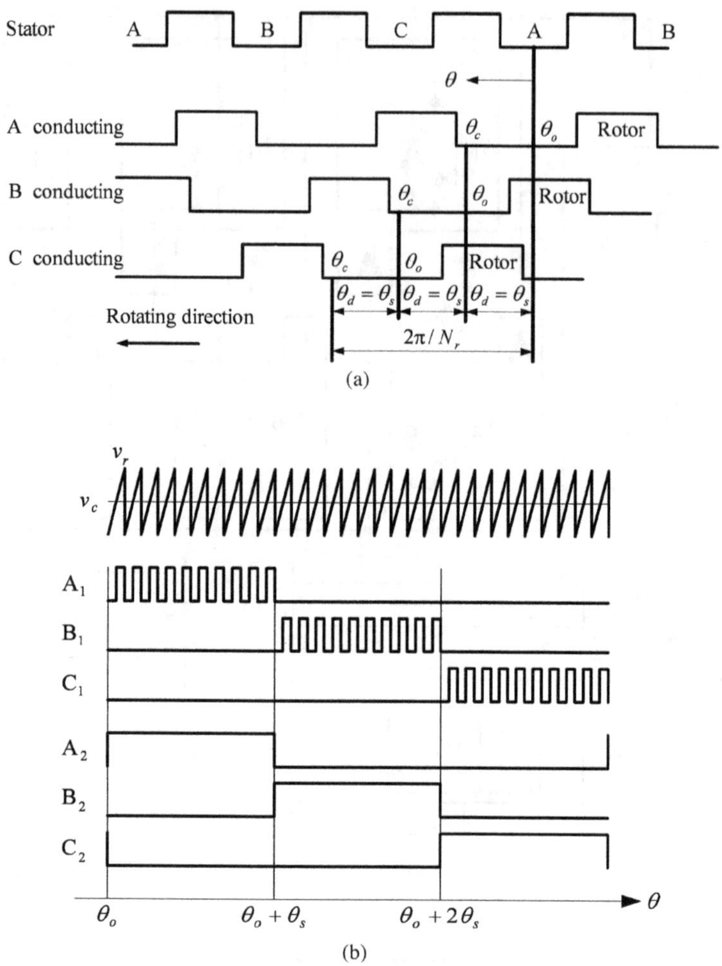

Figure 5.2 Operation principle of voltage-controlled SR drive. (a) Rotor positions. (b) Switching signals

$$T_e = -\frac{\partial}{\partial \theta}\int_0^{\psi_1} i_1(\theta,\psi)\mathrm{d}\psi - \frac{\partial}{\partial \theta}\int_0^{\psi_2} i_2(\theta-\theta_s,\psi)\mathrm{d}\psi \tag{5.4}$$

$$u_k = \begin{cases} V_s\varepsilon(v_r(\theta)-v_c) & (\theta \bmod 2\theta_s) \in [\theta_o, \theta_o+\theta_s) \\ -V_s\varepsilon(\psi_k) & (\theta \bmod 2\theta_s) \in [\theta_o+\theta_s, \theta_o+2\theta_s) \end{cases} \tag{5.5}$$

where u_k is the phase voltage, V_s is the DC supply voltage, R is the phase resistance, B is the viscous damping, J is the load inertia, T_e is the electromagnetic torque, T_l is the load torque, and ε is the unit step function. Because of its high nonlinearity, $\psi_k(\theta, i_k)$ is approximated by a series of two-dimensional quadratic Lagrange interpolation functions of θ and i_k. These two-dimensional grids are formulated by using the manufacturer's design data or by employing a finite element analysis of the SR motor. This approach can significantly reduce the complexity in the calculation of flux linkages. Moreover, $i_k(\theta, \psi_k)$ can similarly be obtained by the numerical inversion of the relationship of $\psi_k(\theta, i_k)$. Since the current is more easily measured than the flux linkage, it is chosen as the output variable. By defining the state vector as $\mathbf{X} = (\theta, \omega, \psi_1, \psi_2)^{\mathrm{T}}$ and the output vector as $\mathbf{Y} = (\theta, \omega, i)^{\mathrm{T}}$, where $i = i_1 + i_2$, the system equation given by (5.3) can be rewritten as:

$$\begin{cases} \dot{\mathbf{X}} = f_t(\mathbf{X}) \\ \mathbf{Y} = M_t(\mathbf{X}) \end{cases} \tag{5.6}$$

5.1.2 Analysis

In order to construct the Poincaré map, a hyperplane $\Sigma \in \Re^3$ is defined as:

$$\Sigma := \{\mathbf{X} : [(\theta - \theta_o) \bmod \theta_s] = 0\} \tag{5.7}$$

The trajectory of $\mathbf{X}$ under observation repeatedly passes through Σ when θ increases monotonically. Thus, the sequence of Σ crossing defines a Poincaré map P: $\Re^3 \times \Re^3$ as given by:

$$(\omega, \psi_1, \psi_2)^{\mathrm{T}}_{n+1} = P((\omega, \psi_1, \psi_2)^{\mathrm{T}}_n) \tag{5.8}$$

Actually, the solution of this map, the so-called orbit (Parker and Chua, 1989), is a sequence of samples at the turn-on angle of each phase winding. In order to avoid the calculation of Σ crossing, the rotor displacement θ, rather than time t, is selected as the independent variable of the system equation given by (5.3). The next crossing of the plane $\theta_{n+1} = \theta_o + (n+1)\,\theta_s$ can be directly calculated by integrating from $\theta_n = \theta_o + n\theta_s$ to θ_{n+1}. To make θ an independent variable, (5.3) is expressed as:

$$\begin{cases} \dfrac{\mathrm{d}\omega}{\mathrm{d}\theta} = (-B\omega + T_e(\theta, \psi_1, \psi_2, \theta_s) - T_l)/(J\omega) \\[2ex] \dfrac{\mathrm{d}\psi_1}{\mathrm{d}\theta} = (-Ri_1(\theta, \psi_1) + u_1(\theta))/\omega \\[2ex] \dfrac{\mathrm{d}\psi_2}{\mathrm{d}\theta} = (-Ri_2(\theta - \theta_s, \psi_2) + u_2(\theta - \theta_s))/\omega \end{cases} \tag{5.9}$$

Since u_k is piecewise continuous in terms of θ, it is much easier to discover the discontinuity points of u_k in (5.9) than in (5.3). Accordingly, (5.9) can be separately solved within different continuous intervals. Hence, it is much more efficient to use (5.9) than (5.3) to calculate the Poincaré map. By redefining the state vector as $\mathbf{X}(\theta) = (\omega, \psi_1, \psi_2)^{\mathrm{T}}$ and the output vector as $\mathbf{Y}(\theta) = (\omega, i)^{\mathrm{T}}$, (5.9) can be rewritten as:

$$\begin{cases} \dot{\mathbf{X}} = f_\theta(\mathbf{X}, \theta) \\ \mathbf{Y} = M_\theta(\mathbf{X}) \end{cases} \tag{5.10}$$

The Poincaré map (5.8) can also be rewritten as:

$$\begin{cases} \mathbf{X}_{n+1} = P(\mathbf{X}_n) \\ \mathbf{Y}_n = M_\theta(\mathbf{X}_n) \end{cases} \tag{5.11}$$

Although f_θ is piecewise continuous, the solution of (5.10) is continuous and hence P is also continuous. It should be noted that the map M_θ in (5.10) is a noninvertible map within the whole set of the solution. For example, if ψ_1 and ψ_2 are simultaneously nonzero, $\mathbf{X}$ is a multivalued function of $\mathbf{Y}$. However, the map M_θ in (5.11) is homeomorphism, namely both M_θ and M_θ^{-1} are continuous, because one of ψ_1 and ψ_2 of $\mathbf{X}_n$ is zero. It results in a new map Q: $\mathbf{Y}_{n+1} = Q(\mathbf{Y}_n)$ which is defined as $Q = M_\theta \cdot P \cdot M_\theta^{-1}$. Thus, the maps P and Q are topologically conjugate (Wiggins, 1990), where the corresponding orbits $\{\mathbf{X}_n\}$ and $\{\mathbf{Y}_n\}$ have the same dynamics although they represent different physical variables of the SR drive system. For example, if $\mathbf{X}^*$ is a fixed point of P, then $\mathbf{Y}^* = M_\theta(\mathbf{X}^*)$ is also a fixed point of Q. The eigenvalues of their Jacobian matrices $DP(\mathbf{X}^*)$ and $DQ(\mathbf{Y}^*)$ are identical. Since the orbit $\{\mathbf{X}_n\}$ represents the system states, it is useful for the analysis of periodic and chaotic orbits. On the other hand, it is convenient to use the orbit $\{\mathbf{Y}_n\}$ to illustrate the trajectories and waveforms since the rotor speed and armature current are easy to observe both in simulation and in experiments.

The Jacobian matrix $DP(\mathbf{X}_n)$ of the Poincaré map P with respect to $\mathbf{X}_n$ is the solution $\mathbf{Z}_{n+1} = \mathbf{Z}(\theta_{n+1}) \in \Re^3 \times \Re^3$ of the variational equation of its underlying system as given by:

$$\dot{\mathbf{Z}}(\theta) = \frac{\partial f_\theta(\mathbf{X}(\theta), \theta)}{\partial \mathbf{X}} \mathbf{Z}(\theta), \quad \mathbf{Z}(\theta_n) = \mathbf{Z}_n \tag{5.12}$$

Considering winding 1 to be inactivated, it yields:

$$\frac{\partial f_\theta}{\partial \mathbf{X}} = \begin{pmatrix} \dfrac{-T_e + T_l}{J\omega^2} & \dfrac{1}{J\omega}\dfrac{\partial i_1}{\partial \theta} & \dfrac{1}{J\omega}\dfrac{\partial i_2}{\partial \theta} \\ \dfrac{Ri_1 - u_1}{\omega^2} & -\dfrac{\delta_\theta(\theta_\psi - \theta)}{\omega} - \dfrac{R}{\omega}\dfrac{\partial i_1}{\partial \psi_1} & 0 \\ \dfrac{1}{\omega}\dfrac{\partial u_2}{\partial \omega} + \dfrac{Ri_2 - u_2}{\omega^2} & 0 & -\dfrac{R}{\omega}\dfrac{\partial i_2}{\partial \psi_2} \end{pmatrix} \tag{5.13}$$

where $\dfrac{\partial u_2}{\partial \omega} = \dfrac{gV_s\delta_\theta(\theta - \theta_\omega)}{g(-B\omega(\theta_\omega) + T_e(\theta_\omega) - T_l)/(J\omega(\theta_\omega)) - (v_u - v_l)/\theta_T}$; both $\dfrac{\partial i_k}{\partial \theta}$ and $\dfrac{\partial i_k}{\partial \psi_k}$ can be obtained by directly differentiating $i_k(\theta, \psi_k)$; δ_θ is the Dirac delta function that is the derivative of ε; θ_ψ and θ_ω are the discontinuity points of u_1 and u_2, which can be obtained by solving $\psi_1(\theta_\psi) = 0$ and $v_r(\theta_\omega) - v_c(\theta_\omega) = 0$, respectively.

For an arbitrary point $\mathbf{X}_n$, the corresponding initial value $\mathbf{Z}_n$ of the variational equation given by (5.12) is usually an identity matrix. However, since $\psi_2(\theta_n) = 0$, it gives $\partial \mathbf{X}_n / \partial \psi_2 = (0 \quad 0 \quad 0)^{\mathrm{T}}$. Hence, the third column of the identity matrix $\mathbf{Z}_n$ should be replaced by a zero vector. Actually, the period of ψ_1 and ψ_2 of $\mathbf{X}$ is always $2\theta_s$ rather than θ_s. In order to attain the period-1 orbit of P, winding 1 always represents the inactivated winding at each iteration of P, with the result that ψ_1 and ψ_2 must exchange their values after each iteration of P. Thus, the fixed point $\mathbf{X}^*$ of P and its Jacobian matrix are defined as:

$$\mathbf{X}^* = \mathbf{C}P(\mathbf{X}^*) \tag{5.14}$$

$$\mathbf{J}_1 = \mathbf{C}DP(\mathbf{X}^*) \tag{5.15}$$

where $\mathbf{C} = \begin{pmatrix} 1 & 0 & 0 \\ 0 & 0 & 1 \\ 0 & 1 & 0 \end{pmatrix}$.

The fixed point $\mathbf{X}^*$ of P can also be located by using a Newton–Raphson algorithm, as given by:

$$\mathbf{X}^{(i+1)} = \mathbf{X}^{(i)} - (\mathbf{C}DP(\mathbf{X}^{(i)}) - \mathbf{1})^{-1}(\mathbf{C}P(\mathbf{X}^{(i)}) - \mathbf{X}^{(i)}) \tag{5.16}$$

where $DP(\mathbf{X}^{(i)})$ can be evaluated from (5.12) and (5.13). By checking the characteristic multipliers, which are the eigenvalues of the Jacobian matrix, the stable region of the period-1 orbit for fundamental operation can readily be obtained.

5.1.3 Simulation

In order to assess the modeling of chaos, a corresponding analysis is carried out based on a practical 3-phase SR drive system that has been designed for an electric vehicle (Zhan, Chan and Chau, 1999). The parameter values are $V_s = 150$ V, $N_s = 12$, $N_r = 8$, $\theta_s = 15°$, $\theta_o = 3.75°$, $R = 0.15\,\Omega$, $B = 0.00075$ Nm/rad/s, $J = 0.025$ kgm^2, $v_u = 5$ V, $v_l = 1$ V, $n_\theta = 10$, $\theta_T = 1.5°$, $\omega^* = 50$ rad/s, and $T_l = 8.6$ Nm.

When $g = 1.3$ V/rad/s, the steady-state behavior of the SR drive system is the fundamental or normal operation. The corresponding simulated waveforms of v_c, v_r, and I as well as the simulated phase-plane trajectory of i versus v_c are shown in Figure 5.3, in which θ is expressed as the integer multiple of θ_s. As shown in Figure 5.3(a), there is no skipping cycle during PWM regulation, namely v_c crosses every v_r. The fluctuation of v_c is also small (from 4.0 to 4.4 V), and the corresponding ω is from 53.1 to 53.4 rad/s. As shown in Figure 5.3(b), the corresponding i is periodic in term of θ_s, and its fluctuation is from 23 to 60 A.

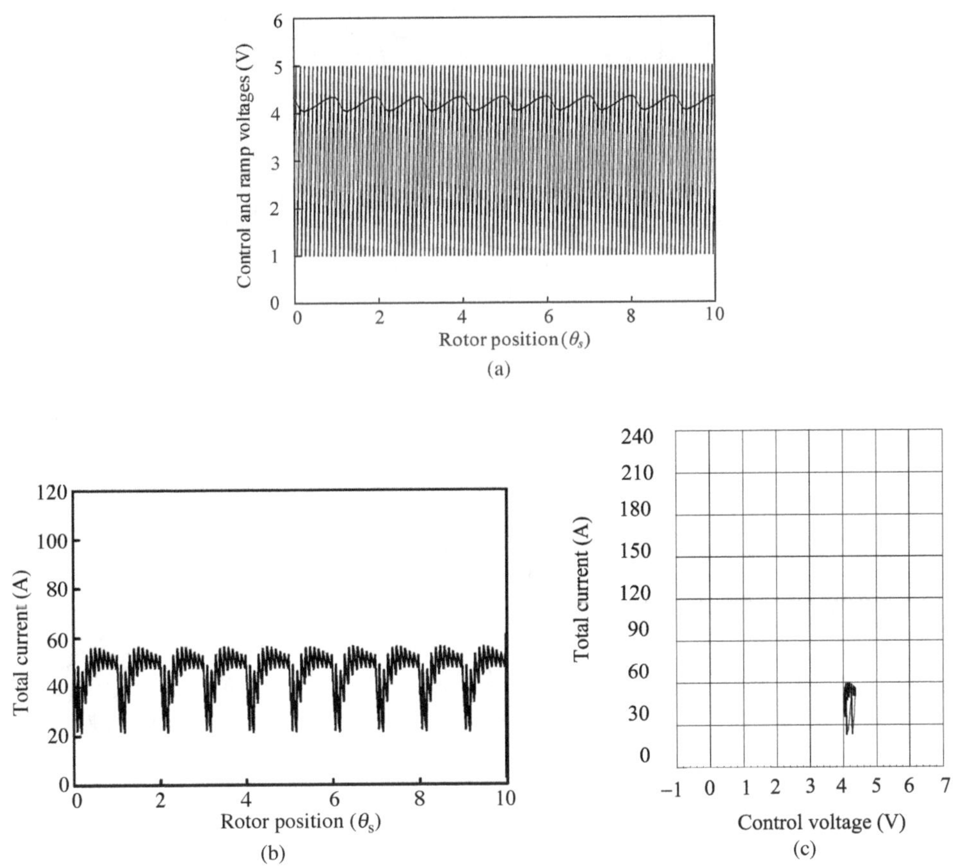

Figure 5.3 Simulated fundamental operation. (a) Control and ramp voltage waveforms. (b) Total current waveform. (c) Trajectory of total current versus control voltage

Since $n_\theta = 10$, i has ten peaks within each θ_s, resulting in the phase-plane trajectory of this periodic solution being a cycle with ten peaks, as shown in Figure 5.3(c).

When $g = 4.8$ V/rad/s, the SR drive system operates in chaos. The simulated chaotic waveforms and trajectory are shown in Figure 5.4. Contrary to the periodic solution, it has skipping cycles within each θ_s (as shown in Figure 5.4(a)) in which v_c is higher than v_r and no intersection occurs. Furthermore, the number of skipping cycles within each θ_s is a random-like variable, with the result that the oscillating magnitudes of v_c (being the same shape of ω) and i are all fluctuating as shown in Figures 5.4(a) and (b), respectively. As expected, Figure 5.4(c) shows that the trajectory of i versus v_c is a phase portrait with a random-like but bounded feature. It can be found that even though the fluctuation of ω is still acceptable (from 50.6 to 51.6 rad/s), the fluctuation of v_c is severe (from 2.9 to 7.6 V), resulting in a fluctuation of i that is exceptionally large (from 0 to 120 A) which is highly undesirable.

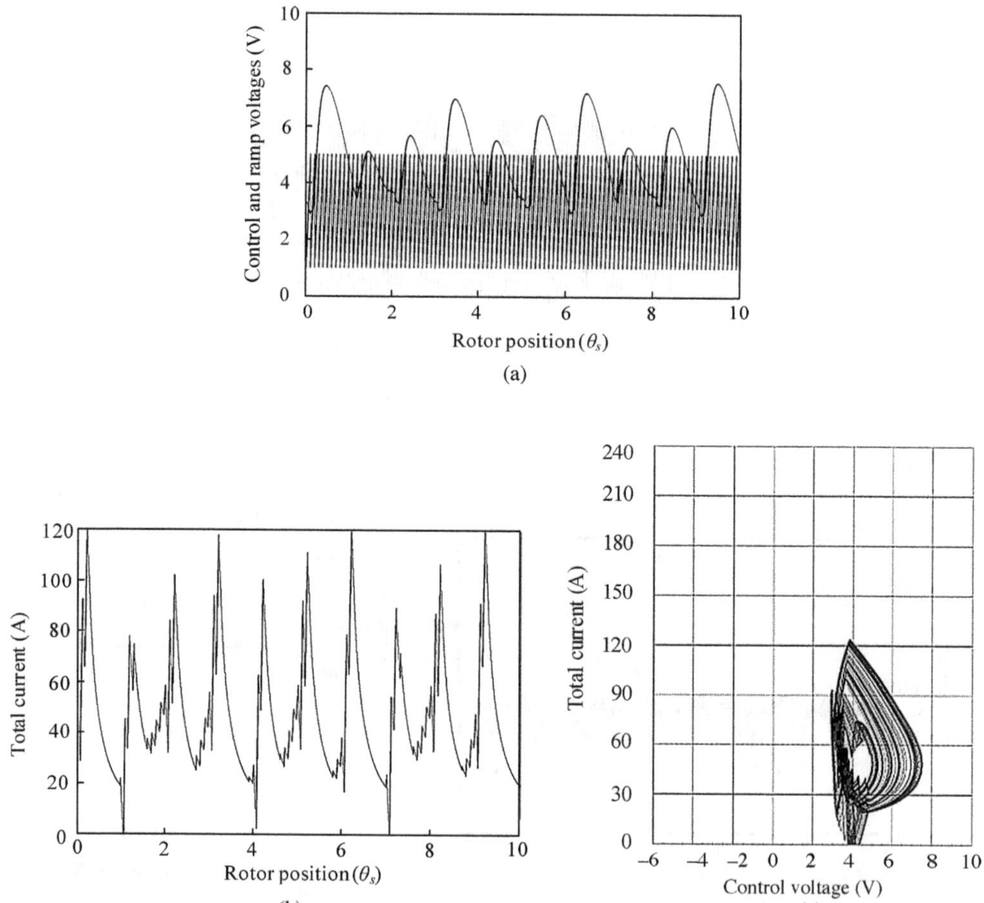

Figure 5.4 Simulated chaotic operation. (a) Control and ramp voltage waveforms. (b) Total current waveform. (c) Trajectory of total current versus control voltage

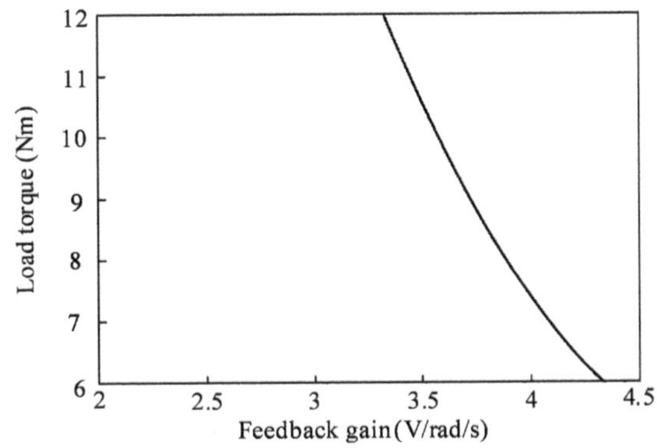

Figure 5.5 Stable and unstable regions for fundamental operation

In order to determine the boundary of the stable fundamental operation, all period-1 orbits are firstly located by using the Newton–Raphson algorithm as given by (5.16). Then, their characteristic multipliers are evaluated by computing the eigenvalues of the corresponding Jacobian matrices obtained from (5.12)–(5.15). By drawing the line where the magnitude of the characteristic multipliers is equal to unity, the stable region of T_l versus g for the fundamental operation can be obtained, as shown in Figure 5.5. If the characteristic multipliers are complex conjugates while their magnitudes are less than unity, the system is still stable, but spirally converges to the fixed point. It should be noted that the unstable region for the fundamental operation involves both subharmonic and chaotic operations. Therefore, the system in Figure 5.5 is highly desirable for the designer of SR drive systems since the fundamental operation is preferred normally.

5.1.4 Experimentation

Based on the same SR drive system, experimentation has been conducted. Since ω has a large DC bias while its variation is relatively small, the speed variation cannot be clearly assessed based on the direct measurement of ω. In contrast, v_c is the amplified speed error which not only exhibits a clear pattern of speed variation, but is also easily measured. Thus, v_c is measured to represent ω. The measured trajectory and waveforms of i and v_c when $g = 1.3$ V/rad/s are shown in Figure 5.6. This illustrates that the SR drive system operates in a period-1 orbit, which is actually a stable fundamental operation. It can be observed that i and v_c are not of exact periodicity, which is due to the inevitable imperfections of the practical SR motor drive, such as the mechanical eccentricity of the drive shaft and the torsional oscillation of the coupler. Also, it can be found that i lies roughly between 20 and 60 A while v_c lies between 4.0 and 4.5 V. By comparing these results with the waveforms and trajectory shown in Figure 5.3, the measured results and the simulation results have good agreement.

Moreover, by selecting $g = 4.8$ V/rad/s, the measured trajectory and waveforms of i and v_c, shown in Figure 5.7, illustrate that the SR drive system operates in chaos. It can be found that the boundaries of i and v_c lie roughly between 0 and 120 A and between 2 and 8 V, respectively. Contrary to the period-1 orbit in which the measured trajectory and waveforms are directly compared with simulation, the chaotic trajectory and waveforms measured in the experiment cannot be compared with the simulation results due to the random-like nature of chaos as well as its dependence on the initial conditions. Nevertheless, it can be found that the measured boundaries of the chaotic trajectory shown in Figure 5.7 resemble the simulation results in Figure 5.4, which is actually a property of chaos.

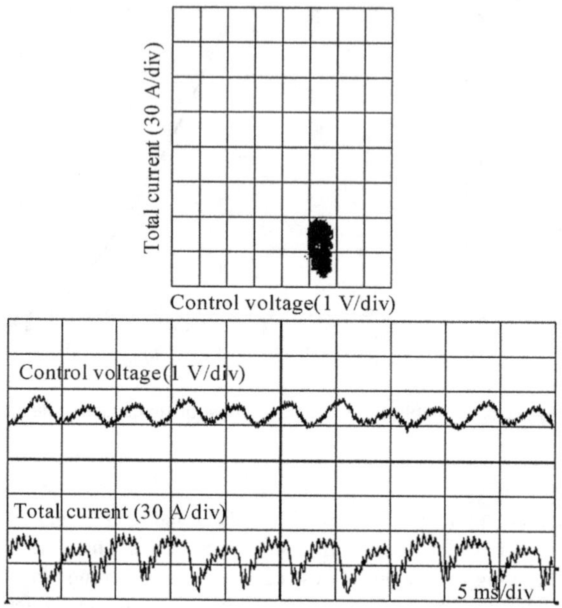

Figure 5.6 Measured trajectory and waveforms of control voltage and total current at fundamental operation

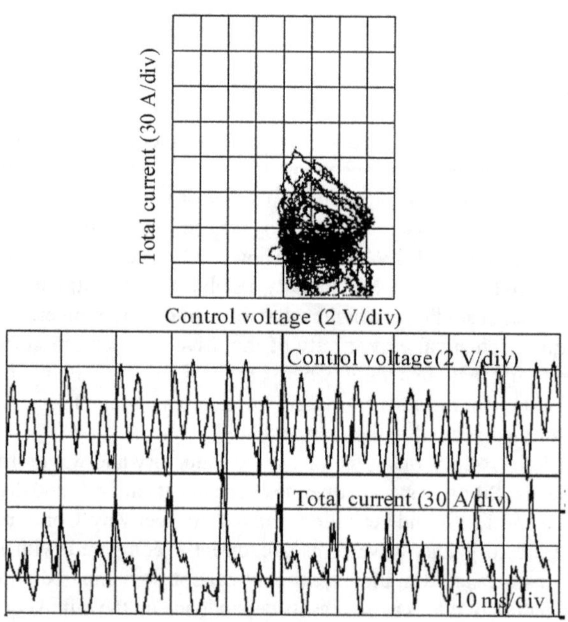

Figure 5.7 Measured trajectory and waveforms of control voltage and total current at chaotic operation

5.2 Current-Controlled Switched Reluctance Drive System

5.2.1 *Modeling*

Figure 5.8 shows a 3-phase SR drive system using current hysteresis regulation, the so-called current-controlled SR drive system. The commutation strategy also uses rotor position feedback to select the turn-on angle θ_o and turn-off angle θ_c of those lower-leg power switches (A_2, B_2, and C_2). In order to simplify the control circuit, the dwell angle $\theta_d = \theta_c - \theta_o$ of each phase winding is selected to be equal to the commutation angle $\theta_s = 2\pi/(mN_r)$, where m is the number of phases and N_r is the number of rotor poles. For each conductive phase winding, the corresponding current control is achieved by applying a current hysteresis controller. The current reference is given by using a proportional speed controller. As the operational amplifier (OA) of the speed controller and the current multiplexer have gains of g and g_i, respectively, the current reference signal $v_c(t)$, current feedback signal $v_i(t)$ and current hysteresis band signal Δv can be expressed as:

$$v_c(t) = g(\omega_{ref} - \omega(t)) \tag{5.17}$$

$$v_i(t) = g_i i_k(t) \tag{5.18}$$

$$\Delta v = \pm g_i \Delta i \tag{5.19}$$

where ω_{ref}, $\omega(t)$, $i_k(t)$, and Δi are the speed reference, the instantaneous speed, the kth phase current, and its hysteresis band, respectively. Then, the current hysteresis controller outputs the signal to turn on or off those upper-leg power switches (A_1, B_1, and C_1), depending on the phase commutation. When $(v_i - v_c) > \Delta v$, the upper-leg switch, being the same phase of the on-state lower-leg switch, is turned off until $(v_i - v_c) < -\Delta v$; then it is turned on. The other phase switches remain off. The switching pattern of each phase winding is illustrated by the waveforms of v_c and v_i shown in Figure 5.8(c).

For the kth phase winding of the SR drive, the voltage u_k depends on the states of its upper-leg and lower-leg power switches. During the interval of dwell angle, the lower-leg switch is always turned on. Hence, when the upper-leg switch is turned on, $u_k = U$, where U is the DC link voltage; otherwise $u_k = 0$, which is caused by one of the freewheeling diodes. During the other interval $u_k = -U$, which is caused by two freewheeling diodes because both the upper-leg and lower-leg switches are turned off. Since the current i_k is always unipolar for the converter topology shown in Figure 5.8, u_k can be considered as zero when i_k, and hence the flux linkage ψ_k becomes zero in this interval. Thus, for an m-phase SR drive whose phase current is continuously conducted within l commutation periods, the number of simultaneously nonzero voltage equations can be reduced from m to l. Consequently, the dimensionality of the state equations is reduced from $(m+2)$ to $(l+2)$. Since the dwell angle is equal to the commutation angle, only two adjacent phase windings have currents at the same time and hence $l=2$ as shown in Figure 5.8(c). For the sake of clarity and simplicity, m phase windings conducted in turn are represented by two virtual phase windings (winding 1 and winding 2) activated alternately. When the winding is conducted, it is said to be activated, otherwise it is said to be inactivated.

Rather than using i_k, ψ_k is also chosen as the state variable in the current-controlled SR drive system so that the system differential equation does not involve the calculation of $\partial\psi_k(\theta, i_k)/\partial\theta$ and $\partial\psi_k(\theta, i_k)/\partial i_k$ (Stephenson and Corda, 1979). When winding 2 is lagging behind winding 1 by θ_s, the

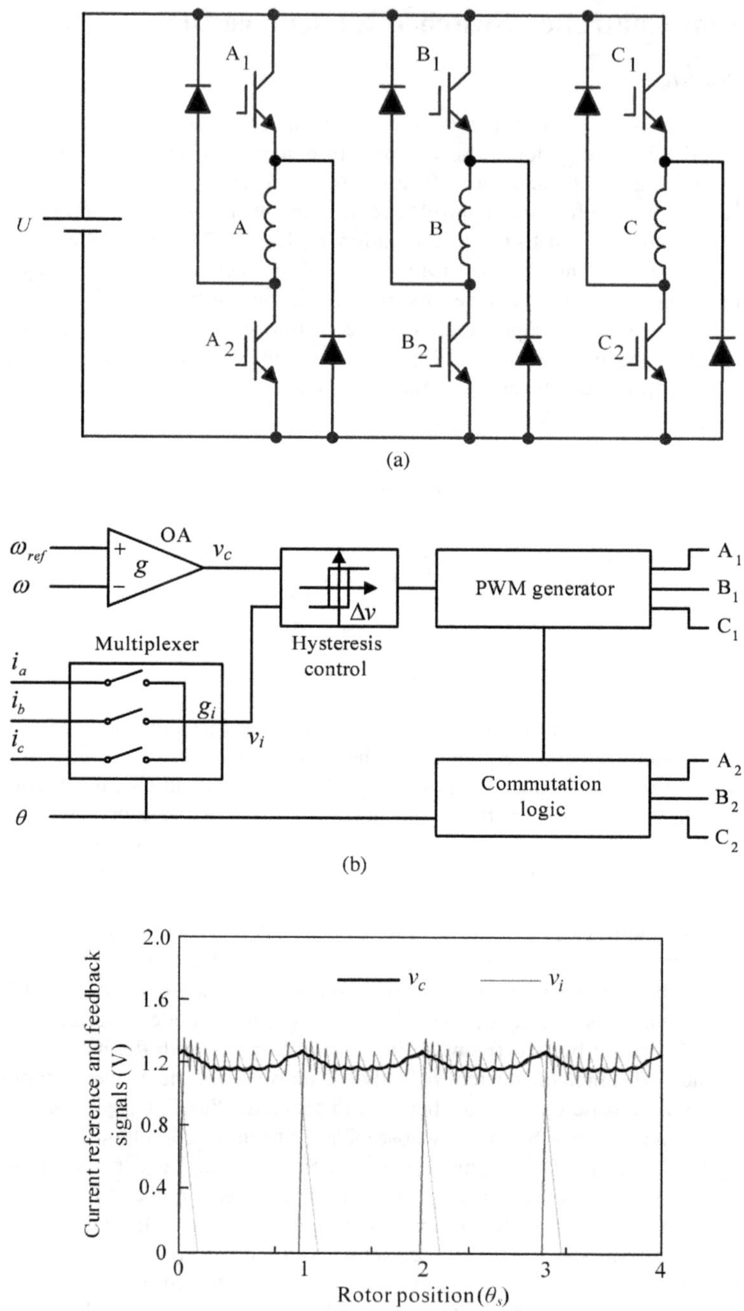

Figure 5.8 Current-controlled SR drive system. (a) Power circuitry. (b) Control circuitry. (c) Current reference and feedback signals

system equation of the SR drive can be expressed as:

$$\begin{cases} \dfrac{d\theta}{dt} = \omega \\ \dfrac{d\omega}{dt} = (-B\omega + T_e(\theta, \psi_1, \psi_2, \theta_s) - T_l)/J \\ \dfrac{d\psi_1}{dt} = -Ri_1(\theta, \psi_1) + u_1(\theta) \\ \dfrac{d\psi_2}{dt} = -Ri_2(\theta - \theta_s, \psi_2) + u_2(\theta - \theta_s) \end{cases} \tag{5.20}$$

$$T_e = -\frac{\partial}{\partial\theta}\int_0^{\psi_1} i_k(\theta, \psi)d\psi - \frac{\partial}{\partial\theta}\int_0^{\psi_2} i_k(\theta - \theta_s, \psi)d\psi \tag{5.21}$$

where R is the phase resistance, B is the viscous damping, J is the load inertia, T_e is the electromagnetic torque, and T_l is the load torque.

The nonlinear phase flux linkage $\psi_k(\theta, i_k)$ of the SR motor can be approximated by a series of two-dimensional quadratic Lagrange interpolation functions. Based on a finite element analysis of the SR motor, its two-dimensional grid of tabulated values $\psi_k(\theta, i_k)$ can be obtained. The flux linkage at any operating point can be numerically calculated by using the two-dimensional quadratic polynomials constructed by nine tabulated values near the operating point. These local quadratic interpolations considerably reduce the complexity of the flux model, and ensure the continuity of its first derivatives. Thus, $i_k(\theta, \psi_k)$ can be obtained by using numerical inversion of $\psi_k(\theta, i_k)$. By applying numerical integration and differentiation to $i_k(\theta, \psi_k)$, T_e can be attained by the use of (5.21).

As the phase current, rather than the flux linkage, can be directly measured in a practical SR drive, the current is chosen as the output variable. By defining the state vector as $\mathbf{X} = (\theta, \omega, \psi_1, \psi_2)^{\mathrm{T}}$ and the output vector as $\mathbf{Y} = (\theta, \omega, i)$, where $i = i_1 + i_2$, (5.21) can be rewritten as:

$$\begin{cases} \dot{\mathbf{X}} = f_t(\mathbf{X}) \\ \mathbf{Y} = M_t(\mathbf{X}) \end{cases} \tag{5.22}$$

5.2.2 Analysis

A Poincaré map is used to analyze the chaotic behavior of this current-controlled SR drive system, and a sequence of samples of the continuous-time solution is thereby produced. The steady-state behavior of the Poincaré map, known as the orbit, corresponds to the steady-state waveform of the continuous-time dynamical system.

In order to construct the Poincaré map, a hyperplane $\Sigma \in \Re^3$ is defined as:

$$\Sigma := \{\mathbf{X} : [(\theta - \theta_o) \bmod \theta_s] = 0\} \tag{5.23}$$

The trajectory of $\mathbf{X}$ under observation repeatedly passes through Σ when θ increases monotonically. The sequence of Σ crossings defines a Poincaré map $P: \Re^3 \times \Re^3$ as given by:

$$(\omega, \psi_1, \psi_2)_{n+1}^{\mathrm{T}} = P((\omega, \psi_1, \psi_2)_n^{\mathrm{T}}) \tag{5.24}$$

Actually, the orbit of this map is a sequence of samples at the beginning of the dwell angle of each phase winding. In order to avoid calculating the sequence of Σ crossing, the rotor displacement θ, rather than time t, is selected as the independent variable of the system equation given by (5.20). The crossing of the plane $\theta_{n+1} = \theta_o + (n+1)\theta_s$ can be directly calculated by integrating from $\theta_n = \theta_o + n\theta_s$ to θ_{n+1}. To make θ an independent variable, (5.20) is expressed as:

$$\begin{cases} \dfrac{\mathrm{d}\omega}{\mathrm{d}\theta} = (-B\omega + T_e(\theta, \psi_1, \psi_2, \theta_s) - T_L)/(J\omega) \\ \dfrac{\mathrm{d}\psi_1}{\mathrm{d}\theta} = (-Ri_1(\theta, \psi_1) + u_1(\theta))/\omega \\ \dfrac{\mathrm{d}\psi_2}{\mathrm{d}\theta} = (-Ri_2(\theta - \theta_s, \psi_2) + u_2(\theta - \theta_s))/\omega \end{cases} \tag{5.25}$$

By redefining the state vector as $\mathbf{X}(\theta) = (\omega, \psi_1, \psi_2)^{\mathrm{T}}$ and the output vector as $\mathbf{Y}(\theta) = (\omega, i)^{\mathrm{T}}$, (5.25) can be rewritten as:

$$\begin{cases} \dot{\mathbf{X}} = f_\theta(\mathbf{X}, \theta) \\ \mathbf{Y} = M_\theta(\mathbf{X}) \end{cases} \tag{5.26}$$

The Poincaré map (5.24) can also be rewritten as:

$$\begin{cases} \mathbf{X}_{n+1} = P(\mathbf{X}_n) \\ \mathbf{Y}_n = M_\theta(\mathbf{X}_n) \end{cases} \tag{5.27}$$

Although f_θ is piecewise continuous, the solution of (5.26) is continuous and hence P is also continuous. Similar to Section 5.1, the map M_θ in (5.27) is a noninvertible map within the whole set of the solution. For example, if ψ_1 and ψ_2 are simultaneously nonzero, $\mathbf{X}$ is a multivalued function of $\mathbf{Y}$. Fortunately, the map M_θ in (5.27) is a homeomorphism, namely M_θ and M_θ^{-1} are continuous, because one of ψ_1 and ψ_2 of $\mathbf{X}_n$ is zero. Thus, the orbits $\{\mathbf{X}_n\}$ and $\{\mathbf{Y}_n\}$ are topologically equivalent. It follows that they have the same dynamics although they represent different physical variables of the SR drive system. The orbit $\{\mathbf{X}_n\}$ is used to locate the periodic and chaotic orbits, whereas the orbit $\{\mathbf{Y}_n\}$ is used to depict the trajectories and bifurcation diagrams.

Due to the nonlinearity of the flux linkage, the period of the fundamental operation (period-1 orbit) is referred to as the commutation period θ_s. However, the period of ψ_1 and ψ_2 of $\mathbf{X}$ is always $2\theta_s$ instead of θ_s. In order to attain the period-1 orbit of P, winding 1 always stands for the inactivated winding at each iteration of P, with the result that ψ_1 and ψ_2 must exchange their values after each iteration of P. Thus, the fixed point $\mathbf{X}^*$ of the period-1 orbit and the cyclic points $\{\mathbf{X}_1^*, \ldots, \mathbf{X}_p^*\}$ of the period-p orbit are also defined as:

$$\mathbf{X}^* = \mathbf{C}P(\mathbf{X}^*) \tag{5.28}$$

$$\mathbf{X}_{k+1}^* = \mathbf{C}P(\mathbf{X}_k^*) \quad (k = 1, \ldots, p-1), \quad \mathbf{X}_1^* = \mathbf{C}P(\mathbf{X}_p^*) \tag{5.29}$$

where the transformation matrix $\mathbf{C}$ is given by:

$$\mathbf{C} = \begin{pmatrix} 1 & 0 & 0 \\ 0 & 0 & 1 \\ 0 & 1 & 0 \end{pmatrix} \tag{5.30}$$

5.2.3 Simulation

Based on the system equation(5.26) and the Poincaré map (5.27), the analysis of the nonlinear dynamics in the SR drive system is carried out by computer simulation. A practical SR drive, originally designed for an electric vehicle (Chan *et al.*, 1996), is used for exemplification. The system setup is shown in Figure 5.8, and the corresponding data are listed in Table 5.1. In the following analysis, the fundamental, subharmonic and chaotic behaviors are simulated by adopting different values of g. Also, a unique property of chaotic behavior, namely a high sensitivity to the initial conditions, is demonstrated. Various bifurcation diagrams with respect to g, T_l, and Δi are simulated. Hence, the route to chaos and the effect of system parameters on the nonlinear dynamics are discussed.

Firstly, when the speed feedback gain is selected as $g = 5$ V/rad/s, the steady-state solution of the SR drive has a normal periodic behavior, the so-called period-1 orbit, which usually exists in the fundamental operation. The corresponding waveforms of ω, v_c, i_1, i_2, and T_e, as well as the phase-plane trajectory of i versus ω, are shown in Figure 5.9, in which θ is expressed as an integer multiple of the stroke angle θ_s. These waveforms consist of two oscillating components – one corresponds to the commutation frequency f_{com} and the other to the PWM frequency f_{PWM}. It can be found that f_{com} is almost constant when the oscillating magnitude of ω is very narrow, as shown in Figure 5.9(a). On the other hand, f_{PWM} gradually decreases with increasing inductance within the commutation period, resulting in irregular waveforms. Due to the nonlinear distribution of electromagnetic torque within the commutation period, the f_{com} oscillating component of ω and v_c dominates the f_{PWM} component, as shown in Figures 5.9(a) and (b). The corresponding extremes of ω and v_c are near the commutation point. Moreover, although the distribution of i_1 (or i_2) is uniform, as shown in Figure 5.9(c), the distribution of T_e is nonuniform, ranging from 10 to 18 Nm, as shown in Figure 5.9(d). The

Table 5.1 Key parameters of SR motor

Number of phases m	3
Stator poles N_s	12
Rotor poles N_r	8
Dwell angle θ_d	15°
Stroke angle θ_s	15°
Turn on angle θ_o	5.5°
DC link voltage U	150 V
Phase resistance R	0.1 Ω
Viscous damping B	0.0005 Nm/rad/s
Load inertia J	0.025 kgm^2
Load torque T_l	12 Nm
Speed reference ω_{ref}	50 rad/s
Current hysteresis band Δi	±5 A
Current feedback gain g_i	0.02 V/A
Speed feedback gain g	10 V/rad/s
PWM frequency f_{PWM}	1–5 kHz

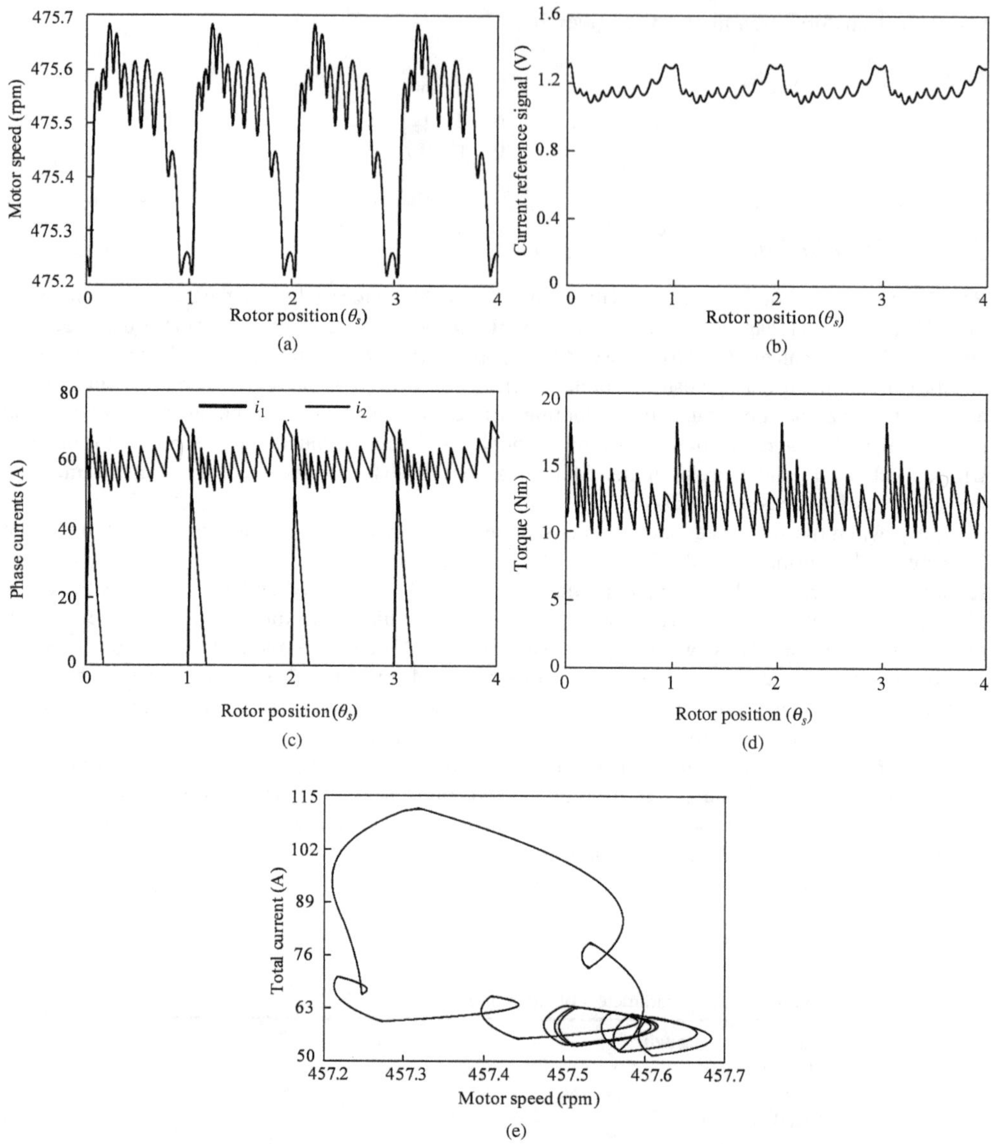

Figure 5.9 Fundamental operation. (a) Motor speed waveform. (b) Current reference waveform. (c) Phase current waveforms. (d) Torque waveform. (e) Trajectory of total current versus speed

phase-plane trajectory of this periodic solution is a cycle embedded with some smaller cycles as shown in Figure 5.9(e). The boundary of ω is from 49.738 to 49.786 rad/s (474.96 to 475.42 rpm), whereas the boundary of i is from 50 to 110 A. Such a high peak current is due to the overlapping of i_1 and i_2 during commutation, even though they are actually confined between 51 and 70 A as shown in Figure 5.9(c).

When the speed feedback gain is changed to $g = 9$ V/rad/s, the steady-state behavior of the SR drive is a period-2 subharmonic solution, as shown in Figure 5.10. It can be found that the waveforms are of period-2 behavior, the so-called period-2 orbit. Comparing Figures 5.10(a) and (5.9)(a), the oscillating magnitudes of ω of period-2 and period-1 orbits are very similar. However, the oscillating magnitude of v_c, as shown in Figure 5.10(b), is larger than that shown in Figure 5.9(b) because of the larger g.

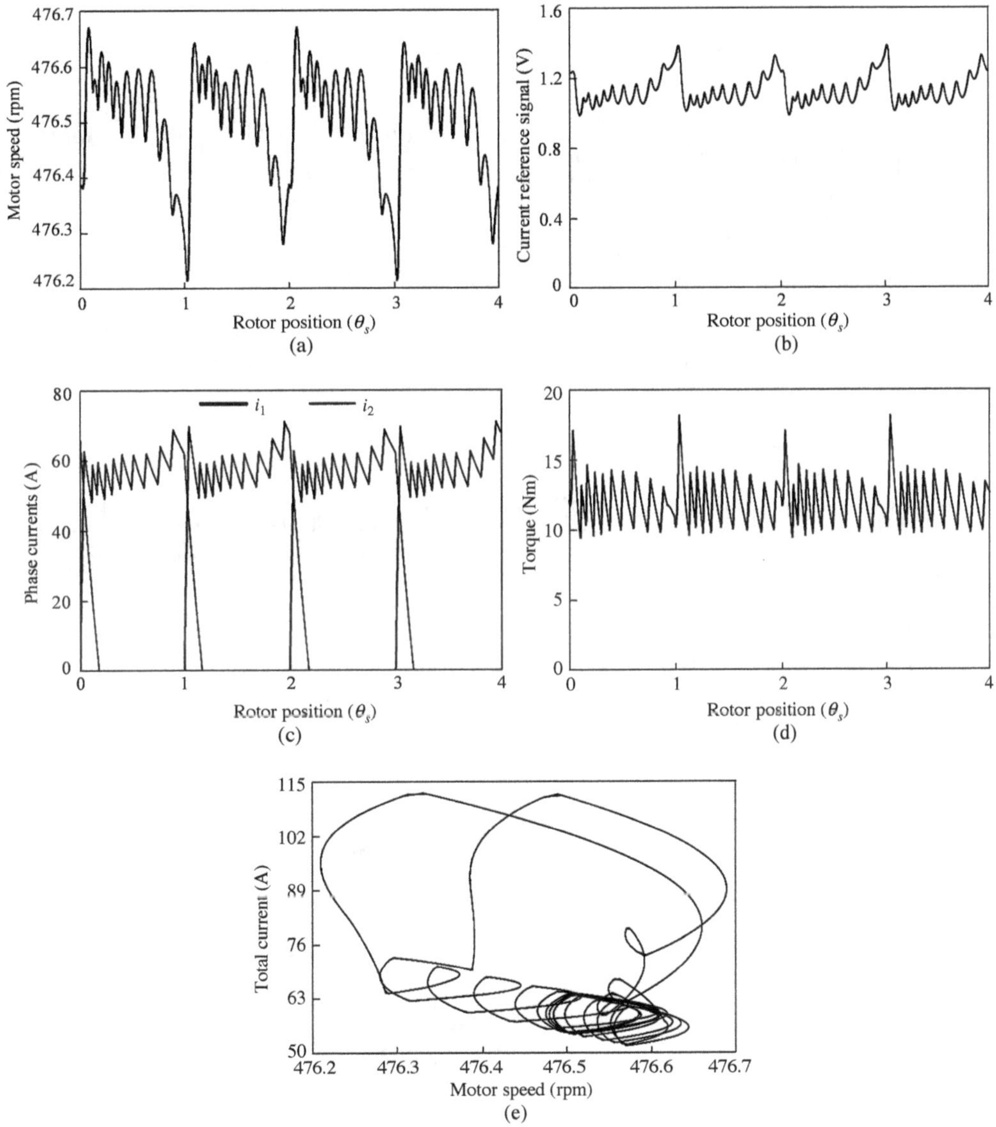

Figure 5.10 Subharmonic operation. (a) Motor speed waveform. (b) Current reference waveform. (c) Phase current waveforms. (d) Torque waveform. (e) Trajectory of total current versus speed

Accordingly, the oscillating magnitudes of i_1 (or i_2) and T_e increase to 50–72 A and 9.5–18 Nm, as shown in Figures 5.10(c) and (d), respectively. The phase-plane trajectory of this period-2 solution is still a cycle embedded with some smaller cycles with boundaries of ω from 49.84 to 49.889 rad/s (475.94 to 476.4 rpm) and i from 50 to 113 A, as shown in Figure 5.10(e).

When the speed feedback gain is further increased to $g = 14$ V/rad/s, the SR drive operates in chaos. The chaotic waveforms and trajectory are shown in Figure 5.11. Similar to the periodic

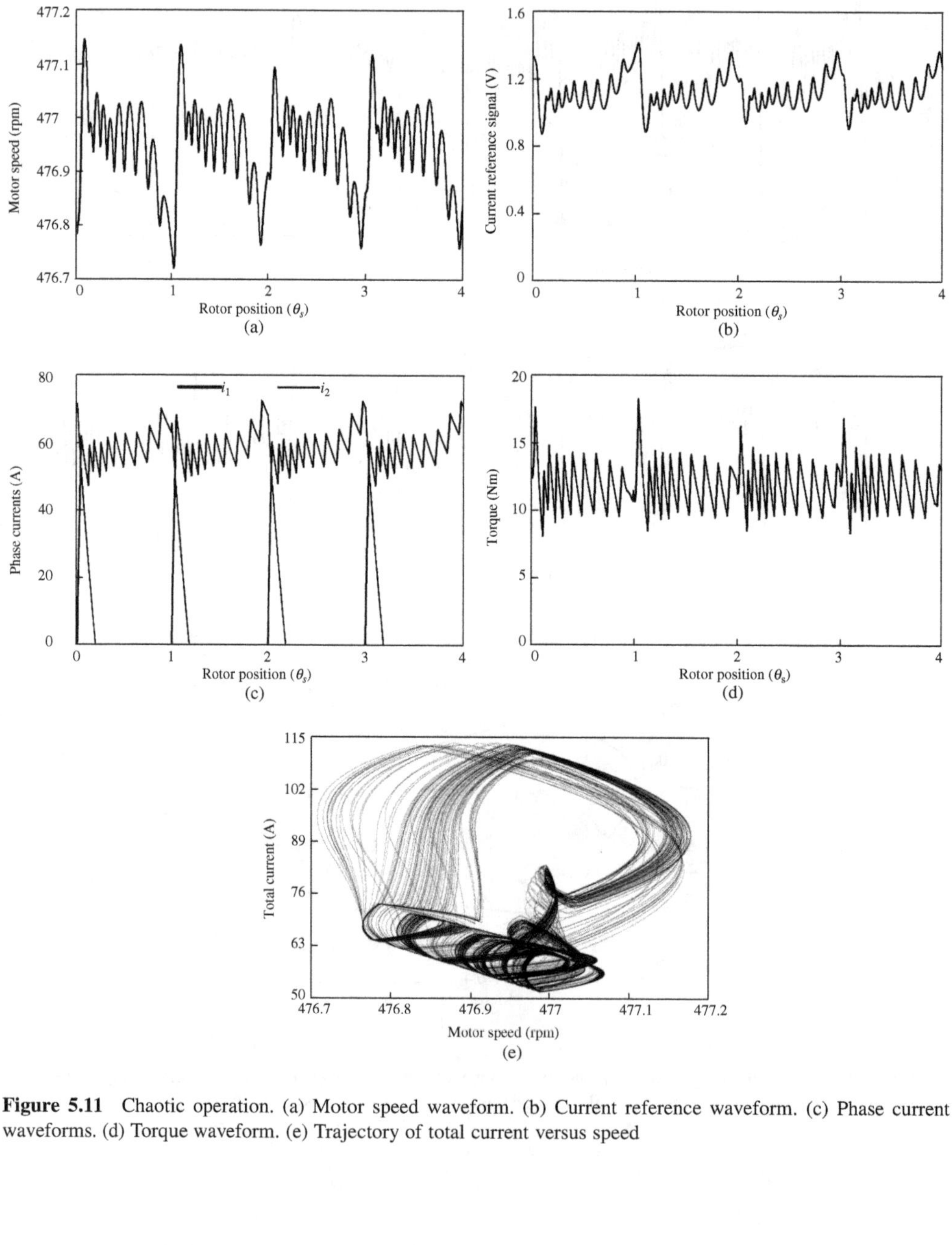

Figure 5.11 Chaotic operation. (a) Motor speed waveform. (b) Current reference waveform. (c) Phase current waveforms. (d) Torque waveform. (e) Trajectory of total current versus speed

(fundamental and subharmonic) operations, the f_{com} oscillating component of chaotic operation still dominates the f_{PWM} component and the extremes of ω and v_c are also near the commutation point, as shown in Figures 5.11(a) and (b). However, these waveforms lose the synchronization of f_{com}, leading to the irregular variation of f_{PWM}. Hence, the corresponding i_1 (or i_2) and T_e exhibit random-like fluctuations with the boundaries of 48–72 A and 9–18 Nm, as shown in Figures 5.11(c) and (d), respectively. As shown in Figure 5.11(e), the trajectory of i versus ω is a random-like bounded phase portrait with boundaries of ω from 49.896 to 49.938 rad/s (476.47 to 476.87 rpm) and i from 50 to 114 A. By comparing Figures 5.9–5.11, it can be found that the chaotic waveform of T_e exhibits a random-like variation of the peaky values even though the maximum peaky value is almost identical to that of the fundamental and subharmonic waveforms. This leads to a reduction of the torque ripples that contribute to acoustic noise. It also explains why the oscillating magnitude of ω in chaotic operation is less than that in fundamental and subharmonic operations, even though the oscillating magnitude of v_c in chaotic operation is slightly larger than that in fundamental and subharmonic operations because of the larger g. This discovery implies that the chaotic behavior of a current-controlled SR drive system may be beneficially utilized to reduce electromagnetic interference and acoustic noise.

5.2.4 *Phenomena*

A key property of chaotic behavior is the divergence of nearby trajectories. This property can be used to distinguish between aperiodic behavior due to chaos and that due to external noise. The divergence of nearby trajectories results in a sensitive dependence on the initial conditions. On the other hand, the chaotic behavior can be realized by its chaotic orbit of the Poincaré map. This chaotic orbit has the twin properties of stretching and folding. The stretching property means that the orbits lying initially close together in a region are mapped into separate orbits, whereas the folding property means that the orbits outside this region are returned to its interior. The above properties can be illustrated by the error of a current reference signal Δv_c, which is a linear function of the speed error $\Delta\omega$ as given by (5.17). By using (5.27), the variation of Δv_c with an initial difference of 0.0028 V (equivalent to 0.0002 rad/s) is shown in Figure 5.12. It can be found that

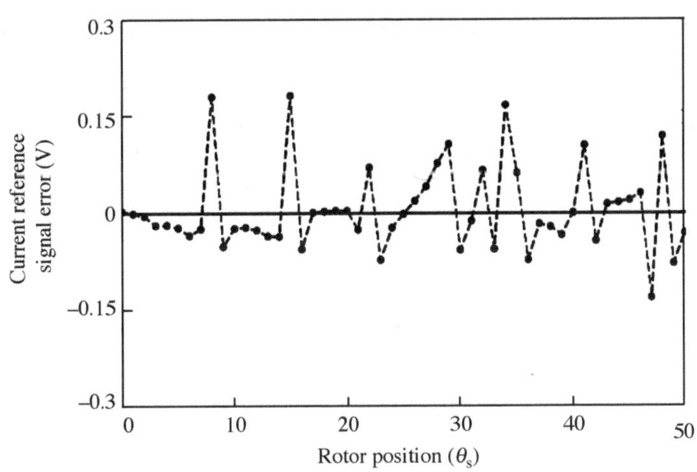

Figure 5.12 Waveform of current reference signal error

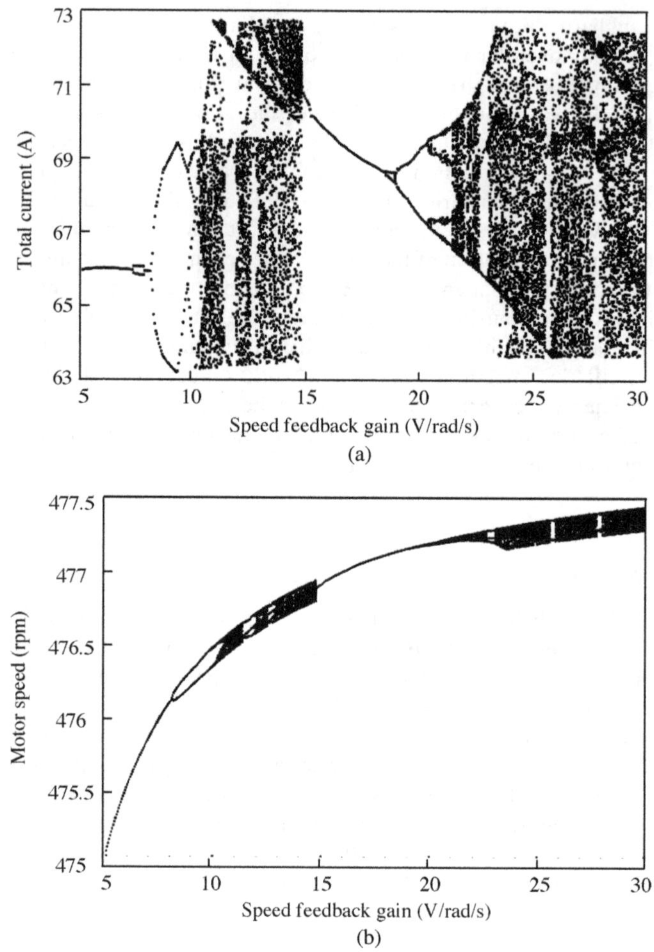

Figure 5.13 Bifurcation diagrams with respect to speed feedback gain. (a) Total current. (b) Motor speed

Δv_c is very sensitive to its initial value of Δv_c, indicating the property of divergence of nearby trajectories. Increasingly, it shows that both stretching and folding properties occur irregularly. It is the stretching property that renders the chaotic behavior long-term unpredictable, while the folding property keeps the chaotic behavior bounded. Hence, the chaotic behavior is characterized by a random-like bounded orbit.

By computing the Poincaré map in (5.27), the bifurcation diagrams of i and ω with respect to g, T_l, and Δi are shown in Figures 5.13–5.15, respectively. These bifurcation diagrams can depict the fundamental, subharmonic and chaotic orbits at a glance, hence illustrating how to route to chaos. While calculating the bifurcation diagrams, only the bifurcation parameter varies and the other parameters are fixed.

As shown in Figure 5.13, both i and ω bifurcate from a period-1 orbit to a period-2 orbit at $g = 7.5$ V/rad/s, then to a period-1 orbit again at $g = 7.9$ V/rad/s, then back to a period-2 orbit at $g = 8$ V/rad/s, then to a period-4 orbit at $g = 9.6$ V/rad/s, and then to chaos at $g = 10.2$ V/rad/s. After

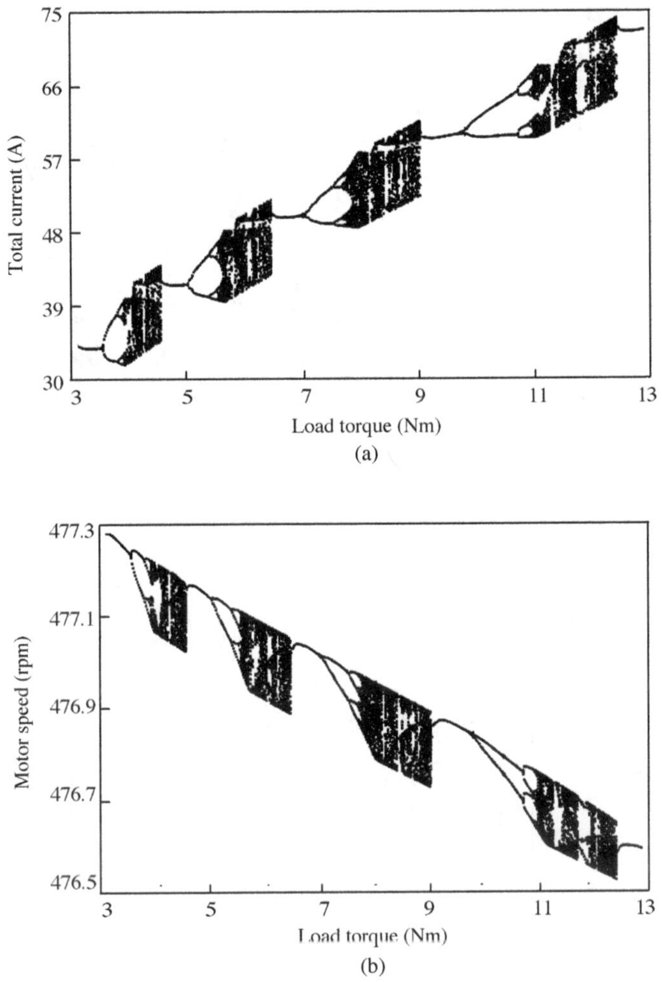

Figure 5.14 Bifurcation diagrams with respect to load torque. (a) Total current. (b) Motor speed

chaos, the system comes back to a period-1 orbit at $g = 14.9$ V/rad/s, and then to chaos again via a similar route. It can be also found that there are period-3 orbits within the chaotic band, the so-called period-3 window. As shown in Figure 5.14, both i and ω deviate from a period-1 orbit to a period-2 orbit at $T_l = 3.5$ Nm, then to a period-4 orbit at $T_l = 3.7$ Nm, and then to chaos at $T_l = 3.82$ Nm. After this chaotic band, both i and ω return to a period-1 orbit again at $T_l = 4.5$ Nm, and then to chaos again via a period-doubling route. It is interesting to note that this route to chaos appears again and again. As shown in Figure 5.15, both i and ω with $g = 14$ V/rad/s also have a period-doubling route to chaos, which repeats throughout the range of Δi. It should be noted that the reduction of Δi (which is constrained by the switching frequency of power devices) can reduce the fluctuation of i and ω, but is unable to avoid the occurrence of chaos.

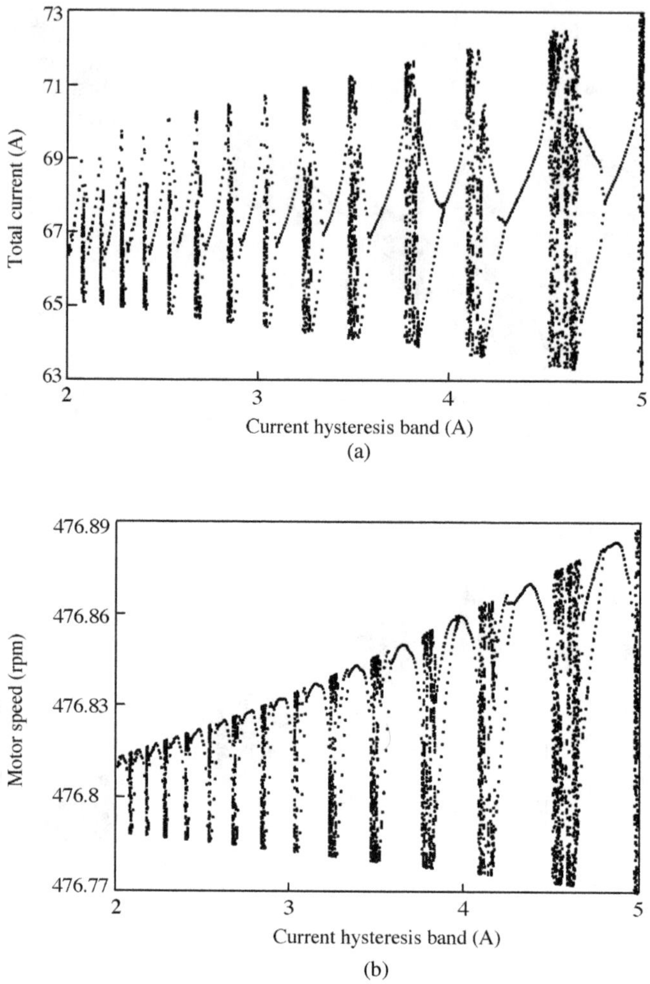

Figure 5.15 Bifurcation diagrams with respect to current hysteresis band. (a) Total current. (b) Motor speed

References

Chan, C.C., Jiang, Q., Zhan, Y.J., and Chau, K.T. (1996) A high-performance switched reluctance drive for P-star EV project. Proceedings of International Electric Vehicle Symposium, pp. 78–83.

Chen, J.H., Chau, K.T., Chan, C.C., and Jiang, Q. (2002) Subharmonic and chaos in switched reluctance motor drives. *IEEE Transactions on Energy Conversion*, **17**, 73–78.

De Castro, M.R., Robert, B.G.M., and Goeldel, C. (2008) Bifurcations and chaotic dynamics in a linear switched reluctance motor. Proceedings of International Power Electronics and Motion Control Conference, pp. 2126–2133.

Lawrenson, P.J. (1992) A brief status review of switched reluctance drives. *EPE Journal*, **2**, 133–144.

Miller, T.J.E. (1993) *Switched Reluctance Motors and Their Control*, Magna Physics Publishing and Clarendon Press, Oxford.

Parker, T.S. and Chua, L.O. (1989) *Practical Numerical Algorithms for Chaotic Systems*, Springer-Verlag, New York.

Stephenson, J.M. and Corda, J. (1979) Computation of torque and current in doubly salient reluctance motors from nonlinear magnetisation data. *IEE Proceedings - B*, **126**, 393–396.

Wiggins, S. (1990) *Introduction to Applied Nonlinear Dynamical Systems and Chaos*, Springer-Verlag, New York.

Zhan, Y.J., Chan, C.C., and Chau, K.T. (1999) A novel sliding mode observer for indirect position sensing of switched reluctance motor drives. *IEEE Transactions on Industrial Electronics*, **46**, 390–397.

Zhusubaliyev, Z.T. and Mosekilde, E. (2003) *Bifurcations and Chaos in Piecewise-Smooth Dynamical Systems*, World Scientific Publishing Company, Singapore.

Part Three

Control of Chaos in Electric Drive Systems

6

Stabilization of Chaos in Electric Drive Systems

Chaos is observed as an unpredictable phenomenon due to its sensitivity to initial states. It is a kind of steady-state but locally unstable behavior, and exhibits irregular properties. Such random-like phenomenon is normally regarded as unstable operation which results in additional loss, and therefore is a harmful behavior. Various control methods have been proposed to stabilize the chaotic behavior, such as the Ott–Grebogi–Yorke (OGY) method (Ott, Grebogi, and Yorke, 1990; Hunt, 1991), the time-delay feedback method (Pyragas, 1992), the non-feedback method (Rajasekar, Murali, and Lakshmanan, 1997; Ramesh and Narayanan, 1999), the proportional feedback method (Jackson and Grosu, 1995; Casas and Grebogi, 1997), the nonlinear control method (Khovanov *et al.*, 2000; Tian, 1999), the adaptive control method (Boccaletti, Farini, and Arecchi, 1997; Liao and Lin, 1999), the neutral networks method (Hirasawa *et al.*, 2000; Poznyak, Yu, and Sanchez, 1999), and the fuzzy control method (Tanaka, Ikeda, and Wang, 1998). Some of them have also been proposed to stabilize the chaotic behavior in electric drive systems.

In this chapter, various control approaches, including the time-delay feedback control, the nonlinear feedback control, the backstepping control, the dynamic surface control and the sliding mode control, are introduced to stabilize the chaos that occurs in both DC and AC drive systems.

6.1 Stabilization of Chaos in DC Drive System

6.1.1 Modeling

As shown in Figure 6.1, the time-delay feedback control method is used to stabilize chaos in a voltage-controlled DC drive system. Time-delay feedback control has some definite advantages: it does not desire *a priori* analytical knowledge of the system dynamics; it does not require a reference signal corresponding to the desired unstable periodic orbit (UPO); and it does not need fast sampling or a computer analysis of the state of the system. Also, the corresponding perturbation is small when the delayed time is close to the period of the desired UPO (Kittel, Parisi, and Pyragas, 1995).

The speed control of the DC drive is implemented by constant-frequency pulse width modulation (PWM). The ramp voltage signal v_r, which functions to generate the PWM signal, is represented by:

$$v_r = v_l + (v_u - v_l)(t \bmod T) \tag{6.1}$$

Chaos in Electric Drive Systems: Analysis, Control and Application, First Edition. K.T. Chau and Zheng Wang.
© 2011 John Wiley & Sons (Asia) Pte Ltd. Published 2011 by John Wiley & Sons (Asia) Pte Ltd.

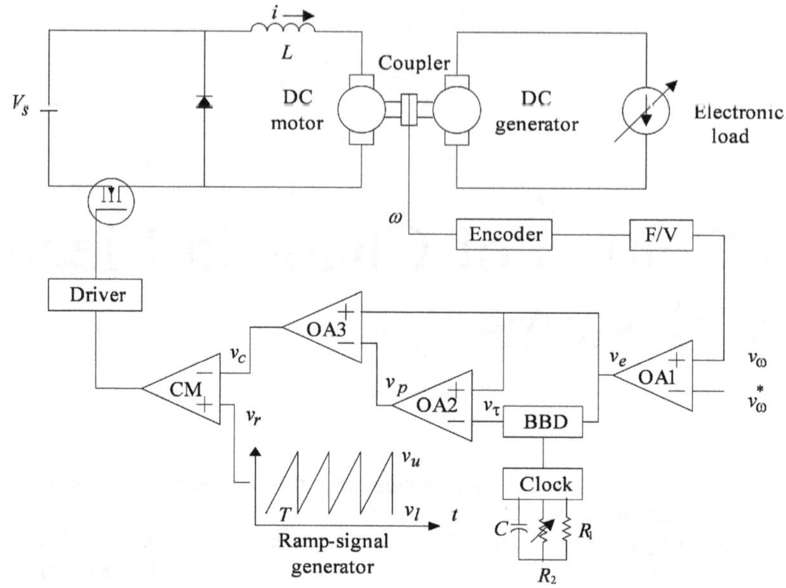

Figure 6.1 Voltage-controlled DC drive system with time-delay feedback control

where v_l, v_u, and T are the lower limit, upper limit and period of the sawtooth wave. On the other hand, the speed error signal v_e is given by:

$$v_e = g(\omega - \omega^*) \tag{6.2}$$

where g is the speed feedback gain, which is actually the overall gain of the speed encoder, frequency-to-voltage (F/V) converter, and operational amplifier 1 (OA1), ω is the actual speed, and ω^* is the reference speed. This speed error is chosen as the feedback variable. By using the bucket-brigade delay (BBD) line, the corresponding delayed signal v_τ for the time-delay feedback control is described as:

$$v_\tau = v_e(t - \tau) \tag{6.3}$$

where τ is the time delay. When the desired orbit is the embedded unstable period-p orbit, τ is normally chosen to be pT. Thus, the feedback control signal v_c can be represented as:

$$v_c = v_e - \eta(v_e - v_\tau) \tag{6.4}$$

where η is the delayed feedback gain, which is actually the gain in the operational amplifier 2 (OA2). The operational amplifier 3 (OA3) simply adopts a unity gain. Then, v_c and v_r are fed to the comparator (CM), and thus generate the PWM signal to turn on and off the power switch S. When v_c is larger than v_r, S is turned off and the diode D conducts. Otherwise, S is turned on and D is turned off. The inductor L is connected in series with the DC motor to ensure a continuous conduction mode of operation. So, the voltage-controlled DC drive system with a time-delay feedback control can be modeled as:

$$\begin{cases} \dot{\mathbf{X}}(t) = \mathbf{A}\mathbf{X}(t) + \mathbf{E}(t) \\ h(t) = v_r(t) - [\mathbf{G}_v(\mathbf{X}(t) - \mathbf{E}_c) - \eta\mathbf{G}_v(\mathbf{X}(t) - \mathbf{X}(t - pT))] \end{cases} \tag{6.5}$$

It should be noted that $h(t)$ comprises a delayed component, and the dynamical equation in (6.5) is actually a time-delayed differential equation. The delayed feedback only exists in the switching condition $h(t)$. As discussed in Chapter 3, the corresponding Poincaré map with multiple switching pulses can be represented by:

$$\begin{cases} \mathbf{X}(nT+\delta_0 T) = \mathbf{X}_n \\ \mathbf{X}(nT+\delta_i T) = -\mathbf{A}^{-1}\mathbf{E}(nT+\delta_{i-1}^+ T)+\Phi(\delta_i T-\delta_{i-1} T) \\ \qquad \times(\mathbf{X}(nT+\delta_{i-1}T)+\mathbf{A}^{-1}\mathbf{E}(nT+\delta_{i-1}^+ T)) \quad (i=1,\cdots,m) \\ \mathbf{X}_{n+1} = -\mathbf{A}^{-1}\mathbf{E}(nT+\delta_m^+ T)+\Phi(T-\delta_m T)(\mathbf{X}(nT+\delta_m T) \\ \qquad +\mathbf{A}^{-1}\mathbf{E}(nT+\delta_m^+ T)) \end{cases} \tag{6.6}$$

The switching points $\delta_1,\cdots,\delta_m$ can be determined by using the switching condition described in (6.5):

$$\begin{aligned} h_i(\delta_i) = v_r(\delta_i T)-[\mathbf{G}_v(\mathbf{X}(nT+\delta_i T)-\mathbf{E}_c)-\eta\mathbf{G}_v(\mathbf{X}(nT+\delta_i T) \\ -\mathbf{X}((n-p)T+\delta_i T))] = 0 \qquad (i=1,\cdots,m) \end{aligned} \tag{6.7}$$

Then, the switching points between the interval $(n-p)T$ and $(n-p+1)T$ are given by:

$$\begin{cases} t=(n-p)T+\delta'_j T \\ 0=\delta'_0<\delta'_1<\cdots<\delta'_{m'}<\delta'_{m'+1}=1 \quad (j=1,\cdots,m') \end{cases} \tag{6.8}$$

The corresponding system states in (6.7) are given by:

$$\begin{aligned} \mathbf{X}((n-p)T+\delta_i T) = -\mathbf{A}^{-1}\mathbf{E}((n-p)T+\delta_j'^{+} T)+\Phi(\delta_i T-\delta'_j T) \\ \times(\mathbf{X}((n-p)T+\delta'_j T)+\mathbf{A}^{-1}\mathbf{E}((n-p)T+\delta_j'^{+} T)) \\ (i=1,\cdots,m;\ j=\{j\in\{1,\cdots,m'\}|\delta_i\in[\delta'_j,\delta'_{j+1}]\}) \end{aligned} \tag{6.9}$$

Thus, the switching points can be calculated by using (6.6), (6.7), and (6.9), which yields:

$$h_i(\mathbf{X}_{n-p},\delta'_j,\mathbf{X}_n,\delta_{i-1},\delta_i)=0 \quad (i=1,\cdots,m;\ j=\{j\in\{1,\cdots,m'\}|\delta_i\in[\delta'_j,\delta'_{j+1}]\}) \tag{6.10}$$

Given the values of $\mathbf{X}_{n-p},\delta'_1,\cdots,\delta'_m$, the values of $\mathbf{X}_n,\delta_1,\cdots,\delta_m$ can be calculated by using (6.10).

For a practical voltage-controlled DC drive system, there is only a single switching pulse. So, the corresponding Poincaré map can be represented by:

$$\begin{aligned} \mathbf{X}_{n+1} = -\mathbf{A}^{-1}\mathbf{E}_0+\Phi(T)(\mathbf{X}_n+\mathbf{A}^{-1}\mathbf{E}_0) \\ +(1-\Phi(T-\delta_1 T))\mathbf{A}^{-1}(\mathbf{E}_0-\mathbf{E}_1) \end{aligned} \tag{6.11}$$

When $\delta_1\in(0,\delta'_1]$, the instant of switching action in (6.10) is given by:

$$\begin{aligned} h_I(\mathbf{X}_{n-p},\mathbf{X}_n,\delta_1) = v_r(\delta_1 T)-\mathbf{G}_v[-\mathbf{A}^{-1}\mathbf{E}_0+\Phi(\delta_1 T)(\mathbf{X}_n+\mathbf{A}^{-1}\mathbf{E}_0) \\ -\eta\Phi(\delta_1 T)(\mathbf{X}_n-\mathbf{X}_{n-p})-\mathbf{E}_c] = 0 \end{aligned} \tag{6.12}$$

When $\delta_1 \in [\delta_1', 1)$, the instant of switching action in (6.10) is given by:

$$\begin{aligned} h_r(\mathbf{X}_{n-p}, \delta_1', \mathbf{X}_n, \delta_1) = {} & v_r(\delta_1 T) - \mathbf{G}_v[-\mathbf{A}^{-1}\mathbf{E}_0 + \Phi(\delta_1 T)(\mathbf{X}_n + \mathbf{A}^{-1}\mathbf{E}_0) \\ & -\eta[\Phi(\delta_1 T)(\mathbf{X}_n - \mathbf{X}_{n-p}) - (1 - \Phi(\delta_1 T - \delta_1' T)) \\ & \times \mathbf{A}^{-1}(\mathbf{E}_0 - \mathbf{E}_1)] - \mathbf{E}_c] = 0 \end{aligned} \tag{6.13}$$

Since $\Phi(\delta_1' T)$ can be represented by $\mathbf{X}_{n-p}$ and $\mathbf{X}_{n-p+1}$, (6.13) is rewritten as:

$$\begin{aligned} h_r(\mathbf{X}_{n-p+1}, \mathbf{X}_n, \delta_1) = {} & v_r(\delta_1 T) - \mathbf{G}_v[-\mathbf{A}^{-1}\mathbf{E}_0 + \Phi(\delta_1 T)(\mathbf{X}_n + \mathbf{A}^{-1}\mathbf{E}_0) \\ & -\eta[\Phi(\delta_1 T)(\mathbf{X}_n + \mathbf{A}^{-1}\mathbf{E}_0) \\ & -\Phi(\delta_1 T - T)(\mathbf{X}_{n-p+1} + \mathbf{A}^{-1}\mathbf{E}_0) \\ & -(1 - \Phi(\delta_1 T - T))\mathbf{A}^{-1}(\mathbf{E}_0 - \mathbf{E}_1)] - \mathbf{E}_c] = 0 \end{aligned} \tag{6.14}$$

It should be noted that h_l is the switching conditions when δ_1 is on the left side of δ_1', and h_r is the switching conditions when δ_1 is on the right side of δ_1'. The stability of control can be described by (6.11), (6.12), and (6.14). Since τ is normally chosen to be the period T, namely $p = 1$, the corresponding Poincaré map can be simplified as:

$$\begin{aligned} \mathbf{X}_{n+1} = {} & -\mathbf{A}^{-1}\mathbf{E}_0 + \Phi(T)(\mathbf{X}_n + \mathbf{A}^{-1}\mathbf{E}_0) \\ & +(1 - \Phi(T - \delta_1 T))\mathbf{A}^{-1}(\mathbf{E}_0 - \mathbf{E}_1) \end{aligned} \tag{6.15}$$

$$\begin{aligned} h_l(\mathbf{X}_{n-1}, \mathbf{X}_n, \delta_1) = {} & v_r(\delta_1 T) - \mathbf{G}_v[-\mathbf{A}^{-1}\mathbf{E}_0 + \Phi(\delta_1 T)(\mathbf{X}_n + \mathbf{A}^{-1}\mathbf{E}_0) \\ & -\eta\Phi(\delta_1 T)(\mathbf{X}_n - \mathbf{X}_{n-1}) - \mathbf{E}_c] = 0 \end{aligned} \tag{6.16}$$

$$\begin{aligned} h_r(\mathbf{X}_n, \delta_1) = {} & v_r(\delta_1 T) - \mathbf{G}_v[-\mathbf{A}^{-1}\mathbf{E}_0 + \Phi(\delta_1 T)(\mathbf{X}_n + \mathbf{A}^{-1}\mathbf{E}_0) \\ & -\eta[\Phi(\delta_1 T)(1 - \Phi(-T))(\mathbf{X}_n + \mathbf{A}^{-1}\mathbf{E}_0) \\ & -(1 - \Phi(\delta_1 T - T))\mathbf{A}^{-1}(\mathbf{E}_0 - \mathbf{E}_1)] - \mathbf{E}_c] = 0 \end{aligned} \tag{6.17}$$

When $\delta_1 \in [\delta_1', 1)$, δ_1 is an explicit function of $\mathbf{X}_n$; and when $\delta_1 \in (0, \delta_1']$, δ_1 is an implicit function of $\mathbf{X}_n$ and $\mathbf{X}_{n-1}$. Let $\mathbf{Z}_n = \mathbf{X}_{n-1}$, the Poincaré map can be rewritten as:

$$\begin{cases} \mathbf{X}_{n+1} = P_l(\mathbf{X}_n, \delta_1(\mathbf{X}_n, \mathbf{Z}_n)) \\ \mathbf{Z}_{n+1} = \mathbf{X}_n \end{cases} \quad \delta_1 \in (0, \delta_1'] \tag{6.18}$$

$$\mathbf{X}_{n+1} = P_r(\mathbf{X}_n, \delta_1(\mathbf{X}_n)) \quad \delta_1 \in [\delta_1', 1) \tag{6.19}$$

where P_l and P_r are the Poincaré maps corresponding to the conditions that $\delta_1 \in (0, \delta_1']$ and $\delta_1 \in [\delta_1', 1)$, respectively.

6.1.2 Analysis

The Poincaré map with the time-delay feedback control is different from that of the system without control. By choosing a suitable delay feedback gain η, the chaotic behavior can be stabilized into periodic behavior. The effective range for stabilization is called as the stable domain.

6.1.2.1 Fundamental Operation

The fundamental operation of the DC drive system corresponds to the period-1 orbit, namely $\mathbf{X}_n = \mathbf{Z}_n = \mathbf{X}^*$. So, the delay component in (6.16) and (6.17) become zero. Namely, δ_1 and $\mathbf{X}^*$ are independent of η. Therefore, the fixed point of period-1 orbit can be written as:

$$\begin{cases} \mathbf{X}^* & = -\mathbf{A}^{-1}\mathbf{E}_0 + (1-\Phi(T))^{-1}(1-\Phi(T-\delta_1 T))\mathbf{A}^{-1}(\mathbf{E}_0-\mathbf{E}_1) \\ h(\delta_1) & = v_r(\delta_1 T)-\mathbf{G}_v[-\mathbf{A}^{-1}\mathbf{E}_0+\Phi(\delta_1 T)(1-\Phi(T))^{-1} \\ & \quad \times(1-\Phi(T-\delta_1 T))\mathbf{A}^{-1}(\mathbf{E}_0-\mathbf{E}_1)-\mathbf{E}_c] = 0 \end{cases} \tag{6.20}$$

The characteristic multiplier of (6.20) is also independent of η. According to the implicit theorem, the Jacobian matrices of (6.18) and (6.19) are given by:

$$\mathbf{J}_l = \begin{pmatrix} \mathbf{C}_1 & \mathbf{C}_2 \\ \mathbf{1} & \mathbf{0} \end{pmatrix} \tag{6.21}$$

$$\begin{aligned} \mathbf{J}_r &= \frac{\partial P_r}{\partial \mathbf{X}_n} - \frac{\partial P_r}{\partial \delta_1}\left(\frac{\partial h_r}{\partial \delta_1}\right)^{-1}\frac{\partial h_r}{\partial \mathbf{X}_n} \\ &= \Phi(T-\delta_1 T)(\mathbf{1}-d_r(1-\eta)\mathbf{S}-d_r\eta\mathbf{S}\Phi(-T))\Phi(\delta_1 T) \end{aligned} \tag{6.22}$$

where

$$\begin{aligned} \mathbf{C}_1 &= \frac{\partial P_l}{\partial \mathbf{X}_n} - \frac{\partial P_l}{\partial \delta_1}\left(\frac{\partial h_l}{\partial \delta_1}\right)^{-1}\frac{\partial h_l}{\partial \mathbf{X}_n} \\ &= \Phi(T-\delta_1 T)(1-d_l(1-\eta)\mathbf{S})\Phi(\delta_1 T) \end{aligned} \tag{6.23}$$

$$\mathbf{C}_2 = -\frac{\partial P_l}{\partial \delta_1}\left(\frac{\partial h_l}{\partial \delta_1}\right)^{-1}\frac{\partial h_l}{\partial \mathbf{Z}_n} = \Phi(T-\delta_1 T)(-d_l\eta\mathbf{S})\Phi(\delta_1 T) \tag{6.24}$$

$$\mathbf{S} = (\mathbf{E}_1-\mathbf{E}_2)\mathbf{G}_v \tag{6.25}$$

$$\begin{aligned} d_l = {} & [(v_u-v_l)/T-\mathbf{G}_v[\Phi(\delta_1 T)(\mathbf{A}\mathbf{X}_n+\mathbf{E}_0) \\ & -\eta\Phi(\delta_1 T)\mathbf{A}(\mathbf{X}_n-\mathbf{Z}_n)]]^{-1} \end{aligned} \tag{6.26}$$

$$d_r = [(v_u - v_l)/T - \mathbf{G}_v[\boldsymbol{\Phi}(\delta_1 T)(\mathbf{AX}_n + \mathbf{E}_0) \\ -\eta\boldsymbol{\Phi}(\delta_1 T)(\mathbf{AX}_n + \mathbf{E}_0 - \boldsymbol{\Phi}(-T)(\mathbf{AX}_n + \mathbf{E}_1))]]^{-1} \tag{6.27}$$

Since $\mathbf{X}_n = \mathbf{Z}_n = \mathbf{X}^*$, $d_l = d_r = d^*$ can be represented by:

$$d^* = [(v_u - v_l)/T - \mathbf{G}_v\boldsymbol{\Phi}(\delta_1 T)(\mathbf{AX}^* + \mathbf{E}_0)]^{-1} \tag{6.28}$$

Theorem 6.1 *When the eigenvalues of a matrix $\mathbf{S} \in \Re^{n\times n}$ are zero, the eigenvalues of the matrix $a\mathbf{S} + b\mathbf{SS}(a, b \in \Re^1)$ are also zero.*

Proof When $\mathbf{J}_s$ is the Jordan normalization form of $\mathbf{S}$, it results in $\mathbf{S} = \mathbf{Q}^{-1}\mathbf{J}_s\mathbf{Q}$. Thus, it yields $a\mathbf{S} + b\mathbf{SS} = \mathbf{Q}^{-1}(a\mathbf{J}_s + b\mathbf{J}_s\mathbf{J}_s)\mathbf{Q}$. When the eigenvalues of $\mathbf{S}$ are equal to zero, $\mathbf{J}_s$ can be described as $\mathbf{J}_s = \begin{pmatrix} 0 & \cdots & & 0 \\ 1 & 0 & & \\ \vdots & \ddots & \ddots & \vdots \\ 0 & \cdots & 1 & 0 \end{pmatrix}$.

Hence, it deduces that $\mathbf{J}_s\mathbf{J}_s = 0$, and the eigenvalues of $a\mathbf{S} + b\mathbf{SS}$ are also zero.

Theorem 6.2 *When the two eigenvalues of $\mathbf{S} \in \Re^{2\times 2}$ in (6.25) are zero, two of the four eigenvalues of $\mathbf{J}_l \in \Re^{4\times 4}$ are zero, and the other two eigenvalues of $\mathbf{J}_l \in \Re^{4\times 4}$ are equal to those of $\mathbf{J}_r \in \Re^{2\times 2}$. These two eigenvalues are also equal to the eigenvalues of the matrix $\mathbf{J} \in \Re^{2\times 2}$, which is expressed as:*

$$\mathbf{J} = \boldsymbol{\Phi}(T)(1 - d^*(1-\eta)\mathbf{S} - d^*\eta\mathbf{S}\boldsymbol{\Phi}(-T)) \tag{6.29}$$

Proof Given $\mathbf{Q} = \begin{pmatrix} \mathbf{1} & \mathbf{0} \\ \mathbf{C}_1^{-1} & \mathbf{1} \end{pmatrix}$ and $\mathbf{Q}^{-1} = \begin{pmatrix} \mathbf{1} & \mathbf{0} \\ -\mathbf{C}_1^{-1} & \mathbf{1} \end{pmatrix}$, yields $\mathbf{Q}^{-1}\mathbf{J}_l\mathbf{Q} = \begin{pmatrix} \mathbf{C}_1 + \mathbf{C}_2\mathbf{C}_1^{-1} & \mathbf{C}_2 \\ -\mathbf{C}_1^{-1}\mathbf{C}_2\mathbf{C}_1^{-1} & -\mathbf{C}_1^{-1}\mathbf{C}_2 \end{pmatrix}$.

From (6.23) and (6.24), it can be deduced that $\mathbf{C}_1^{-1}\mathbf{C}_2\mathbf{C}_1^{-1} = 0$. The eigenvalues of $\mathbf{J}_l$ are the same as those of $\mathbf{Q}^{-1}\mathbf{J}_l\mathbf{Q}$. Two of the eigenvalues of $\mathbf{J}_l$ are the eigenvalues of $\mathbf{C}_1 + \mathbf{C}_2\mathbf{C}_1^{-1}$, and the other two are the eigenvalues of $-\mathbf{C}_1^{-1}\mathbf{C}_2$. Thus, from (6.23)–(6.26), $-\mathbf{C}_1^{-1}\mathbf{C}_2$ can be described as:

$$-\mathbf{C}_1^{-1}\mathbf{C}_2 = \boldsymbol{\Phi}(-\delta_1 T)(d_l\eta\mathbf{S} + d_l^2(1-\eta)\eta\mathbf{SS})\boldsymbol{\Phi}(\delta_1 T) \tag{6.30}$$

According to Theorem 1, when the eigenvalues of $\mathbf{S}$ are all equal to zero, the eigenvalues of $a\mathbf{S} + b\mathbf{SS}$ are zero. When $a = d_l\eta$ and $b = d_l^2(1-\eta)\eta$, the eigenvalues of $-\mathbf{C}_1^{-1}\mathbf{C}_2$ are zero. From (6.23)–(6.28), $\mathbf{C}_1 + \mathbf{C}_2\mathbf{C}_1^{-1}$ can be described as:

$$\mathbf{C}_1 + \mathbf{C}_2\mathbf{C}_1^{-1} = \mathbf{J}_r \\ = \boldsymbol{\Phi}(-\delta_1 T)\boldsymbol{\Phi}(T)(1 - d^*(1-\eta)\mathbf{S} - d^*\eta\mathbf{S}\boldsymbol{\Phi}(-T))\boldsymbol{\Phi}(\delta_1 T) \tag{6.31}$$

So, the eigenvalues of $\mathbf{C}_1 + \mathbf{C}_2\mathbf{C}_1^{-1}$ or $\mathbf{J}_r$ are equal to the eigenvalues of $\mathbf{J}$ in (6.29).

Because

$$\mathbf{S} = (\mathbf{E}_1 - \mathbf{E}_0)\mathbf{G}_v = \begin{pmatrix} 0 & 0 \\ (V_s/L)g & 0 \end{pmatrix},$$

the eigenvalues of $\mathbf{S}$ are equal to zero. Thus, the aforementioned theorems can be applied to a voltage-controlled DC drive system. The characteristic multipliers of a period-1 orbit can be calculated by the eigenvalues of $\mathbf{J}$ in (6.29). By calculating the characteristic multipliers, the stable domain of the control parameters – namely the time-delay feedback gain versus various system parameters for effective stabilization – can be determined. It should be noted that the calculation of the eigenvalues of $\mathbf{J}$ is easier than those for $\mathbf{J}_l$ and $\mathbf{J}_r$. In particular, the dimension of $\mathbf{J}_l$ is twice that of $\mathbf{J}$.

6.1.2.2 Subharmonic Operation

The stable domain for the period-p ($p > 1$) is illustrated by the period-2 operation. When $\delta_1 \geq \delta_2$, a second-order Poincaré map of the drive system can be obtained by using (6.18) and (6.19):

$$\begin{cases} \mathbf{X}_{n+1} = P_r(\mathbf{X}_n, \delta_1(\mathbf{X}_n)) \\ \mathbf{Z}_{n+1} = \mathbf{X}_n \end{cases} \delta_1 \in [\delta_1', 1) \tag{6.32}$$

$$\begin{cases} \mathbf{X}_{n+2} = P_l(\mathbf{X}_{n+1}, \delta_1(\mathbf{X}_{n+1}, \mathbf{Z}_{n+1})) \\ \mathbf{Z}_{n+2} = \mathbf{X}_{n+1} \end{cases} \delta_2 \in (0, \delta_2'] \tag{6.33}$$

Contrary to the fixed point of the period-1 orbit, the fixed point of the period-2 orbit is dependent on η, which can be represented by:

$$\begin{cases} \mathbf{X}_1^* = -\mathbf{A}^{-1}\mathbf{E}_0 + (1-\Phi(2T))^{-1}[\Phi(T)(1-\Phi(T-\delta_1 T)) \\ \qquad + (1-\Phi(T-\delta_2 T))]\mathbf{A}^{-1}(\mathbf{E}_0-\mathbf{E}_1) \\ \mathbf{X}_2^* = -\mathbf{A}^{-1}\mathbf{E}_0 + (1-\Phi(2T))^{-1}[(1-\Phi(T-\delta_1 T)) \\ \qquad + \Phi(T)(1-\Phi(T-\delta_2 T))]\mathbf{A}^{-1}(\mathbf{E}_0-\mathbf{E}_1) \end{cases} \tag{6.34}$$

$$\begin{cases} h_1(\delta_1, \delta_2) = v_r(\delta_1 T) - \mathbf{G}_v[-\mathbf{A}^{-1}\mathbf{E}_0 + \Phi(\delta_1 T)(1-\eta(1-\Phi(T)))(\mathbf{X}_1^* \\ \qquad + \mathbf{A}^{-1}\mathbf{E}_0) + \eta(1-\Phi(\delta_1 T - T))\mathbf{A}^{-1}(\mathbf{E}_0-\mathbf{E}_1) - \mathbf{E}_c] = 0 \\ h_2(\delta_1, \delta_2) = v_r(\delta_2 T) - \mathbf{G}_v[-\mathbf{A}^{-1}\mathbf{E}_0 + \Phi(\delta_2 T)(\mathbf{X}_2^* + \mathbf{A}^{-1}\mathbf{E}_0) \\ \qquad - \eta(1-\Phi(\delta_2 T)(\mathbf{X}_2^* - \mathbf{X}_1^*) - \mathbf{E}_c] = 0 \end{cases} \tag{6.35}$$

Substituting (6.34) into (6.35), δ_1 and δ_2 can be deduced by solving (6.35), and the period-2 orbit $\{\mathbf{X}_1^*, \mathbf{X}_2^*\}$ can be obtained from (6.34). Similar to the period-1 orbit, the characteristic multipliers for the period-2 orbit are the eigenvalues of the Jacobian matrix $\mathbf{J}_2 = \partial\mathbf{X}_{n+2}/\partial\mathbf{X}_n$ which can be deduced from the equation:

$$\mathbf{J}_2 = \frac{\partial P_l(\mathbf{X}_{n+1}, \delta_1(\mathbf{X}_{n+1}, \mathbf{Z}_{n+1}))}{\partial \mathbf{X}_{n+1}} \frac{\partial P_r(\mathbf{X}_n, \delta_1(\mathbf{X}_n))}{\partial \mathbf{X}_n} + \frac{\partial P_l(\mathbf{X}_{n+1}, \delta(\mathbf{X}_{n+1}, \mathbf{Z}_{n+1}))}{\partial \mathbf{Z}_{n+1}}$$
$$= \Phi(T-\delta_2 T)(1-d_2^*(1-\eta)\mathbf{S})\Phi(\delta_2 T)\Phi(T-\delta_1 T)(1-d_1^*(1-\eta)\mathbf{S} - d_1^*\eta\mathbf{S}\Phi(-T)) + \Phi(T-\delta_2 T)(1-d_2^*\eta\mathbf{S})\Phi(\delta_2 T) \quad (6.36)$$

where

$$d_1^* = [(v_u - v_l)/T - \mathbf{G}_v[\Phi(\delta_1 T)(\mathbf{A}\mathbf{X}_1^* + \mathbf{E}_0) - \eta\Phi(\delta_1 T) \times (\mathbf{A}\mathbf{X}_1^* + \mathbf{E}_0 - \Phi(-T)(\mathbf{A}\mathbf{X}_1^* + \mathbf{E}_1))]]^{-1} \quad (6.37)$$

$$d_2^* = [(v_u - v_l)/T - \mathbf{G}_v\Phi(\delta_2 T)[(\mathbf{A}\mathbf{X}_n + \mathbf{E}_0) - \eta\mathbf{A}(\mathbf{X}_2^* - \mathbf{X}_1^*)]]^{-1} \quad (6.38)$$

By calculating the characteristic multipliers, the corresponding stable domain of the control parameters can be obtained.

6.1.3 Simulation

In order to validate the aforementioned stabilization of chaos, a computer simulation is carried out. The parameters adopted for the simulation are based on a practical voltage-controlled DC drive system, namely $R = 4.1\,\Omega$, $K_E = 0.1356$ V/rad/s, $K_T = 0.1324$ Nm/A, $J = 0.000557$ kgm^2, $B = 0.000275$ Nm/rad/s, $L = 28$ mH, $T_l = 0.5$ Nm, $v_u = 2.2$ V, $v_l = 0$ V, $T = 6.667$ ms, and $v_\omega^* = 7$ V. The delayed feedback gain η, speed feedback gain g, and supply voltage V_s are the control parameters for this simulation.

Based on the above stability analysis, the stable domain of η versus g under $V_s = 60$ V for period-1 and period-2 behaviors is depicted in Figure 6.2(a), while the stable domain of η versus V_s under

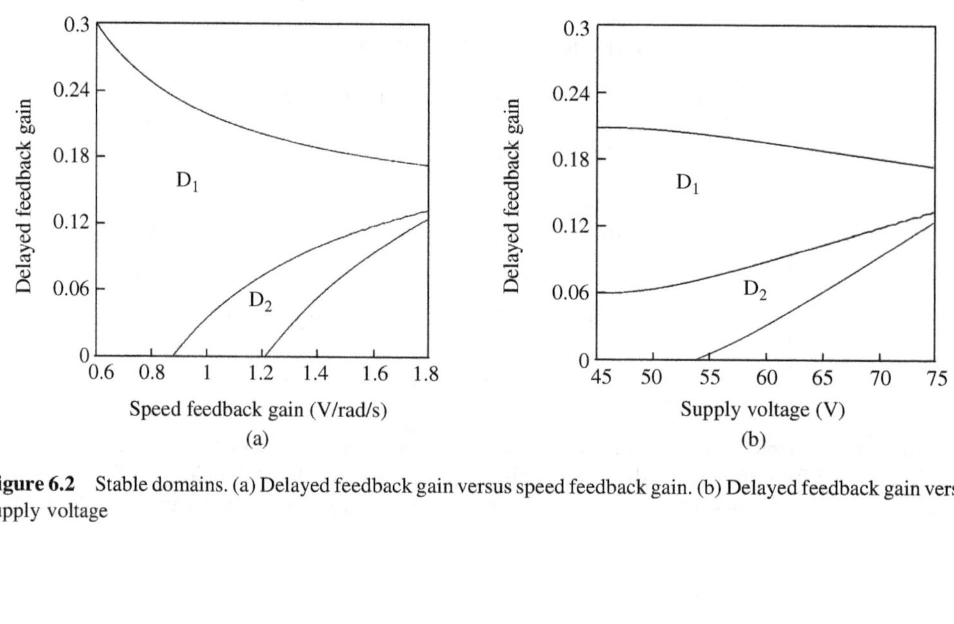

Figure 6.2 Stable domains. (a) Delayed feedback gain versus speed feedback gain. (b) Delayed feedback gain versus supply voltage

$g = 1.4$ V/rad/s is depicted in Figure 6.2(b). From Figure 6.2(a), it can be seen that when $g < 0.8$ V/rad/s the stable domain of period-1 motion D_1 embraces the case $\eta = 0$. This means that when $g < 0.8$ V/rad/s, a DC drive system without a time-delay feedback control can operate stably. Similar to the period-1 orbit, a DC drive system without a time-delay feedback control can also exhibit a stable period-2 orbit within the domain D_2 when 0.8 V/rad/s $< g <$ 1.15 V/rad/s. It also can be observed that, with the increase of g and V_s, the range of η for D_1 and D_2 become narrower. This feature indicates that with an increase of g and V_s, the corresponding stable domain becomes narrower and chaos is more prone to occur. The same phenomena can be observed in Figure 6.2(b).

In order to illustrate the stabilization of chaos, the trajectory of armature current i versus feedback control signal v_c is used for analysis. This feedback control signal is actually used to describe the motor speed. When $\eta = 0$, $V_s = 60$ V and $g = 1.6$ V/rad/s, the system operates in a chaotic mode, and the trajectory of i versus v_c is as plotted in Figure 6.3(a). It can be found that the trajectory exhibits a random-like but bounded behavior, with boundaries of $v_c \in [0\text{ V}, 3.3\text{ V}]$ and $i \in [1.3\text{ A}, 6.8\text{ A}]$. When η is set as 0.15 and 0.11, the chaotic behavior can be stabilized into a period-1 orbit and a period-2 orbit, as shown in Figures 6.3(b) and 6.3(c), which are in good agreement with the stable domains shown in Figure 6.2(a). The corresponding boundaries of the period-1 trajectory are $v_c \in [0.66\text{ V}, 1.9\text{ V}]$ and $i \in [2.2\text{ A}, 5.8\text{ A}]$, and those of the period-2 are $v_c \in [-0.5\text{ V}, 3.6\text{ V}]$ and $i \in [1.4\text{ A}, 7.4\text{ A}]$.

6.1.4 *Experimentation*

According to Figure 6.1, an experimental DC drive system is prototyped. In principle, there are three main subsystems, namely a power electronic DC chopper, a motor-generator set, and an analog electronic controller. The DC chopper, consisting of a DC power supply V_s, a power MOSFET switch IRFI640G, a power diode BYW29E200, and an inductor L, functions to regulate the input power flowing into the drive. The motor-generator set includes a DC motor, a DC generator, a coupler, and an electronic load, where the mechanical load torque is electronically controlled by the current sink of the electronic load. The electronic controller involves simple hardware, namely an encoder M57962L, a F/V converter LM331, three op-amps (OA1, OA2, and OA3) LM833, a bucket-brigade delay (BBD) line MN3004 and its clock MN3101, a ramp-signal generator, a comparator (CM) LM311, and a MOSFET driver DS0026.

Based on the encoder and the F/V converter, the motor speed ω is converted into an analog signal v_ω (with gain γ) which is then compared with the command speed signal v_ω^* to produce the error signal v_e via OA1 with gain α. Hence, the speed feedback gain g equals $\alpha\gamma$. According to the principle of delayed self-controlling feedback (Pyragas, 1992), v_e and its delayed version $v_\tau = v_e(t-\tau)$ are fed into OA2 with gain β to produce the perturbation signal v_p which is then compared with v_e to generate the desired control signal v_c via OA3 with gain unity. Finally, v_c is compared with the ramp signal v_r (with period T and upper and lower bound voltages v_u and v_l) via the CM to produce the PWM switching signal for driving the power MOSFET. The core of this controller is the BBD line and the associated clock. By tuning the clock frequency via its externally connected R_1–R_2–C network, the BBD line can allow for a time delay varying from 2.56 to 25.6 ms. In the controller, the time delay τ is set to the switching period T which in fact corresponds to the fundamental operation (the period-1 orbit).

Based on the same conditions for computer simulation, the measured chaotic, period-1, and period-2 phase portraits are shown in Figure 6.4. It can be found that the chaotic orbit has boundaries of $v_c \in [-0.3\text{ V}, 3.5\text{ V}]$ and $i \in [1.1\text{ A}, 7.2\text{ A}]$, the period-1 orbit has boundaries of $v_c \in [-0.5\text{ V}, 2\text{ V}]$ and $i \in [2.1\text{ A}, 6.7\text{ A}]$, and the period-2 orbit has boundaries of $v_c \in [-1\text{ V}, 4\text{ V}]$ and $i \in [1.1\text{ A}, 7.5\text{ A}]$. Comparing Figures 6.4 and 6.3, the measured results are in good agreement with the simulation results. Nevertheless, the boundaries of those phase portraits still have some discrepancies, and the stabilized orbits are slightly shaking, which is due to some inevitable imperfections in the DC drive system, such as the uneven contacts of the DC commutator, the torsional oscillation of the coupler, and the phase distortion of the BBD line.

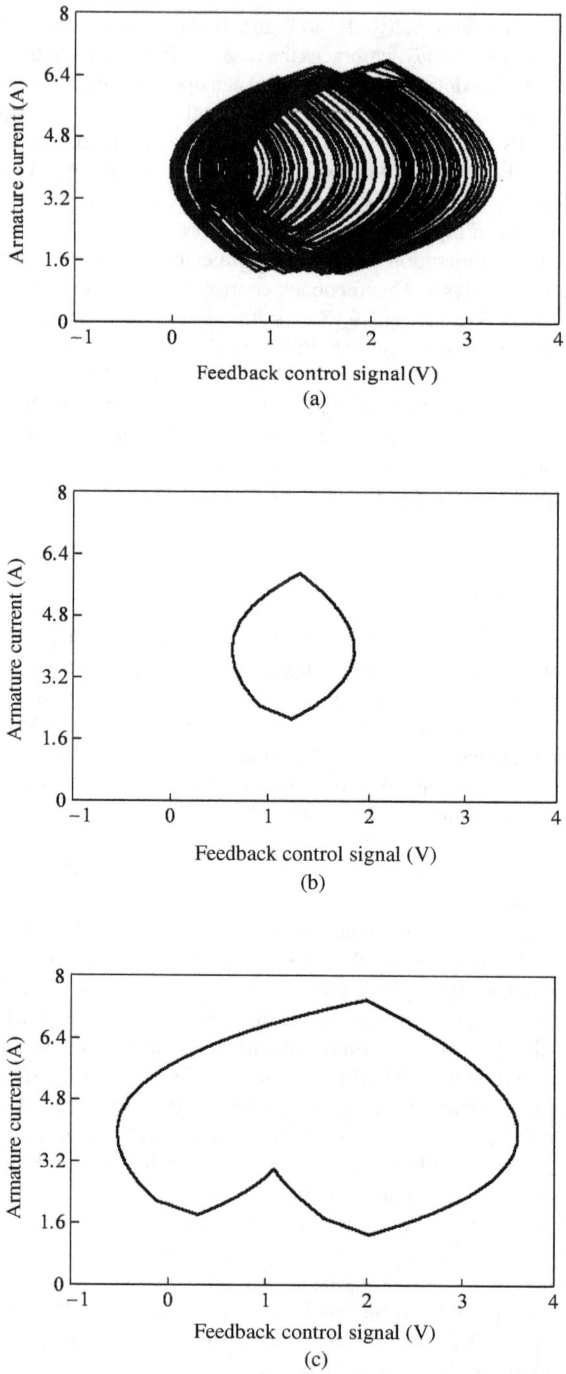

Figure 6.3 Simulated phase portraits. (a) Chaotic orbit. (b) Period-1 orbit. (c) Period-2 orbit

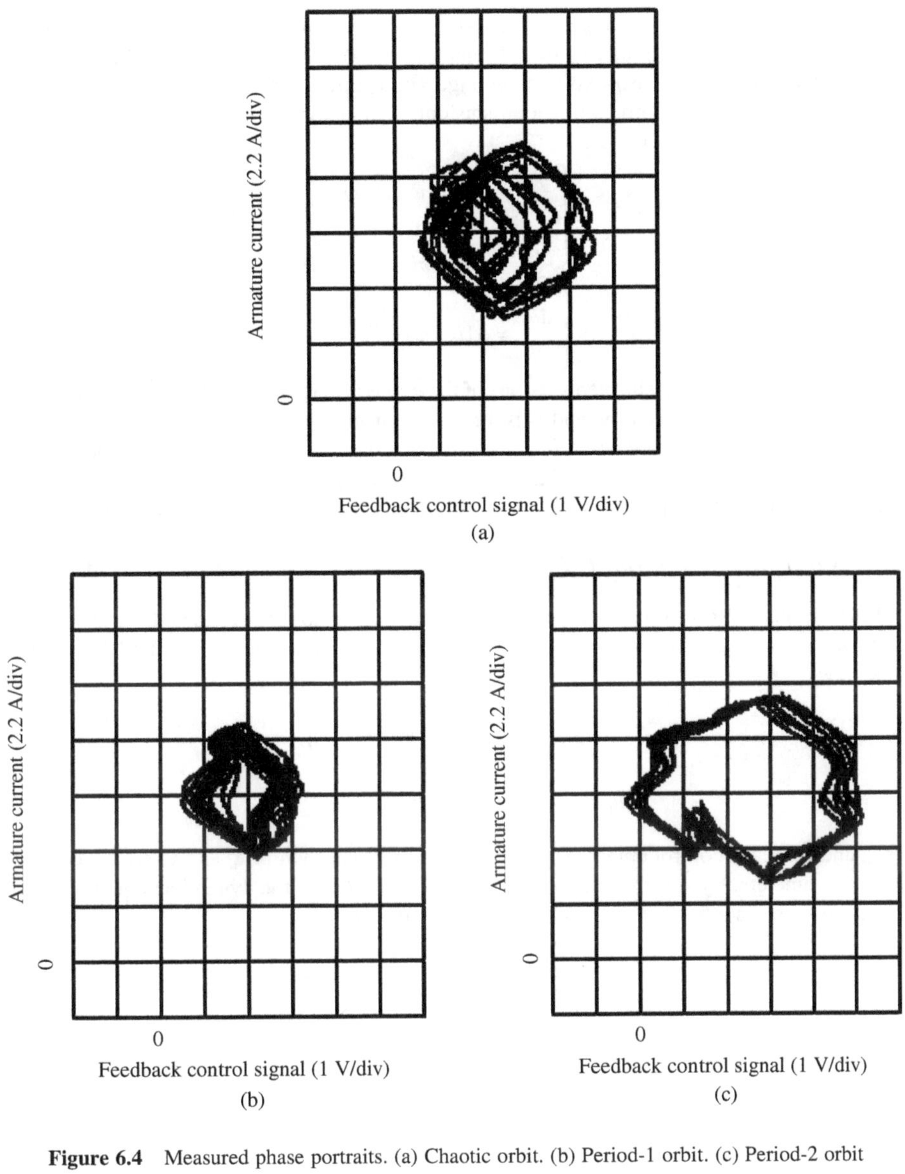

Figure 6.4 Measured phase portraits. (a) Chaotic orbit. (b) Period-1 orbit. (c) Period-2 orbit

6.2 Stabilization of Chaos in AC Drive System

As discussed in Chapter 4, chaotic behavior can be observed in the permanent magnet synchronous motor (PMSM) drive system under some operating conditions. By defining

$$\alpha = (LB)/(JR), \quad \eta = L/R, \quad \bar{u}_q = (n_p^2\eta^2\psi_f u_q)/(LB),$$

$$\bar{u}_d = (n_p^2\eta^2\psi_f u_d)/(LB) + (\beta LB + n_p^2\psi_f^2\eta)/(LB) \text{ and } \bar{T}_l = (n_p\eta^2 T_l)/J,$$

where L is the armature winding inductance, R is the armature winding resistance, B is the rotor viscous friction coefficient, J is the rotor inertia, n_p is the number of PM pole-pairs, ψ_f is the PM flux linkage, u_d is the d-axis input voltage, u_q is the q-axis input voltage, and T_l is the load torque, the dynamical equation of the PMSM drive can be transformed into a dimensionless model which is in the form of the well-known Lorenz system (Hemati, 1994):

$$\begin{cases} \dfrac{d\bar{i}_d}{dt} = -\bar{i}_d + \bar{\omega}\bar{i}_q + \bar{u}_d \\ \dfrac{d\bar{i}_q}{dt} = -\bar{i}_q - \bar{\omega}\bar{i}_d + \beta\bar{\omega} + \bar{u}_q \\ \dfrac{d\bar{\omega}}{dt} = \alpha(\bar{i}_q - \bar{\omega}) - \bar{T}_l \end{cases} \tag{6.39}$$

where $\bar{i}_d$, $\bar{i}_q$, and $\bar{\omega}$ are the transformed versions of the d-axis armature current i_d, q-axis armature current i_q, and motor speed ω, respectively, and β is a free parameter.

This section presents an overview of four control methods to stabilize the chaotic motion in a PMSM drive system. Namely, the nonlinear feedback control, backstepping control, dynamic surface control, and sliding mode control are employed to stabilize the chaotic motor speed to the desired value.

6.2.1 Nonlinear Feedback Control

A nonlinear feedback control has been proposed to stabilize chaos in a PMSM drive system (Ren and Liu, 2006). This nonlinear feedback control uses $\bar{u}_d$ and $\bar{u}_q$ as the manipulated variables. The control law is governed by:

$$\bar{u}_d = \bar{i}_d - \bar{\omega}\bar{i}_q - k_d(\bar{i}_d - \bar{i}_d^*) + \frac{d\bar{i}_d^*}{dt} \tag{6.40}$$

$$\bar{u}_q = \bar{i}_q + \bar{\omega}\bar{i}_d - \beta\bar{\omega} - k_q(\bar{i}_q - \bar{i}_q^*) + \frac{d\bar{i}_q^*}{dt} \tag{6.41}$$

where $\bar{i}_d^*$ and $\bar{i}_q^*$ are the control objectives of $\bar{i}_d$ and $\bar{i}_q$ respectively, while k_d and k_q are the corresponding controller parameters. By substituting (6.40) and (6.41) into (6.39), the dynamical equation of a PMSM drive system with a nonlinear feedback control is rewritten as:

$$\begin{cases} \dfrac{d\bar{i}_d}{dt} = -k_d(\bar{i}_d - \bar{i}_d^*) + \dfrac{d\bar{i}_d^*}{dt} \\ \dfrac{d\bar{i}_q}{dt} = -k_q(\bar{i}_q - \bar{i}_q^*) + \dfrac{d\bar{i}_q^*}{dt} \\ \dfrac{d\bar{\omega}}{dt} = \alpha(\bar{i}_q - \bar{\omega}) - \bar{T}_l \end{cases} \tag{6.42}$$

Since $\bar{i}_d^*$ and $\bar{i}_q^*$ are normally constant parameters, this yields $d\bar{i}_d^*/dt = 0$ and $d\bar{i}_q^*/dt = 0$. Thus, the control law can be reduced as:

$$\bar{u}_d = \bar{i}_d - \bar{\omega}\bar{i}_q - k_d(\bar{i}_d - \bar{i}_d^*) \tag{6.43}$$

$$\bar{u}_q = \bar{i}_q + \bar{\omega}\bar{i}_d - \beta\bar{\omega} - k_q(\bar{i}_q - \bar{i}_q^*) \tag{6.44}$$

The desired speed reference $\bar{\omega}^*$ is a constant for the PMSM drive system. Therefore, in the controller design, $\bar{i}_q^*$ can be calculated according to the given value of $\bar{\omega}^*$ and $\bar{i}_d^*$ is set based on the requirement of flux. The selection of k_d and k_q determines the responding speed of the system. The manipulated variables $\bar{u}_d$ and $\bar{u}_q$ are externally accessible, while $\bar{i}_d$, $\bar{i}_q$, and $\bar{\omega}$ can be measured in real time. Hence, the controller can be physically realized.

In order to assess the performance of this nonlinear feedback control, the simulation is carried out with the parameters $\bar{T}_l = 0$, $\alpha = 5.46$, $\beta = 20$. Given $\bar{\omega}^* = 5$, it results in $\bar{i}_q^* = 5$ according to $\bar{i}_q^* = \bar{\omega}^* + T_l/\alpha$, and $\bar{i}_d^* = 3$.

The simulation is carried out with the initial values $\bar{i}_d(0) = 20$, $\bar{i}_q(0) = 0.01$, and $\bar{\omega}(0) = -5$ as well as $k_d = k_q = 4.2$. After startup, $\bar{u}_d$ and $\bar{u}_q$ are set to zero and remain unchanged. At the instant of $t = 25$ s, $\bar{u}_d$ and $\bar{u}_q$ are adjusted in accordance with the control law governed by (6.43) and (6.44). As shown in Figure 6.5, it can be found that the nonlinear feedback control can successfully stabilize both of the armature current components and the motor speed.

6.2.2 Backstepping Control

The backstepping control method has also been designed to stabilize the chaotic motion in a PMSM drive system (Harb, 2004). By setting $\bar{u}_d = 0$ and $\bar{T}_l = 0$, the dynamical equation (6.39) of the PMSM drive is expressed as:

$$\begin{cases} \dfrac{d\bar{i}_d}{dt} = -\bar{i}_d + \bar{\omega}\bar{i}_q \\ \dfrac{d\bar{i}_q}{dt} = -\bar{i}_q - \bar{\omega}\bar{i}_d + \beta\bar{\omega} + \bar{u}_q \\ \dfrac{d\bar{\omega}}{dt} = \alpha(\bar{i}_q - \bar{\omega}) \end{cases} \tag{6.45}$$

The error signals of the corresponding system states are defined as:

$$\begin{cases} e_1 = \bar{i}_d - \bar{i}_d^* \\ e_2 = \bar{i}_q + \xi_1 e_1 \\ e_3 = \bar{\omega} + \xi_2 e_1 + \xi_3 e_2 \end{cases} \tag{6.46}$$

where ξ_1, ξ_2, and ξ_3 are the gains of error signals. Given $\bar{i}_d^* = 0$, and by the use of (6.45), the time derivative of (6.46) can be represented as:

$$\begin{cases} \dfrac{de_1}{dt} = \alpha(e_2 - \xi_1 e_1 - e_1) \\ \dfrac{de_2}{dt} = \beta e_1 - (e_2 - \xi_1 e_1) - e_1(e_3 - \xi_2 e_1 - \xi_3 e_2) + \xi_1 \dfrac{de_1}{dt} + \bar{u}_q \\ \dfrac{de_3}{dt} = -(e_3 - \xi_2 e_1 - \xi_3 e_2) + e_2(e_2 - \xi_1 e_1) + \xi_2 \dfrac{de_1}{dt} + \xi_3 \dfrac{de_2}{dt} \end{cases} \tag{6.47}$$

which constitutes the control law for stabilization.

Based on these error signals, a positive Lyapunov function is constructed as:

$$V = \frac{1}{2}\sum_{i=1}^{3} m_i e_i^2 \tag{6.48}$$

By selecting the control parameters $\xi_1 = \xi_2 = \xi_3 = 0$, $m_1 = m_3 = 0$ and $m_2 = 1$, the system stability is guaranteed, namely $\dot{V} \leq 0$, provided that the control law is given by:

$$\bar{u}_q = -e_2 - \beta e_1 + e_1 e_3 \tag{6.49}$$

Based on the same system to assess the nonlinear feedback control, the stabilization performance of this backstepping control is testified. Also, the control law is applied at the instant of $t = 25$ s. As shown in Figure 6.6, it can be seen that the backstepping control can successfully stabilize the system.

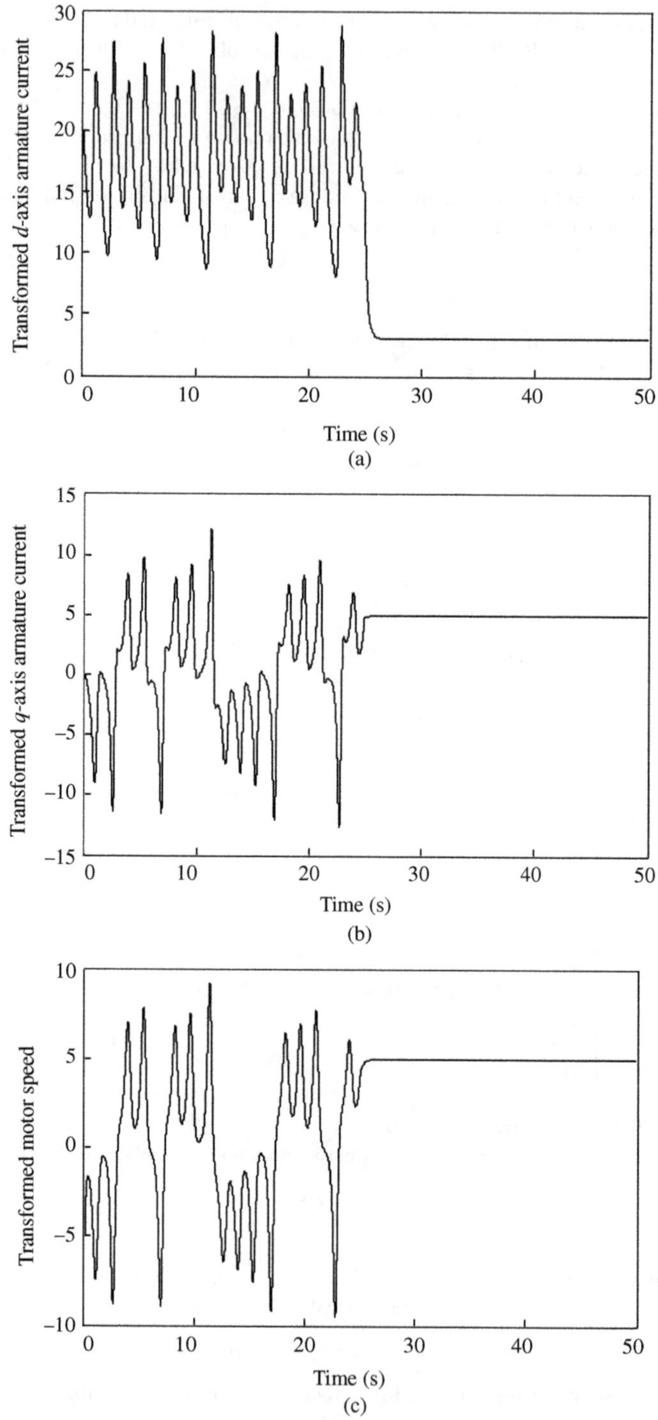

Figure 6.5 Stabilization performance of nonlinear feedback control. (a) Transformed d-axis armature current. (b) Transformed q-axis armature current. (c) Transformed motor speed

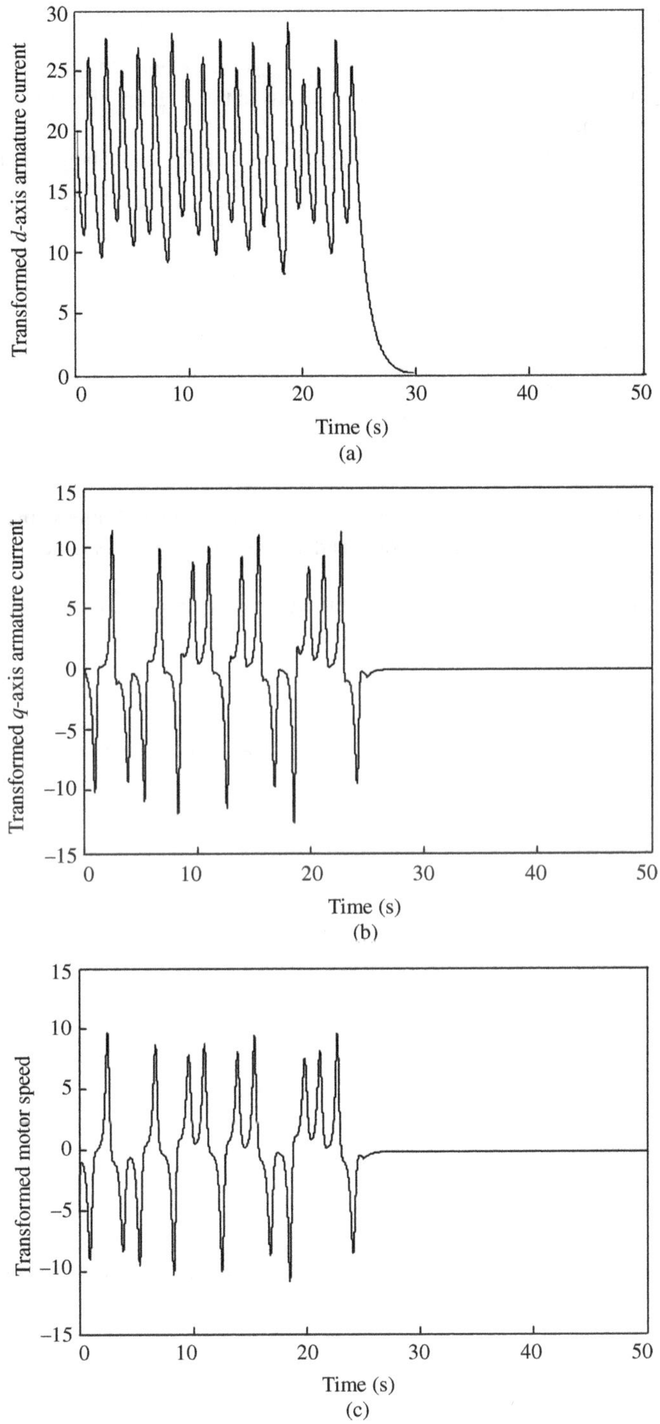

Figure 6.6 Stabilization performance of backstepping control. (a) Transformed d-axis armature current. (b) Transformed q-axis armature current. (c) Transformed motor speed

6.2.3 Dynamic Surface Control

The dynamic surface control has also been used to stabilize chaos in the PMSM drive system. It takes the advantage over the backstepping control because it can avoid the problem of 'explosion of terms' caused by the repeated differentiation of virtual input (Wei *et al.*, 2007).

By using the control variable $\bar{u}_d$ and setting $\bar{u}_q = 0$ in (6.39), the corresponding dynamical equation becomes:

$$\begin{cases} \frac{d\bar{i}_d}{dt} = -\bar{i}_d + \bar{\omega}\bar{i}_q + \bar{u}_d \\ \frac{d\bar{i}_q}{dt} = -\bar{i}_q - \bar{\omega}\bar{i}_d + \beta\bar{\omega} \\ \frac{d\bar{\omega}}{dt} = \alpha(\bar{i}_q - \bar{\omega}) \end{cases} \tag{6.50}$$

The control law of this dynamic surface control includes three steps. Namely, the virtual controllers are designed in the first step and the second step, and the overall control law is designed in the third step (Wei *et al.*, 2007).

Firstly, the dynamic surface S_1 is designed for $\bar{\omega}$ to track the reference value $\bar{\omega}^*$, which is expressed as:

$$S_1 = \bar{\omega} - \bar{\omega}^* \tag{6.51}$$

Hence, by differentiating (6.51) and using (6.50), the dynamics of S_1 is given by:

$$\frac{dS_1}{dt} = \alpha(\bar{i}_q - \bar{\omega}) - \frac{d\bar{\omega}^*}{dt} \tag{6.52}$$

Since $\bar{\omega}^*$ is a constant value and $d\bar{\omega}^*/dt = 0$, (6.52) becomes:

$$\frac{dS_1}{dt} = \alpha(\bar{i}_q - \bar{\omega}) \tag{6.53}$$

So, the first virtual controller is to stabilize S_1 by choosing $\bar{i}_q^{*'}$ as:

$$\bar{i}_q^{*'} = \bar{\omega} - \xi_1 S_1 \tag{6.54}$$

where ξ_1 is the parameter of the first virtual controller. It should be noted that the dynamic surface control eliminates the need for model differentiation by passing $\bar{i}_q^{*'}$ through a first-order filter with a positive time constant τ_2:

$$\tau_2 \frac{d\bar{i}_q^*}{dt} + \bar{i}_q^* = \bar{i}_q^{*'},\ \bar{i}_q^*(0) = \bar{i}_q^{*'}(0) \tag{6.55}$$

where $\bar{i}_q^*$ serves as an estimation of $\bar{i}_q^{*'}$.

Secondly, by using $\bar{i}_q^*$ to supersede $\bar{i}_q^{*'}$, the second surface S_2 is designed as:

$$S_2 = \bar{i}_q - \bar{i}_q^* \tag{6.56}$$

Hence, by differentiating (6.56) and using (6.50), the dynamics of S_2 is given by:

$$\frac{dS_2}{dt} = -\bar{i}_q - \bar{i}_d\bar{\omega} + \beta\bar{\omega} - \frac{d\bar{i}_q^*}{dt} \tag{6.57}$$

So, the second virtual controller is to stabilize S_2 by choosing $\bar{i}_d^{*'}$ as:

$$\bar{i}_d^{*'} = \beta + \frac{\xi_2 S_2}{\bar{\omega}} - \frac{\bar{i}_q}{\bar{\omega}} - \frac{1}{\bar{\omega}}\frac{d\bar{i}_q^*}{dt} \tag{6.58}$$

where ξ_2 is the parameter of the second virtual controller. Consequently, $\bar{i}_d^*$ is deduced by passing $\bar{i}_d^{*'}$ through a first-order filter with a positive time constant τ_3:

$$\tau_3\frac{d\bar{i}_d^*}{dt} + \bar{i}_d^* = \bar{i}_d^{*'},\ \bar{i}_d^*(0) = \bar{i}_d^{*'}(0) \tag{6.59}$$

Thirdly, the third surface S_3 is designed as:

$$S_3 = \bar{i}_d - \bar{i}_d^* \tag{6.60}$$

So, by differentiating (6.60) and using (6.50), the dynamics of S_3 is given by:

$$\frac{dS_3}{dt} = -\bar{i}_d + \bar{i}_q\bar{\omega} - \frac{d\bar{i}_d^*}{dt} + \bar{u}_d \tag{6.61}$$

In order to stabilize S_3, the final control law is designed as:

$$u_d = \bar{i}_d - \bar{i}_q\bar{\omega} + \frac{d\bar{i}_d^*}{dt} - \xi_3 S_3 \tag{6.62}$$

where ξ_3 is the parameter of the final control law.

Based on the same system as mentioned previously, the stabilization performance of this dynamic surface control is assessed. When $\bar{\omega}^* = 5$ is selected and the control is activated at the instant of $t = 25$ s, the responses are as shown in Figure 6.7. This confirms that dynamic surface control can successfully stabilize chaos in a PMSM drive system.

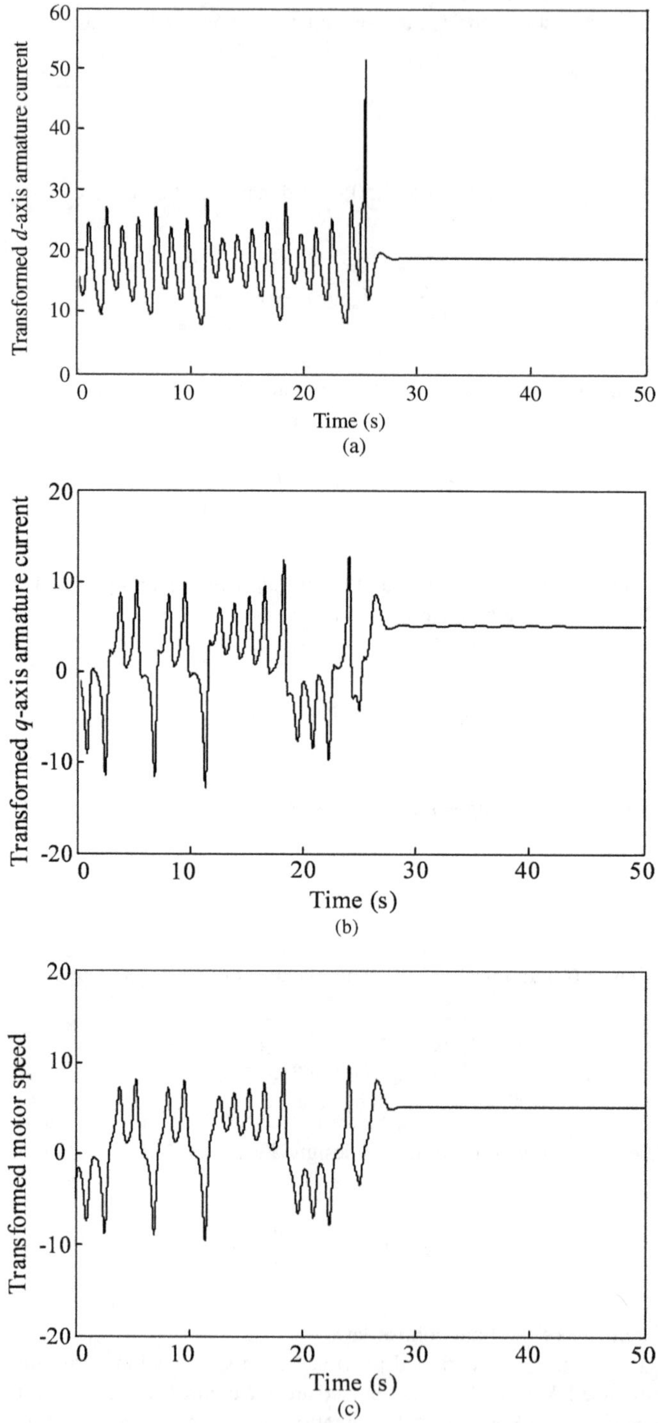

Figure 6.7 Stabilization performance of dynamic surface control. (a) Transformed d-axis armature current. (b) Transformed q-axis armature current. (c) Transformed motor speed

6.2.4 Sliding Mode Control

The sliding mode control has also been developed to stabilize the chaotic behavior in a Lorenz chaotic system, such as a PMSM drive system (Yau and Yan, 2004). By setting $\bar{u}_d = 0$ and $\bar{T}_l = 0$ in (6.39), the dynamical equation of a PMSM drive system can be expressed as:

$$\begin{cases} \frac{d\bar{i}_d}{dt} = -\bar{i}_d + \bar{\omega}\bar{i}_q \\ \frac{d\bar{i}_q}{dt} = -\bar{i}_q - \bar{\omega}\bar{i}_d + \beta\bar{\omega} + \bar{u}_q(u(t)) \\ \frac{d\bar{\omega}}{dt} = \alpha(\bar{i}_q - \bar{\omega}) \end{cases} \tag{6.63}$$

where $\bar{u}_q(u(t))$ is a nonlinear control variable that is a continuous nonlinear function with $\bar{u}_q(0) = 0$ and follows the relationship:

$$\alpha_1 u^2(t) \le u(t)\bar{u}_q(u(t)) \le \alpha_2 u^2(t) \tag{6.64}$$

where α_1 and α_2 are nonzero positive constants. A disturbance term is defined to cancel the following nonlinear term:

$$D(t) = -\bar{\omega}\bar{i}_d \tag{6.65}$$

So, the system dynamics in (6.63) can be expressed as:

$$\begin{cases} \frac{d\bar{i}_d}{dt} = -\bar{i}_d + \bar{\omega}\bar{i}_q \\ \frac{d\bar{i}_q}{dt} = -\bar{i}_q + D(t) + \beta\bar{\omega} + \bar{u}_q(u(t)) \\ \frac{d\bar{\omega}}{dt} = \alpha(\bar{i}_q - \bar{\omega}) \end{cases} \tag{6.66}$$

For analysis, a new system state vector $Z = \begin{bmatrix} z_1 \\ z_2 \end{bmatrix} = P\begin{bmatrix} \bar{\omega} \\ \bar{i}_q \end{bmatrix}$ is defined, where the transformation matrix is $P = \begin{bmatrix} 1/\alpha & 0 \\ -1 & 1 \end{bmatrix}$. The trajectory error states are defined as $e_1 = z_1 - z_1^*$, $e_2 = z_2 - z_2^*$, and $\bar{e} = \bar{i}_d - \bar{i}_d^*$, where the transformed regulation system states are $z_1^* = (1/\alpha)\bar{\omega}^*$ and $z_2^* = -\bar{\omega}^* + \bar{i}_q^*$. Thus, the error state dynamical equations are given by:

$$\begin{cases} \frac{de_1}{dt} = e_2 \\ \frac{de_2}{dt} = \alpha(\beta-1)e_1 - (\alpha+1)e_2 + \alpha(\beta-1)z_1^* + D(t) + \bar{u}_q(u(t)) \\ \frac{d\bar{e}}{dt} = -\bar{e} + F(e_1, e_2) \end{cases} \tag{6.67}$$

where $F(e_1, e_2) = -\bar{i}_d^* + \alpha e_1 e_2 + \alpha e_2 z_1^* + \alpha^2 (e_1 + z_1^*)^2$. So, a sliding surface suitable for the application can be defined as:

$$S = c_1 e_1 + e_2 \tag{6.68}$$

where c_1 is the design parameter. For a sliding mode operation, the necessary and sufficient conditions are given by:

$$S = c_1 e_1 + e_2 = 0 \tag{6.69}$$

$$\frac{dS}{dt} = c_1 \frac{de_1}{dt} + \frac{de_2}{dt} = 0 \tag{6.70}$$

Therefore, the sliding mode dynamics can be obtained as:

$$\frac{de_1}{dt} = -c_1 e_1 \tag{6.71}$$

$$\frac{de_2}{dt} = \alpha(\beta - 1)e_1 - (\alpha + 1)e_2 + \alpha(\beta - 1)z_1^* + D(t) + \bar{u}_q(u(t)) \tag{6.72}$$

$$\frac{d\bar{e}}{dt} = -\bar{e} + F(e_1, e_2) \tag{6.73}$$

When the design parameter $c_1 < 0$, the stability of (6.71) is guaranteed, and e_1 converges to zero. According to (6.69), e_2 is also stable and converges to zero. It has been proved that $\bar{e}$ will converge to zero if e_1 and e_2 converge to zero (Yau and Yan, 2004). Therefore, $(\bar{i}_d, \bar{i}_q, \bar{\omega})$ converges to the desired value $(\bar{i}_d^*, \bar{i}_q^*, \bar{\omega}^*)$ when $t \to \infty$.

The sliding mode control law is designed as:

$$u(t) = -\xi \eta \, \mathrm{sign}(S), \quad \xi > \frac{1}{\alpha_1} \tag{6.74}$$

where

$$\eta = |c_1 e_2| + |\alpha(\beta - 1)e_1 - (\alpha + 1)e_2 + \alpha(\beta - 1)z_1^*| + |\bar{\omega}\bar{i}_d| \tag{6.75}$$

In this way, a reaching condition for the sliding mode $S\dot{S} < 0$ can be guaranteed (Yau and Yan, 2004).

In order to testify the performance of the sliding mode control, the simulation is carried out with the system parameters $\bar{T}_l = 0$, $\alpha = 5.46$, and $\beta = 20$. The design parameter $c_1 = 6$ is selected, while the nonlinear input is defined as:

$$\bar{u}_q(u(t)) = [0.6 + 0.3\sin(u(t))]u(t) \tag{6.76}$$

Thus, the slope of the nonlinear sector $\alpha_1 = 0.3$ and $\alpha_2 = 0.9$ can be obtained, leading to the selection of $\xi = 4$. Based on (6.67), (6.74), and (6.76), the simulated responses of $\bar{i}_d$, $\bar{i}_q$, and $\bar{\omega}$ under $\bar{\omega}^* = 8.5$ are depicted in Figure 6.8 in which the control is activated at the instant of $t = 25$ s. This confirms that a sliding mode control can successfully stabilize chaos in a PMSM drive system.

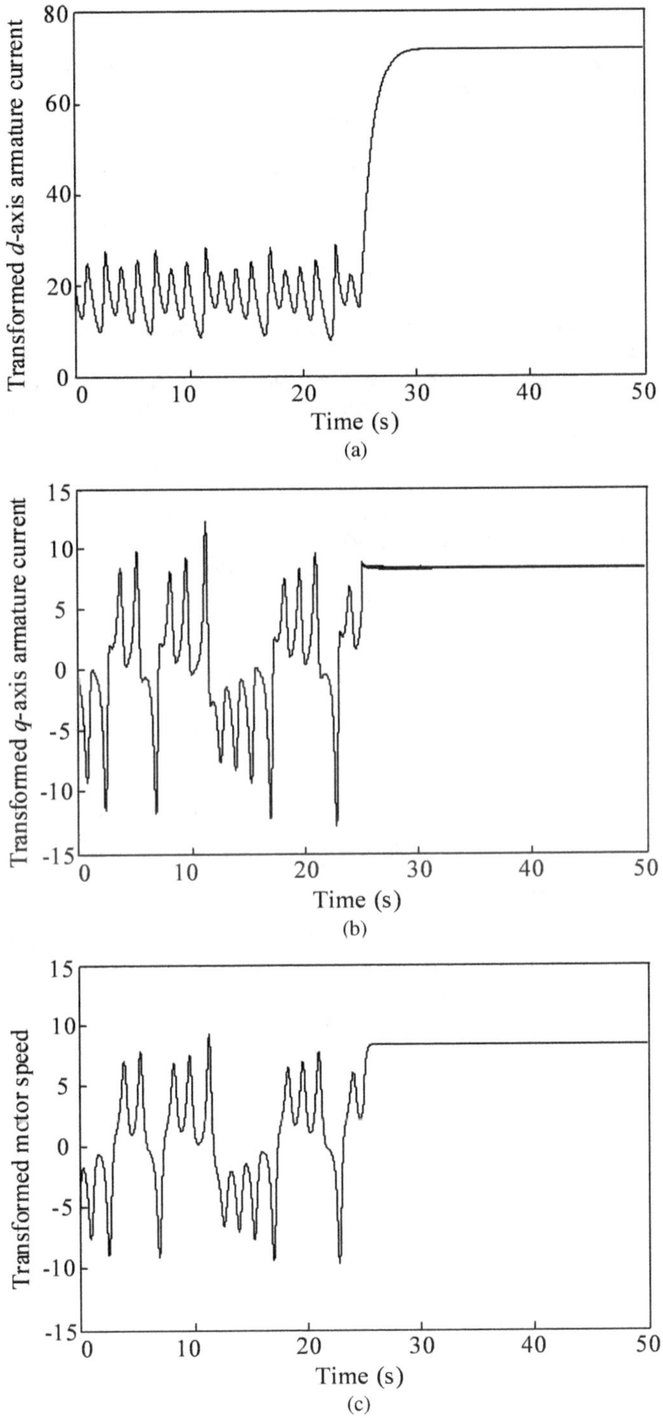

Figure 6.8 Stabilization performance of sliding mode control. (a) Transformed d-axis armature current. (b) Transformed q-axis armature current. (c) Transformed motor speed

References

Boccaletti, S., Farini, A., and Arecchi, F.T. (1997) Adaptive strategies for recognition, control and synchronization of chaos. *Chaos, Solitons and Fractals*, **8**, 1431–1448.

Casas, F. and Grebogi, C. (1997) Control of chaotic impacts. *International Journal of Bifurcation and Chaos*, **7**, 951–955.

Harb, A.M. (2004) Nonlinear chaos control in a permanent magnet reluctance machine. *Chaos, Solitons, and Fractals*, **19**, 1217–1224.

Hemati, N. (1994) Strange attractors in brushless DC motors. *IEEE Transactions on Circuits and Systems - I: Fundamental Theory and Applications*, **41**, 40–45.

Hunt, E.R. (1991) Stabilizing high-period orbits in a chaotic system – the diode resonator. *Physics Review Letters*, **67**, 1953–1955.

Hirasawa, K., Wang, X.F., Murata, J. *et al.* (2000) Universal learning network and its application to chaos control. *Neural Networks*, **13**, 239–253.

Jackson, E.A. and Grosu, I. (1995) An OPCL control of complex dynamic systems. *Physica D*, **85**, 1–9.

Khovanov, I.A., Luchinsky, D.G., Mannella, R., and McClintock, P.V.E. (2000) Fluctuations and the energy-optimal control of chaos. *Physics Review Letters*, **85**, 2100–2103.

Kittel, A., Parisi, J., and Pyragas, K. (1995) Delayed feedback control of chaos by self-adapted delay time. *Physics Letters A*, **198**, 433–436.

Liao, T.L. and Lin, S.H. (1999) Adaptive control and synchronization of Lorenz systems. *Journal of Franklin Institute - Engineering and Applied Mathematics*, **226**, 925–937.

Ott, E., Grebogi, C., and Yorke, J.A. (1990) Controlling chaos. *Physical Review Letters*, **64**, 1196–1199.

Poznyak, A.S., Yu, W., and Sanchez, E.N. (1999) Identification and control of unknown chaotic system via dynamic neural networks. *IEEE Transactions on Circuits and Systems - I: Fundamental Theory and Applications*, **46**, 1491–1495.

Pyragas, K. (1992) Continuous control of chaos by self-controlling feedback. *Physics Letters A*, **170**, 421–428.

Rajasekar, S., Murali, K., and Lakshmanan, M. (1997) Control of chaos by nonfeedback methods in a simple electronic circuit system and the FitzHugh-Nagumo equation. *Chaos, Solitons and Fractals*, **8**, 1545–1558.

Ramesh, M. and Narayanan, S. (1999) Chaos control by nonfeedback methods in the presence of noise. *Chaos, Solitons and Fractals*, **10**, 1473–1489.

Ren, H. and Liu, D. (2006) Nonlinear feedback control of chaos in permanent magnet synchronous motor. *IEEE Transactions on Circuits and Systems - II: Express Briefs*, **53**, 45–50.

Tanaka, K., Ikeda, T., and Wang, H.O. (1998) A unified approach to controlling chaos via arm LMI-based fuzzy control system design. *IEEE Transactions on Circuits and Systems - I: Fundamental Theory and Applications*, **45**, 1021–1040.

Tian, Y.P. (1999) Controlling chaos using invariant manifolds. *International Journal of Bifurcation and Chaos*, **72**, 258–266.

Wei, D.Q., Luo, X.S., Wang, B.H., and Fang, J.Q. (2007) Robust adaptive dynamic surface control of chaos in permanent magnet synchronous motor. *Physics Letters A*, **363**, 71–77.

Yau, H.T. and Yan, J.J. (2004) Design of sliding mode controller for Lorenz chaotic system with nonlinear input. *Chaos, Solitons and Fractals*, **19**, 891–898.

7

Stimulation of Chaos in Electric Drive Systems

Instead of avoiding the occurrence of chaos or stabilizing the chaotic phenomenon in electric drive systems, it is becoming attractive to stimulate the occurrence of chaos and hence to positively utilize chaos. The stimulation of chaos, termed the chaoization, can be classified into two approaches: control-oriented chaoization and design-oriented chaoization. The control-oriented approaches – which offer the definite advantage of flexibility, but need additional control – use control means to stimulate the system exhibiting chaotic behavior. On the other hand, the design-oriented approaches – which offer the definite advantage of simplicity, but lack flexibility – design the system so that it can inherently generate chaotic behavior.

In this chapter, the control-oriented and design-oriented chaoization are both discussed, hence stimulating chaotic operation in various electric drive systems.

7.1 Control-Oriented Chaoization

Many control approaches that have been proposed to stimulate chaos in various nonlinear systems. For the stimulation of chaos in electric drive systems, the available approaches are quite limited, and most of them are still under theoretical development (Ge, Cheng, and Chen, 2004). In this section, several viable control approaches that have been successfully implemented in electric drive systems are discussed, namely time-delay feedback control, proportional time-delay control, and chaotic speed reference control.

7.1.1 Time-Delay Feedback Control of PMDC Drive System

As discussed in Chapter 3, a PMDC motor can be modeled as a single-input/single-output (SISO) linear time-invariant (LTI) system:

$$\frac{d\omega}{dt} + \frac{B}{J}\omega = \frac{T}{J} \tag{7.1}$$

Chaos in Electric Drive Systems: Analysis, Control and Application, First Edition. K.T. Chau and Zheng Wang.
© 2011 John Wiley & Sons (Asia) Pte Ltd. Published 2011 by John Wiley & Sons (Asia) Pte Ltd.

where ω is the motor speed, T is the motor torque, B is the viscous damping coefficient, and J is the load inertia, while T and ω denote the input and output of the system, respectively. By employing the anticontrol of chaos in continuous-time systems (Wang, Chen, and Yu, 2000), the time-delay feedback control can be designed as:

$$T = \mu \xi B f\left(\frac{\omega(t-\tau)}{\xi}\right) \tag{7.2}$$

where μ is the torque parameter, ξ is the speed parameter, and τ is the time-delay parameter. It should be noted that all three parameters are adjustable to achieve the desired chaotic motion (Ye and Chau, 2007). Also, $f(\cdot)$ should be a bounded function to satisfy the condition that the required torque does not exceed the motor torque capability. Substituting (7.2) into (7.1), the time-delay system equation can be formulated as:

$$\frac{d\omega}{dt} + \frac{B}{J}\omega = \frac{\mu \xi B f\left(\frac{\omega(t-\tau)}{\xi}\right)}{J} \tag{7.3}$$

After defining the normalized speed $\Omega = \omega/\xi$, (7.3) is rewritten as:

$$\frac{d\Omega}{dt} + \frac{B}{J}\Omega = \frac{\mu B f(\Omega(t-\tau))}{J} \tag{7.4}$$

For a SISO LTI system, this can be expressed as:

$$y^{(n)}(t) + \alpha_{n-1} y^{(n-1)}(t) + \cdots \alpha_1 y^{(1)}(t) + \alpha_0 y(t) = \beta_0 u(t) \tag{7.5}$$

where $u(t)$ is the input, $y(t)$ is the output, and $\alpha_i (i = 0, \cdots, n-1)$ and β_0 are constants with $\alpha_0 \beta_0 \neq 0$. When the system is incorporated with time-delay feedback control, $u(t)$ and $y(t)$ are related by:

$$u(t) = w(y(t-\tau)) \tag{7.6}$$

where $w(\cdot)$ is a bounded continuous function. Accordingly, a time-delay differential equation can be asymptotically approximated by a difference equation as given by:

$$y(m, \hat{t}) \approx \frac{\beta_0}{\alpha_0} w(y(m-1, \hat{t})), \text{ and } y^{(k)}(m, \hat{t}) \approx 0 \tag{7.7}$$

where $y(m, \hat{t}) = y(m\tau + \hat{t})$ for $m = 0, 1, \cdots$, $k = 1, 2, \cdots, n-1$, τ is sufficiently large, and $\hat{t} : 0 < \hat{t} < \tau$ is large.

For this chaotic motor, $\alpha_0 = B/J$, $\beta_0 = \mu B/J$ and $w(\cdot)$ is represented by $f(\cdot)$. Hence, by using (7.7), (7.4) can be asymptotically approximated by:

$$\Omega(m, \hat{t}) = \mu f(\Omega(m-1, \hat{t})) \tag{7.8}$$

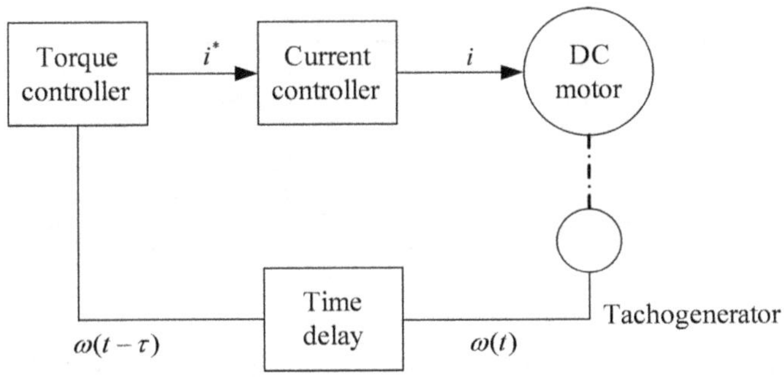

Figure 7.1 Time-delay feedback control of PMDC motor

Since a sine function is a continuous bounded function, it is naturally chosen as $f(\cdot)$. Hence, the system equation can be expressed as:

$$\Omega_m = \mu \sin \Omega_{m-1} \tag{7.9}$$

By writing $\mu = \pi r$ and $\Omega_m = \pi X_{n+1}$, (7.9) can be expressed as:

$$X_{n+1} = r \sin \pi X_n \tag{7.10}$$

which has been proved to exhibit chaotic behavior with certain values of r (Strogatz, 1994). Therefore, the proposed time-delay feedback control scheme given by (7.2) is rewritten as:

$$T = \mu \xi B \sin\left(\frac{\omega(t-\tau)}{\xi}\right) \tag{7.11}$$

which can offer chaotic motion by selecting appropriate values of μ, ξ, and τ.

In order to implement the desired chaotic motor based on the above derivation, both the armature current and the rotor speed of the PMDC motor are used as feedback signals, while the torque command T^* calculated by (7.11) is used to generate the current command i^* for PWM control of the PMDC motor. Figure 7.1 shows the corresponding control system. First, the measured speed feedback is delayed by a preset value of τ. Then, the delayed speed is fed into the torque control block in which proper values of μ and ξ are preset. Hence, it first generates T^* and then i^*. Subsequently, the difference between i^* and the measured current feedback is fed into the current control block in which a simple PI control is adopted. Hence, it generates the desired duty ratio for the full-bridge PWM converter which provides bidirectional current control of the PMDC motor.

Table 7.1 lists some key data of the PMDC motor. Making use of (7.11), bifurcation diagrams with respect to various adjustable parameters are readily deduced. These bifurcation diagrams can illustrate at a glance how the system behavior is affected by varying the parameters. For the sake of illustration, the case of no load torque $T_l = 0$ is adopted for the simulation.

When selecting $\xi = 20$ and $\tau = 1.5$ s, both the speed and current bifurcation diagrams with respect to μ are as shown in Figure 7.2. It can be seen that the motor initially operates at a fixed point (which is equivalent to the normal or so-called period-1 operation) with a small value of μ.

Table 7.1 Key parameters of PMDC motor

Supply voltage U	24 V
Torque constant K	0.05 Nm/A
Armature resistance R	1.1 Ω
Armature inductance L	0.4 mH
Viscous coefficient B	1.0×10^{-4} Nm/rad/s
Rotor inertia J	1.0388×10^{-5} Nm/rad/s^2

Figure 7.3 shows the corresponding motor speed and armature current waveforms when $\mu = 1.2$. With an increase of μ, the motor bifurcates to a period-2 operation, which is equivalent to an abnormal subharmonic operation. Figure 7.4 shows the corresponding speed and current when $\mu = 2.5$. Finally, the motor exhibits chaotic motion when μ is further increased. Figure 7.5 shows

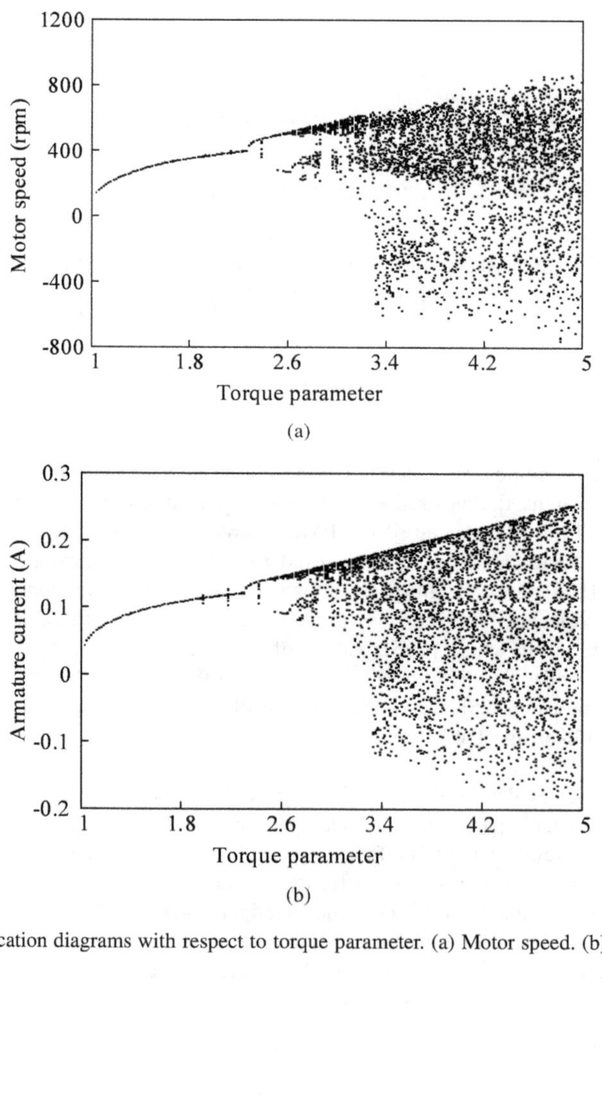

Figure 7.2 Bifurcation diagrams with respect to torque parameter. (a) Motor speed. (b) Armature current

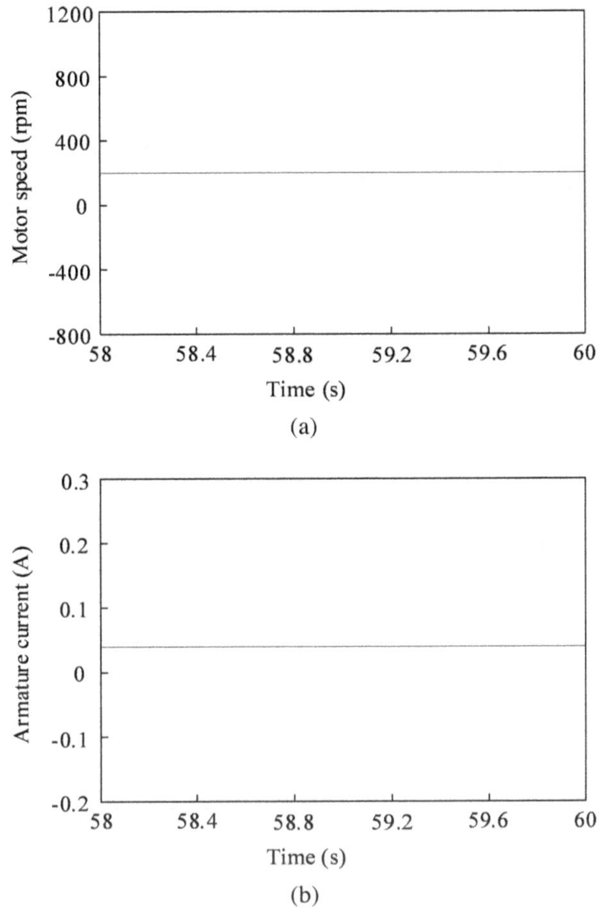

Figure 7.3 Period-1 operation under torque parameter control. (a) Motor speed. (b) Armature current

the corresponding speed and current when $\mu = 5$. When the motor is run in chaotic mode, both the amplitude and direction of the motor speed and armature current change with time and present ergodicity in the range illustrated in Figure 7.2. Therefore, the torque parameter μ is the key to induce chaotic motion from normal operation.

On the other hand, when selecting $\mu = 5$ and $\tau = 1.5$ s, both the speed and current bifurcation diagrams with respect to ξ are shown in Figure 7.6. It can be seen that ξ can be used to almost linearly adjust the speed range of the chaotic motion, which is why it is called the speed parameter. Therefore, the speed parameter ξ serves to adjust the boundary of chaotic motion.

Furthermore, when selecting $\mu = 5$ and $\xi = 20$, the corresponding bifurcation diagrams with respect to τ are as shown in Figure 7.7. It can be observed that the change of τ has little effect on the speed range of the chaotic motion. Nevertheless, this parameter has a significant impact on the system realization. If the parameter is too large, the refreshing rate of the control system will be too slow; if the parameter is too small, it will require too much computational resource. Therefore, the time-delay parameter τ functions to tune the refreshing rate of the chaoization.

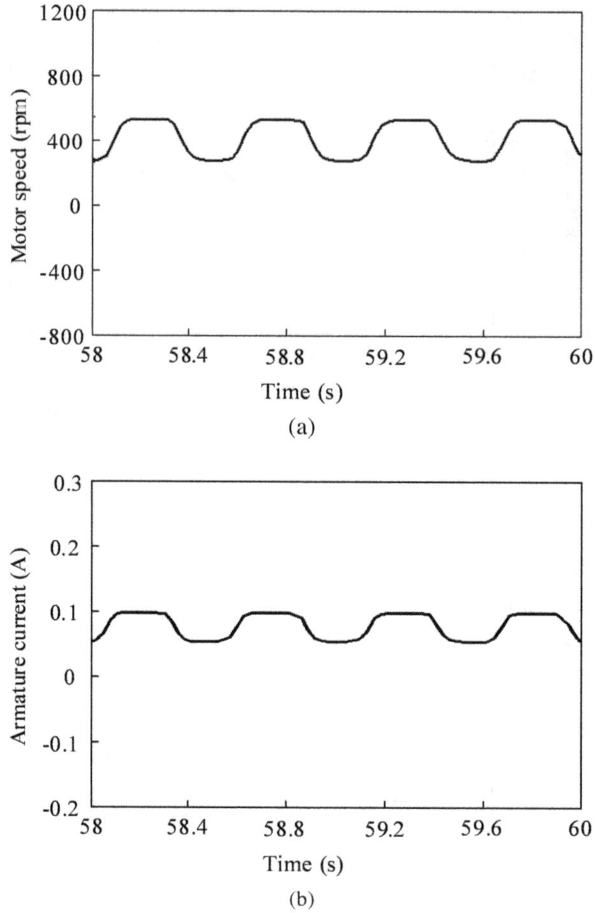

Figure 7.4 Period-2 operation under torque parameter control. (a) Motor speed. (b) Armature current

In order to verify the above analysis, the simulation and experimentation are both performed with the same parameter settings. Firstly, Figure 7.8 shows the simulated waveforms of motor speed and armature current under normal operation ($\mu = 1.2$, $\xi = 200$, $\tau = 1.5$ s), while Figure 7.9 shows the measured normal waveforms of motor speed and armature current under the same conditions. It can be seen that the agreement is very good. Secondly, Figure 7.10 shows the simulated waveforms of motor speed and armature current under chaotic operation ($\mu = 35$, $\xi = 20$, $\tau = 1.5$ s), while Figure 7.11 shows the measured chaotic waveforms of motor speed and armature current under the same conditions. It can be seen that both the simulated and the measured waveforms exhibit chaotic behavior but with different patterns. This is expected because the chaotic behavior is not periodic, so that the period of measurement cannot be the same with that of simulation. Also, chaotic behavior is very sensitive to the initial states. A tiny discrepancy in the initial states will cause a huge difference in chaotic behavior. Nevertheless, it can be found that the measured boundaries of the chaotic waveforms are in good agreement with the simulated ones. This is the actual nature of chaos – random-like but bounded.

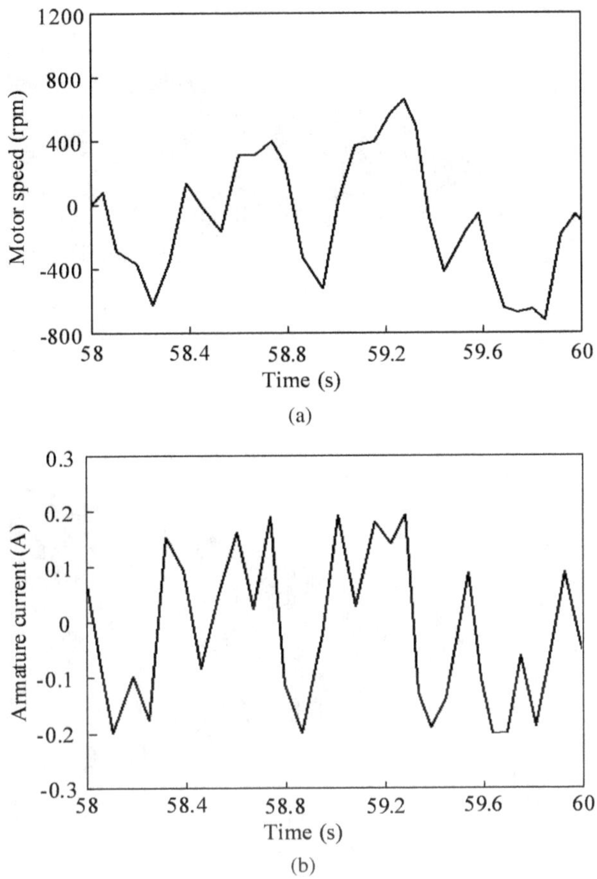

Figure 7.5 Chaotic operation under torque parameter control. (a) Motor speed. (b) Armature current

7.1.2 Time-delay Feedback Control of PM Synchronous Drive System

As discussed in Chapter 4, a 3-phase PM synchronous motor (PMSM) can be transformed into the *d–q* frame. In order to utilize the advantages of vector control, the *d*-axis stator current i_d is set to zero so that the electromagnetic torque T can simply be expressed as:

$$T = \frac{3}{2}\frac{P}{2}\psi_r i_q \tag{7.12}$$

where P is the number of poles, ψ_r is the PM flux linkage in the rotor, and i_q is the q-axis stator current. Since this torque expression resembles that of the PMDC motor, the same time-delay feedback control scheme can be used:

$$T = \mu \, \xi \, B \sin\left(\frac{\omega(t-\tau)}{\xi}\right) \tag{7.13}$$

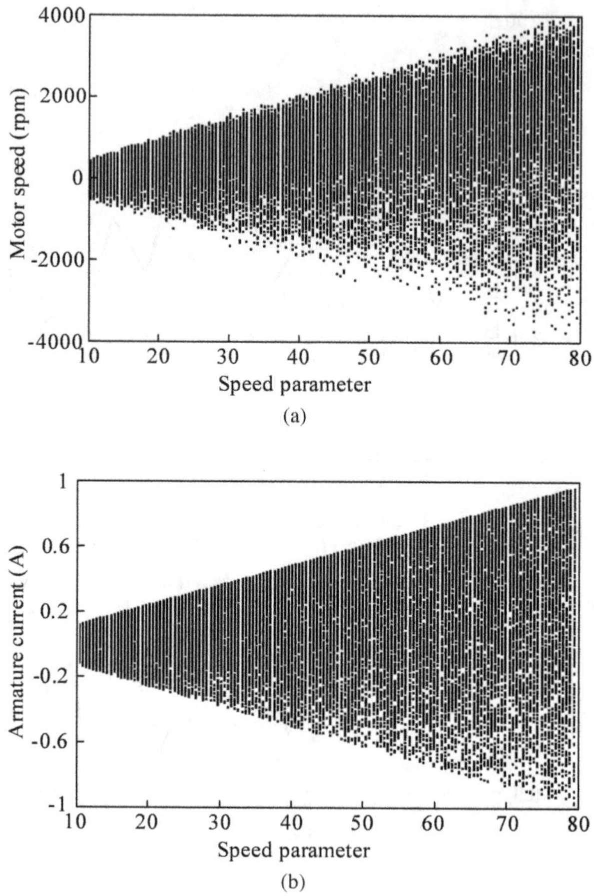

Figure 7.6 Bifurcation diagrams with respect to speed parameter. (a) Motor speed. (b) Armature current

Hence, the torque command T^* calculated by (7.13) is used to generate the torque-component current command i_q^* for the PM synchronous motor (Gao and Chau, 2002). Figure 7.12 shows the corresponding control system.

A practical 3-phase PM synchronous motor is used for exemplification. The corresponding key parameters are $P = 4$, $\psi_r = 0.028$ Wb, $J = 1.44 \times 10^{-5}$ kgm^2, $B = 5.416 \times 10^{-4}$ Nm/rad/s, $L_d = 11.5$ mH, and $L_q = 11.5$ mH. Based on a time-delay feedback, the motor initially operates at a fixed point with a small value of μ. With an increase in μ, periodic motion occurs, and with a further increase in μ, the system exhibits chaotic motion. Figure 7.13 shows the simulated speed and current waveforms under period-2 operation when $\mu = 2.55$, $\xi = 10$, and $\tau = 1$ s, whereas Figure 7.14 shows the measured speed and current waveforms under the same μ and ξ. As expected, the experimental measurement closely matches the simulation waveforms. On the other hand, Figure 7.15 shows the simulated chaotic speed and current waveforms when $\mu = 4$, $\xi = 10$, and $\tau = 1$ s, while Figure 7.16 shows the measured chaotic speed and current waveforms with the same μ and ξ. It can be found that they exhibit chaotic behavior and offer similar boundaries. Notice that chaotic waveforms cannot be directly compared since they are random-like but bounded.

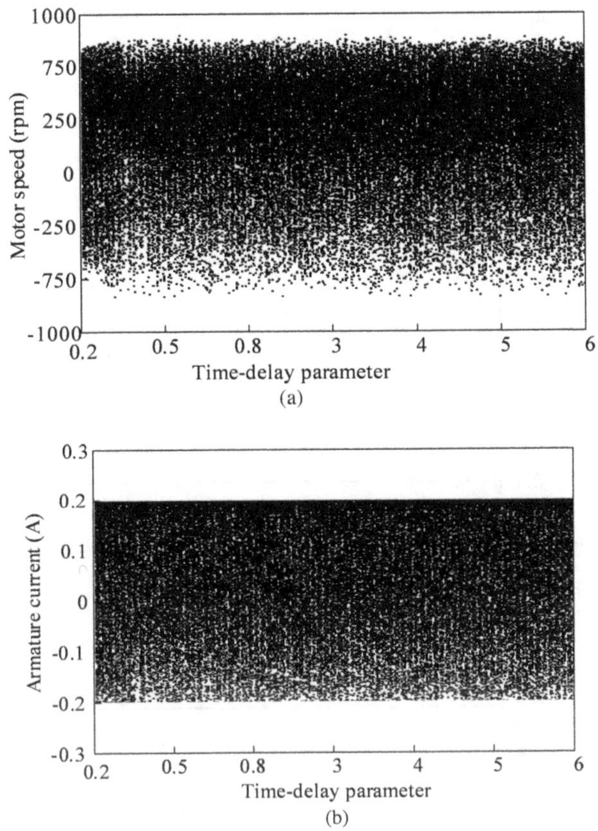

Figure 7.7 Bifurcation diagrams with respect to time-delay parameter. (a) Motor speed. (b) Armature current

7.1.3 Proportional Time-Delay Control of PMDC Drive System

The principle of the proportional-time-delay (PTD) control of a PMDC drive system can be represented as:

$$i^* = \frac{\alpha J}{K_T}\sin[\sigma\omega(t-\tau)] - \frac{\beta J}{K_T}(\omega - \bar{\omega}) \tag{7.14}$$

where i^* is the current reference, σ is the time-delay gain, τ is the time-delay constant, α and β are proportional gains, and $\bar{\omega}$ is the nominal rotational speed. It should be noted that a proportional term $\beta J(\omega - \bar{\omega})/K_T$ is added on the conventional time-delay feedback control term $\alpha J \sin[\sigma\omega(t-\tau)]/K_T$. Thus, this chaoizing approach can offer unidirectional rotation for the PMDC drive. Moreover, it possesses the ability of chaoizing the PMDC drive with a large inertia which is highly desirable for many industrial applications. Figure 7.17 shows the block diagram of this PTD control.

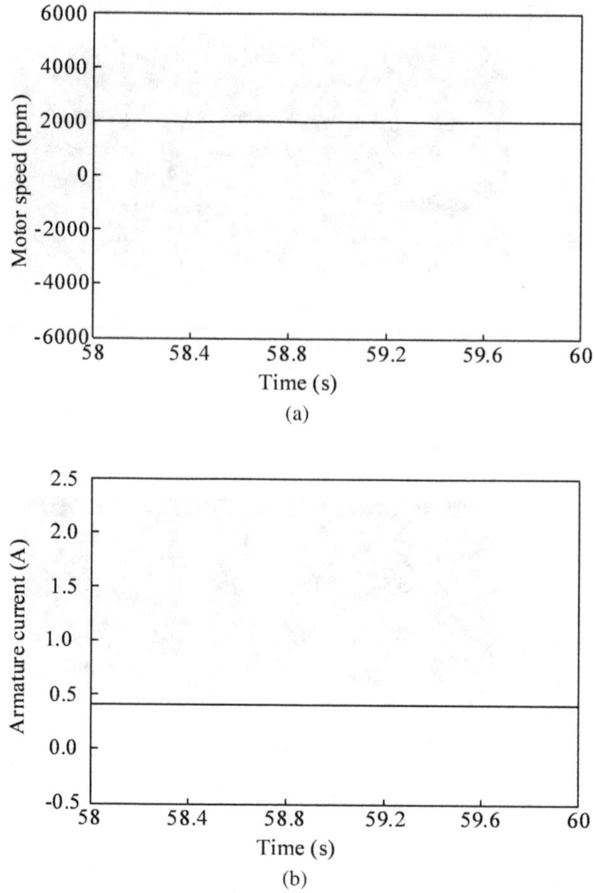

Figure 7.8 Simulated waveforms under normal operation. (a) Motor speed. (b) Armature current

For the PTD control, when i tracks i^* well, the equation of motion can be written as:

$$\frac{\mathrm{d}\omega}{\mathrm{d}t} = -\frac{B+\beta J}{J}\omega + \alpha \sin[\sigma\omega(t-\tau)] + \beta\bar{\omega} - \frac{T_l}{J} \tag{7.15}$$

Accordingly, when $(B+\beta J)/J > 0$, τ is sufficiently large and $\alpha \sin[\sigma\omega(t-\tau)] + \beta\bar{\omega} - T_l/J$ is bounded, the approximated solution of (7.15) can be computed iteratively as (Wang, Chen, and Yu, 2000):

$$\omega_{k+1} = \frac{\alpha J \sin(\sigma\omega_k) + \beta J\,\bar{\omega} - T_l}{B+\beta J} \tag{7.16}$$

A practical PMDC motor is used for exemplification, and its key parameters are listed in Table 7.2. Given $\beta = 10, \sigma = 1, \bar{\omega} = 170\,\mathrm{rad/s}$, and $T_l = 0$, the bifurcation diagram of ω versus α under PTD control can be deduced as depicted in Figure 7.18(a). It can be seen that it is a typical period-doubling route to chaos. Figure 7.18(b) shows the corresponding Lyapunov exponent in which the positive values denote the

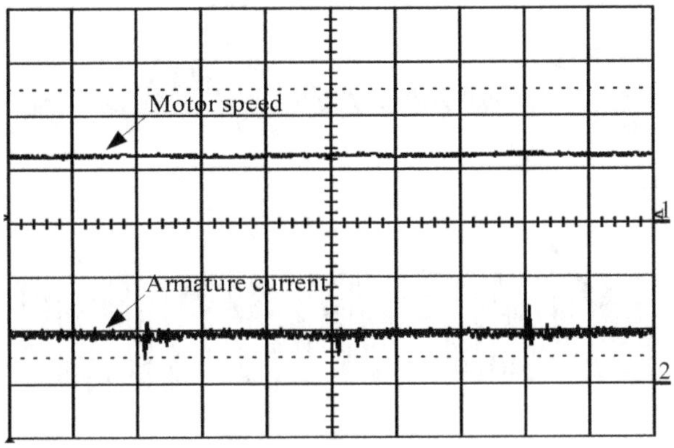

Figure 7.9 Measured waveforms of motor speed and armature current under normal operation (1666.7 rpm/div, 0.5 A/div, 0.2 s/div)

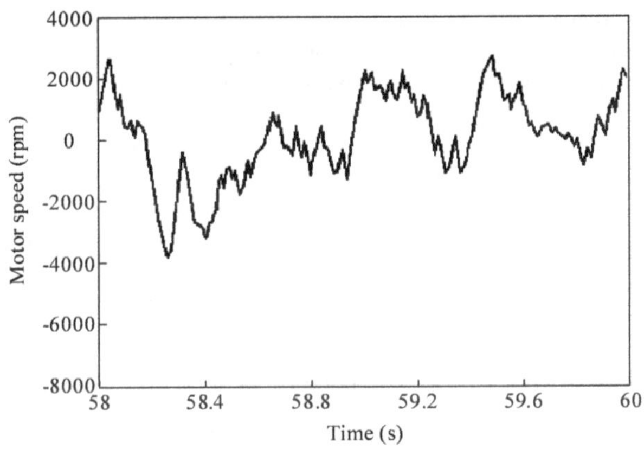

(a)

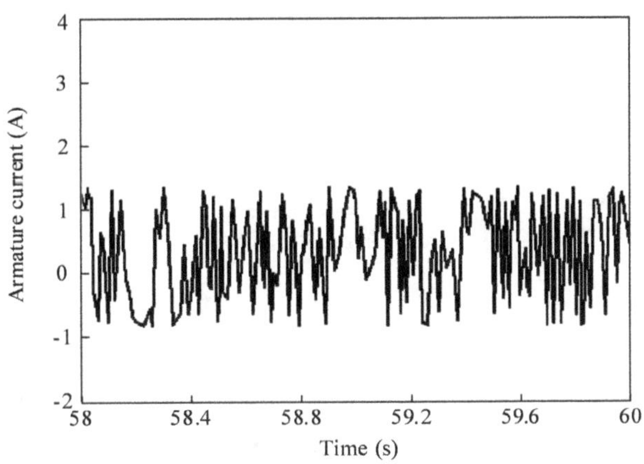

(b)

Figure 7.10 Simulated waveforms under chaotic operation. (a) Motor speed. (b) Armature current

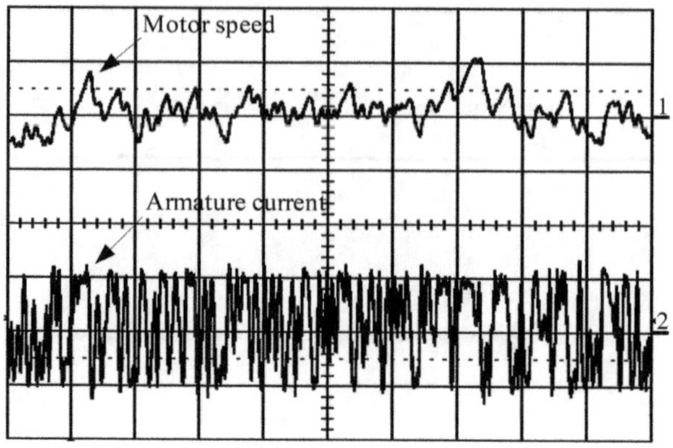

Figure 7.11 Measured waveforms of motor speed and armature current under chaotic operation (1666.7 rpm/div, 1 A/div, 0.2 s/div)

existence of chaos mathematically. Similarly, given $\beta = 10$, $\alpha = 400$, $\bar{\omega} = 170\,\text{rad/s}$, and $T_l = 0$, the bifurcation diagram of ω against σ and the corresponding Lyapunov exponent are shown in Figure 7.19. Also, there is a period-doubling route to chaos with respect to σ.

Both computer simulation and experimentation are used to assess the performance of the chaotic drive system. The selected parameters for chaoization are $T_l = 0$, $\alpha = 400$, $\beta = 10$, $\tau = 1$ s, $\bar{\omega} = 170\,\text{rad/s}$, and $\sigma = 1$. The simulated motor speed and armature current waveforms are shown in Figure 7.20. It can be observed that the waveforms are chaotic, while the motor speed can be maintained at positive values. Under the same operating condition, the measured waveforms are shown in Figure 7.21. It can be seen that the simulated and measured waveforms have the same boundaries although the patterns cannot fully match one another, which is actually the random-like but bounded nature of chaos.

7.1.4 *Chaotic Signal Reference Control of PMDC Drive System*

The chaotic speed reference (CSR) control of a PMDC drive essentially adopts PI control to force the motor speed ω following a chaotic speed reference ω^*. The PI control principle is given by:

$$i^* = K_p(\omega - \omega^*) + K_i \int_0^{+\infty} (\omega - \omega^*)\mathrm{d}t \tag{7.17}$$

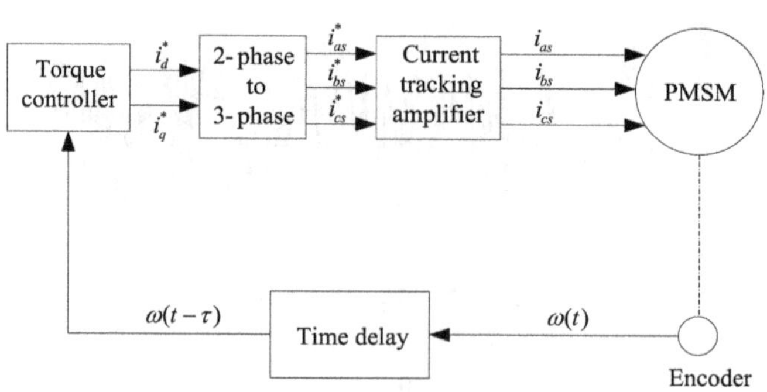

Figure 7.12 Time-delay feedback control of PM synchronous motor

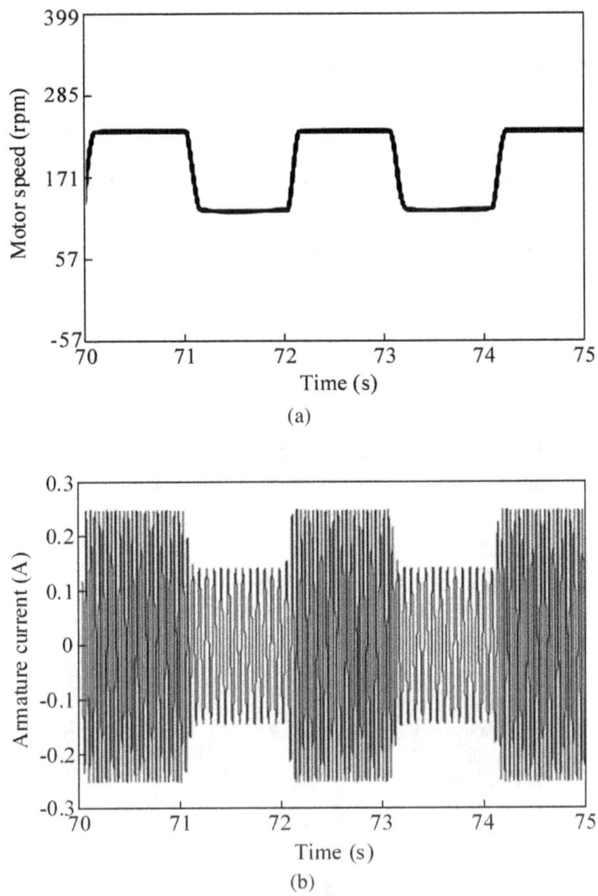

Figure 7.13 Simulated waveforms under period-2 operation. (a) Motor speed. (b) Armature current

where K_p is the proportional coefficient and K_i is the integral coefficient. The chaotic speed reference is generated by the well-known Logistic map which is given by:

$$\omega^* = \tilde{\omega} + \Delta\omega\xi_k \tag{7.18}$$

$$\xi_k = A\xi_{k-1}(1 - \xi_{k-1}) \tag{7.19}$$

where $\tilde{\omega}$ is the base speed, $\Delta\omega$ is the speed boundary, ξ_i is the chaotic series, and A is the control parameter of the Logistic map. Figure 7.22 shows the block diagram of the CSR torque controller.

For the CSR torque control, the value of ω^* is updated according to (7.18) at each interval of 50 ms to enable ω to track ω^* accurately. Thus, the motor speed of the PMDC drive can be expressed as an iterative map:

$$\omega_{k+1} = \frac{-A\omega_k^2}{\Delta\omega} + \frac{A(\Delta\omega + 2\tilde{\omega})\omega_k}{\Delta\omega} + \frac{\tilde{\omega}(\Delta\omega - A\Delta\omega - A\tilde{\omega})}{\Delta\omega} \tag{7.20}$$

Given $\tilde{\omega} = 130\,\text{rad/s}$ and $\Delta\omega = 30\,\text{rad/s}$, Figure 7.23 shows the bifurcation diagram of ω versus A of the PMDC motor using the CSR control, and the corresponding Lyapunov exponent. It can be observed that there is a typical period-doubling route to chaos.

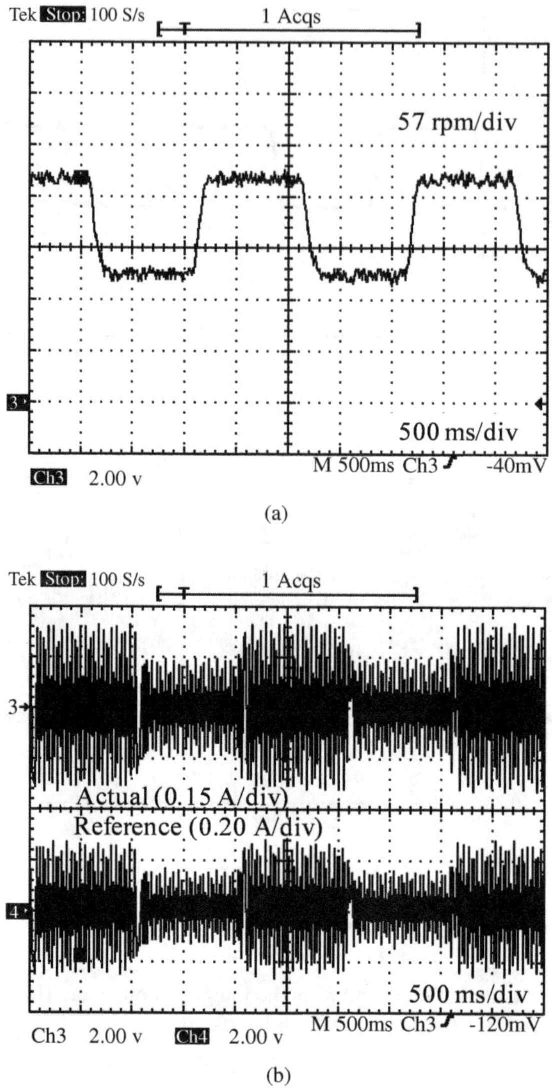

Figure 7.14 Measured waveforms under period-2 operation. (a) Motor speed. (b) Armature current

Both computer simulation and experimentation are used to assess the performance. The selected parameters for chaoization are $\tilde{\omega} = 130\,\text{rad/s}$, $\Delta\omega = 30\,\text{rad/s}$, and $A = 4$. The simulated rotor speed and armature current waveforms are shown in Figure 7.24. Similar to the PTD control, this CSR control can produce the desired chaotic waveforms, namely the rotor speed can be maintained at positive values. Furthermore, by comparing their speed and current waveforms, it can be observed that the PTD control needs lower current magnitudes than the CSR control while the CSR control can offer more accurate speed boundaries than the PTD control. The reason is due to the fact that the PTD control is a model-based method without a closed-loop of speed. Therefore, it is sensitive to motor parameter variation and load torque fluctuation.

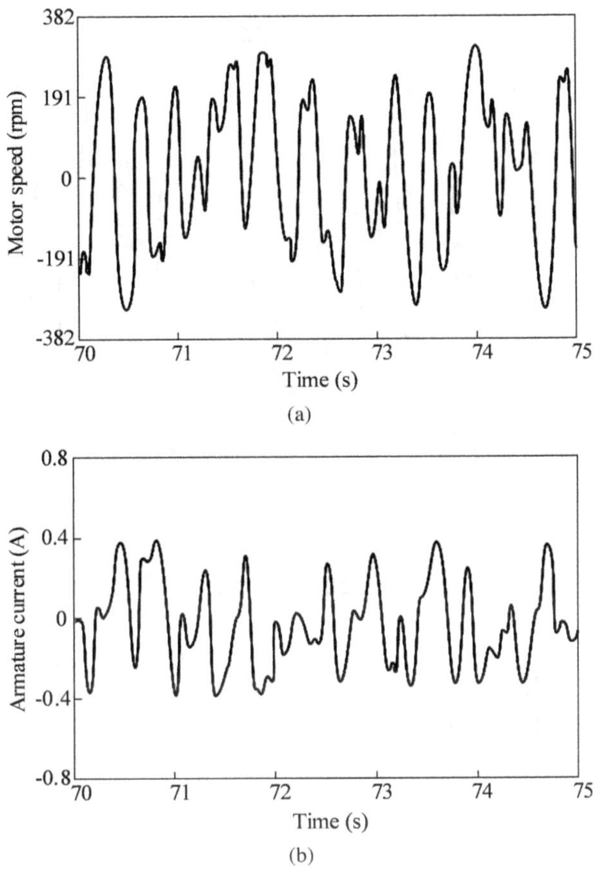

Figure 7.15 Simulated waveforms under chaotic operation. (a) Motor speed. (b) Armature current

Under the same operating condition, the measured waveforms are shown in Figure 7.25. It can be seen that the simulated and measured waveforms have the same boundaries although the patterns cannot fully match one another, which is actually the random-like but bounded nature of chaos. Nevertheless, there are some discrepancies between the simulated and measured current waveforms, especially in the high-frequency region. It is mainly due to the system parasitics and noise in measurement which have been ignored in the simulation model.

7.2 Design-Oriented Chaoization

Instead of relying on control approaches to stimulate chaos, an electric drive system can generate chaotic behavior based on design-oriented approaches. Namely, the electric drive system can spontaneously produce chaotic motion once power on. In this section, two kinds of electric drive systems, namely the doubly salient PM (DSPM) drive and the shaded-pole induction drive, are chaoized by using design-oriented approaches.

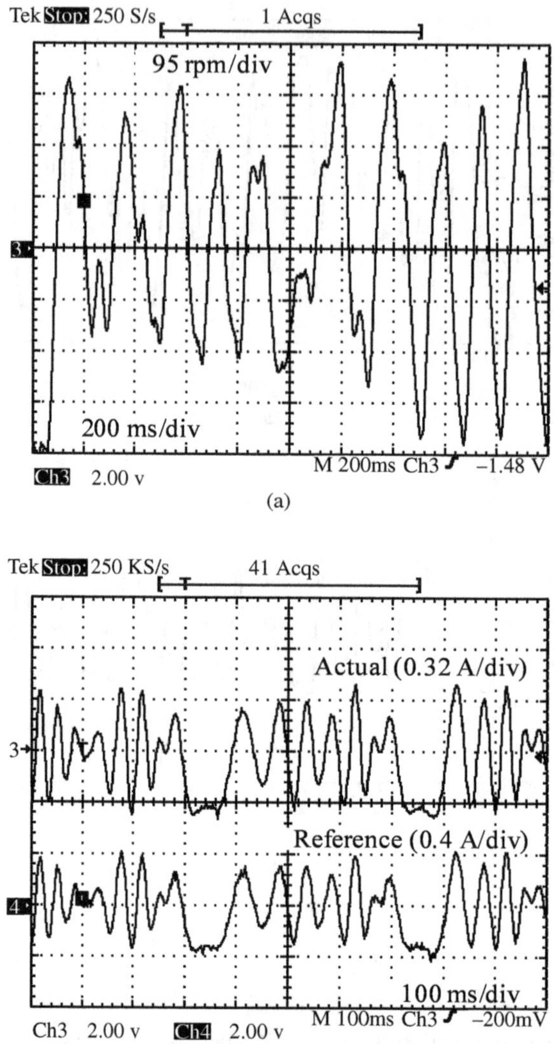

Figure 7.16 Measured waveforms under chaotic operation. (a) Motor speed. (b) Armature current

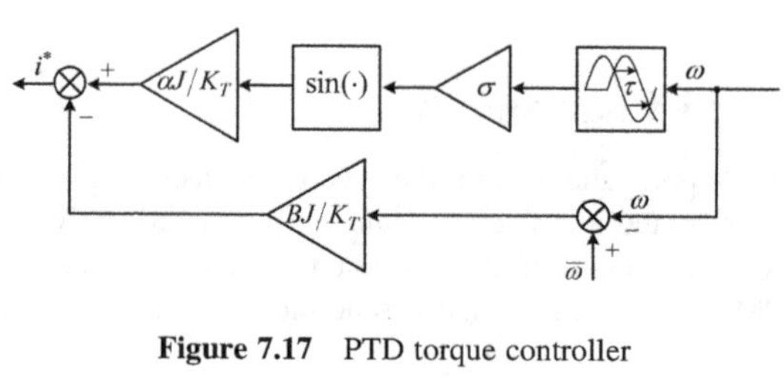

Figure 7.17 PTD torque controller

Table 7.2 Key parameters of PMDC motor

Parameter	Value
Torque constant K_T	0.2286 Nm/A
Back EMF constant K_E	0.2286 V/rad/s
Armature resistance R_a	3.42 Ω
Armature inductance L_a	3.4 mH
Viscous coefficient B	8×10^{-5} Nm/rad/s
Rotor inertia J	1.588×10^{-4} kgm^2

7.2.1 Doubly Salient PM Drive System

The DSPM motor is a new kind of brushless motors (Liu *et al.*, 2008). It combines the advantages of both the switched reluctance motor and the PM brushless motor, hence achieving robust structure, high power density, high efficiency, and high reliability, while being maintenance free (Cheng, Chau, and Chan, 2001). Thus, it is attractive for high-performance applications.

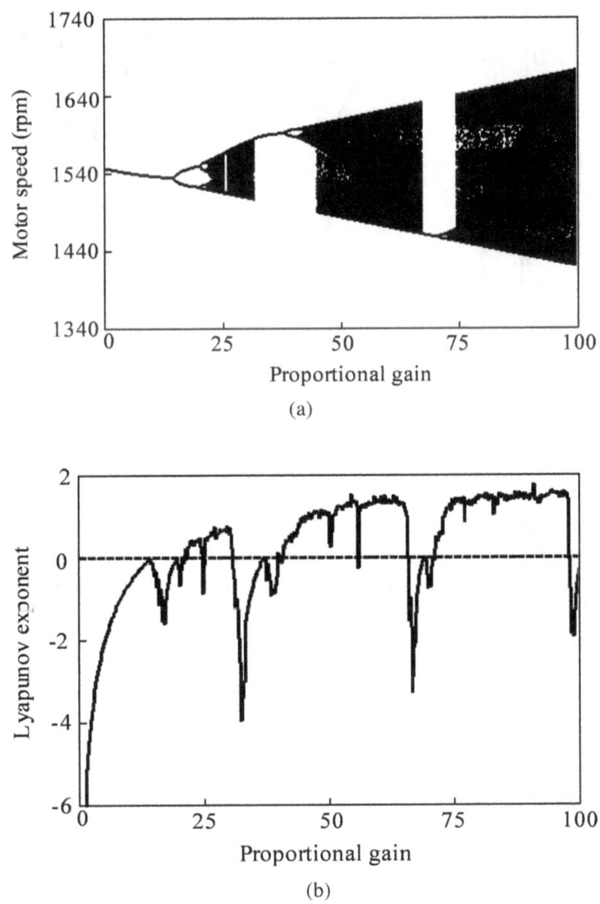

Figure 7.18 Chaotic analysis of PTD with respect to α. (a) Bifurcation diagram. (b) Lyapunov exponent

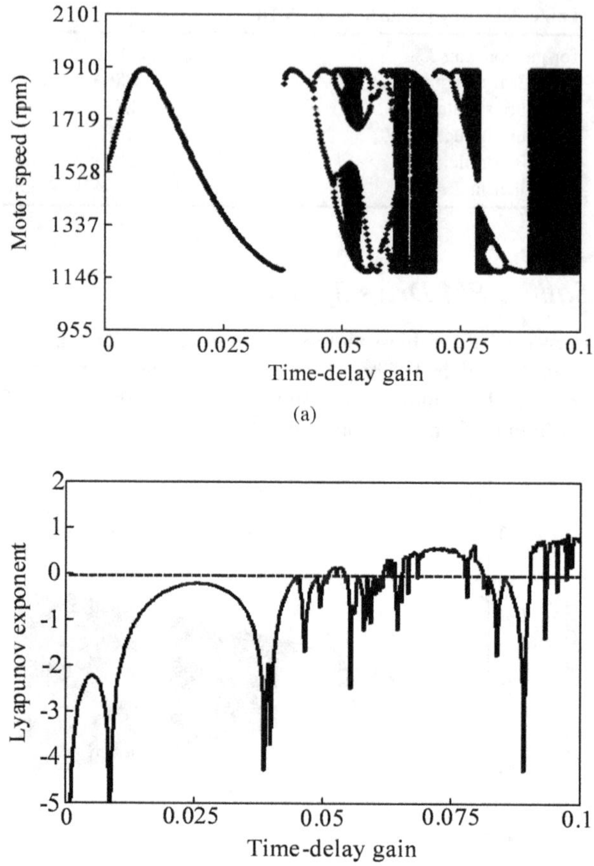

Figure 7.19 Chaotic analysis of PTD with respect to σ. (a) Bifurcation diagram. (b) Lyapunov exponent

A 3-phase 12/8-pole rotor-skewed DSPM motor is adopted for exemplification. As shown in Figure 7.26, it consists of 3-phase armature windings in the stator, 4 pieces of PM material located in the stator, 12 salient poles in the stator, and 8 salient poles in the rotor. With rotor skewing, the PM flux can be expressed as:

$$\psi_{PMx} = \psi_{PM0} + \psi_{PM}\cos(n_p\theta - 2\pi k/3) \tag{7.21}$$

where both $x = a,\ b,\ c$ and $k = 0,\ 1,\ 2$ correspond to the A, B, and C phases, respectively, ψ_{PM0} and ψ_{PM} are the average and amplitude of the PM flux variations, n_p is the number of pole pairs, and θ is the spatial angle with reference to the aligned position of phase A. The mutual inductances of the windings are negligible, whereas the self-inductances can be written as:

$$L_{xx} = L_0 + L_1\cos(n_p\theta - 2\pi k/3) \tag{7.22}$$

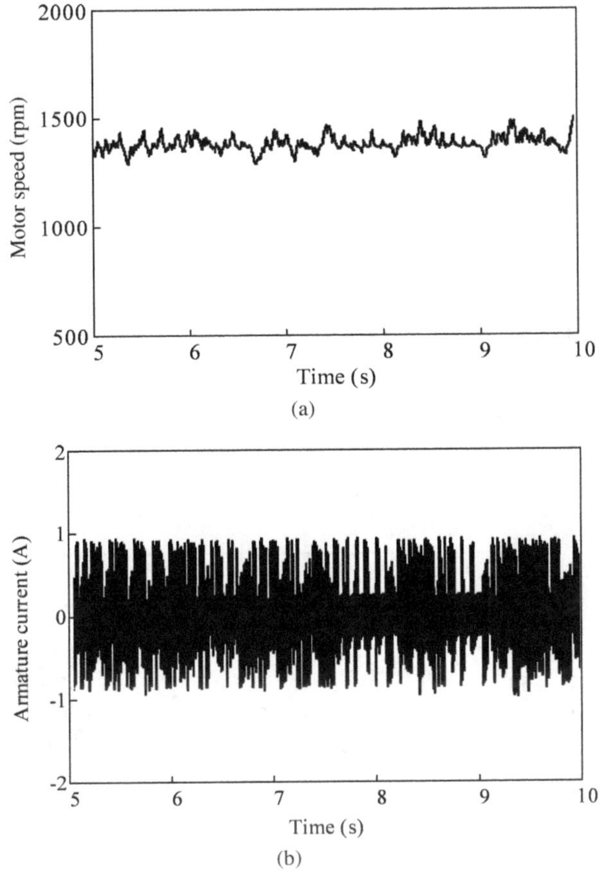

Figure 7.20 Simulated chaotic waveforms using PTD. (a) Motor speed. (b) Armature current

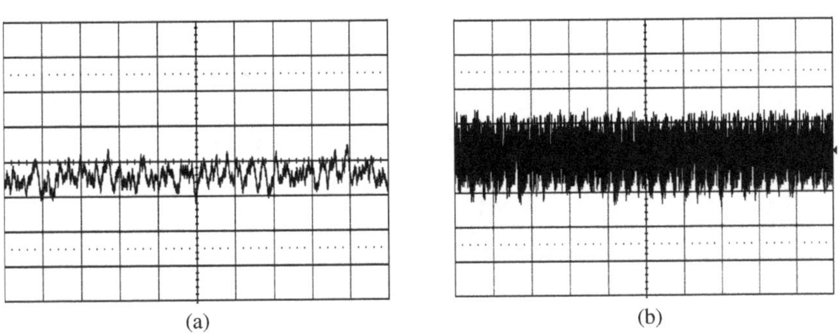

Figure 7.21 Measured chaotic waveforms using PTD. (a) Motor speed (500 ms/div; 191 rpm/div). (b) Armature current (500 ms/div; 0.5 A/div)

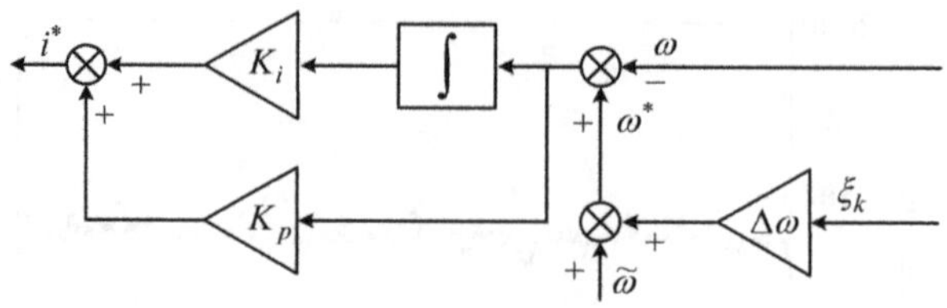

Figure 7.22 CSR torque controller

where L_0 and L_1 are the average and amplitude of self-inductance variations. Hence, the dynamic equations of the DSPM motor fed by 3-phase symmetric sinusoidal voltages are given by:

$$\frac{di_x}{dt} = \frac{1}{L_{xx}}\left(u_x - \frac{dL_{xx}}{dt}i_x - R_s i_x - \frac{d\psi_{PMx}}{dt}\right) \tag{7.23}$$

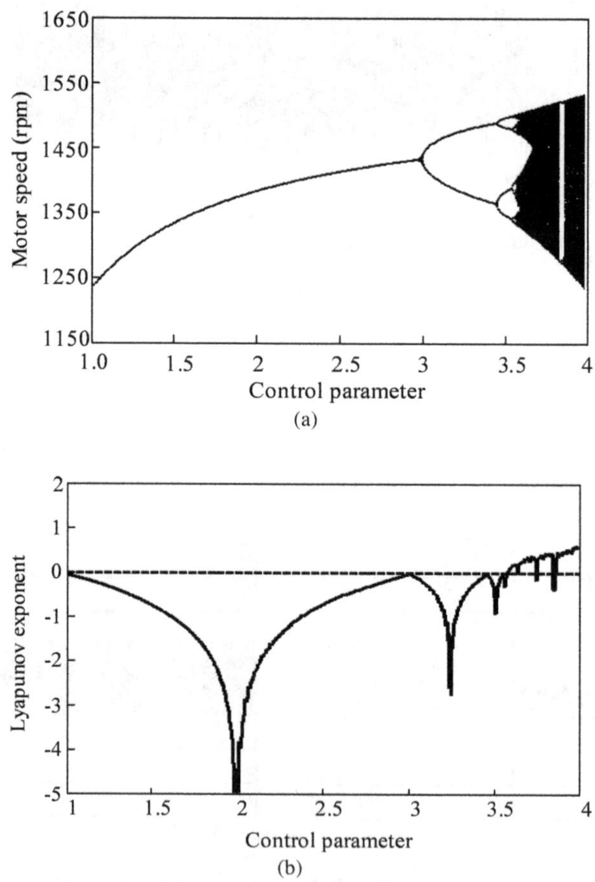

Figure 7.23 Chaotic analysis of CSR. (a) Bifurcation diagram. (b) Lyapunov exponent

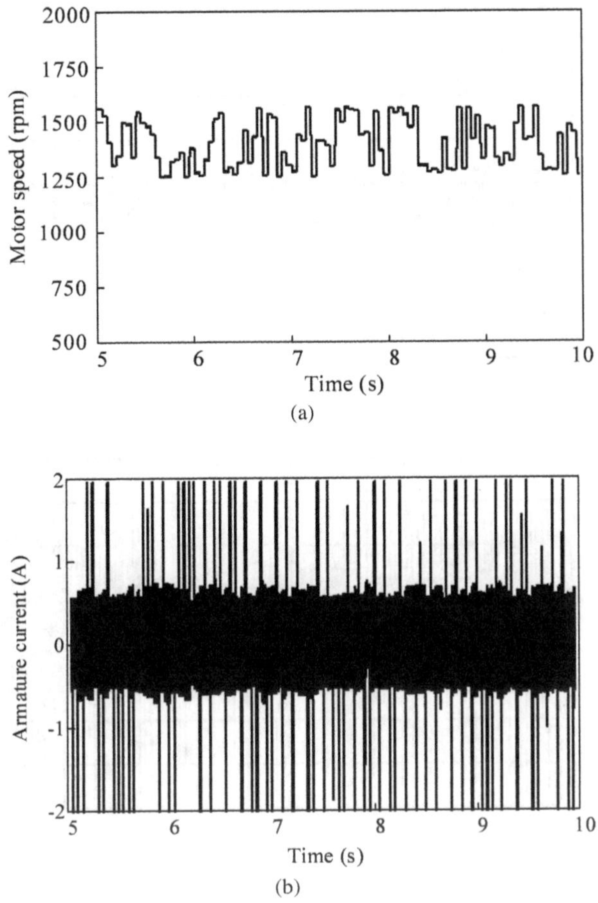

Figure 7.24 Simulated chaotic waveforms using CSR. (a) Motor speed. (b) Armature current

$$\frac{\mathrm{d}\omega}{\mathrm{d}t} = \frac{1}{J}(T_e - T_l - B\omega) \tag{7.24}$$

$$T_e = \sum_{x=a,b,c} \left(\frac{1}{2} i_x^2 \frac{\mathrm{d}L_{xx}}{\mathrm{d}\theta} + i_x \frac{\mathrm{d}\psi_{PMx}}{\mathrm{d}\theta} \right) \tag{7.25}$$

$$u_x = U_x \cos(2\pi f t - 2\pi k/3) \tag{7.26}$$

where u_x is the input phase voltage, U_x is the amplitude of the phase voltage, f is the frequency of the phase voltage, i_x is the phase current, R_s is the winding resistance, ω is the motor speed, T_e is the electromagnetic torque, T_l is the load torque, B is the viscous coefficient, and J is the inertia. Thus, (7.21)–(7.26) form a fourth-order nonautonomous dynamical system.

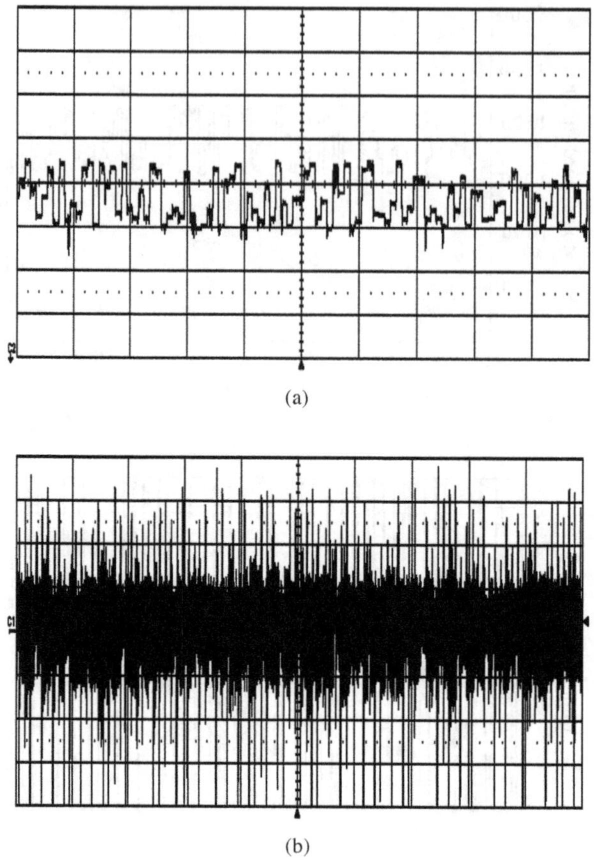

Figure 7.25 Measured chaotic waveforms using CSR. (a) Motor speed (500 ms/div; 191 rpm/div). (b) Armature current (500 ms/div; 0.5 A/div)

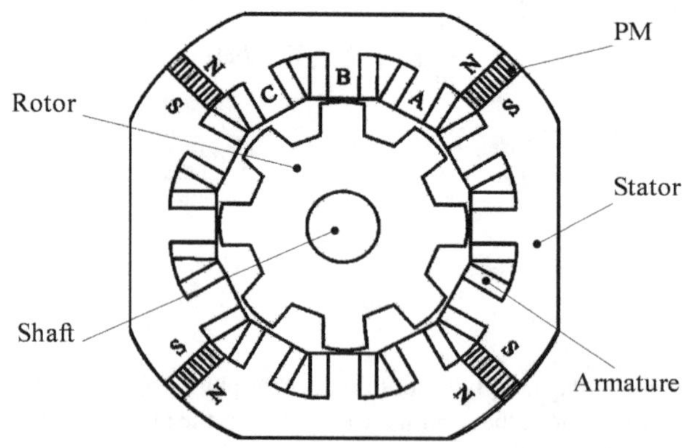

Figure 7.26 Topology of 3-phase 12/8-pole rotor-skewed DSPM motor

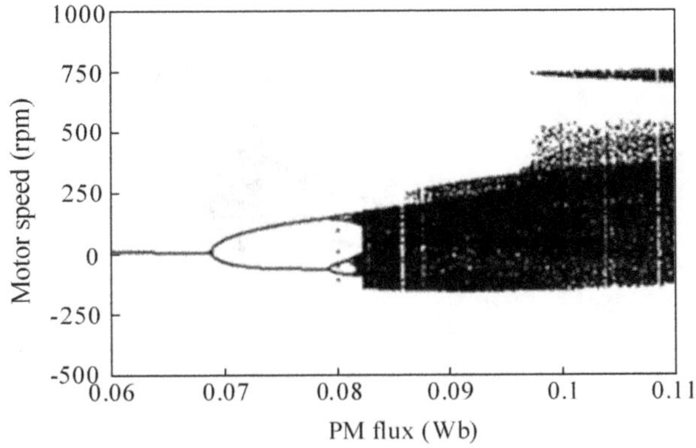

Figure 7.27 Bifurcation diagram of motor speed against PM flux of DSPM drive

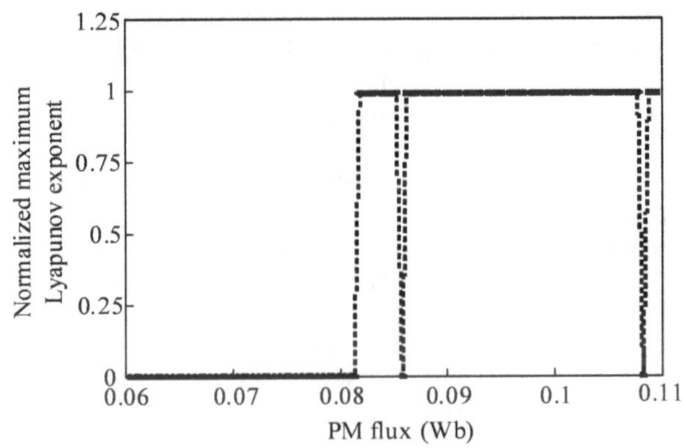

Figure 7.28 Normalized maximum Lyapunov exponent against PM flux of DSPM drive

The system model given by (7.21)–(7.26) illustrates that the variation in the PM flux is an important parameter contributing to the system dynamics. So, the design of PMs – mathematically the value of ψ_{PM} – is used for chaoization of the DSPM drive. By adopting $n_p = 8$, $L_0 = 20\,\text{mH}$, $L_1 = 10\,\text{mH}$, $U_x = 35.4\,\text{V}$, $f = 50\,\text{Hz}$, $R_s = 2.7\,\Omega$, $B = 4 \times 10^{-4}\,\text{Nm/rads}^{-1}$, $T_l = 0$, and $J = 1.35 \times 10^{-3}\,\text{kgm}^2$, the bifurcation diagram of ω with respect to ψ_{PM} can be obtained, as shown in Figure 7.27. It can be seen that the motor speed changes from regular period-1 motion to subharmonic operation at $\psi_{PM} = 0.0695$ Wb, and then bifurcates to chaotic operation at $\psi_{PM} = 0.082$ Wb (the so-called threshold of chaoization). Figure 7.28 shows the corresponding normalized maximum Lyapunov exponent value where the positive values are normalized to be unity while the negative or zero values are normalized to be zero. Thus, a value of unity indicates the existence of chaotic motion. It should be noted that the normalized maximum Lyapunov exponent has a zero value at $\psi_{PM} = 0.086$ Wb, where subharmonic motion occurs according to the bifurcation diagram.

Chaotic motion in the DSPM drive can also be produced by tuning the amplitude of the input voltage. Given $\psi_{PM} = 0.1045$ Wb, the bifurcation diagram of the motor speed against U_x is plotted as shown

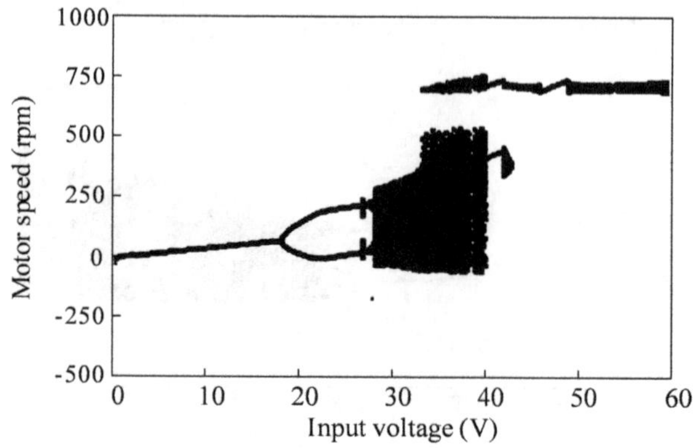

Figure 7.29 Bifurcation diagram of motor speed against input voltage of DSPM drive

in Figure 7.29. Also, there is a critical value of $U_x = 18.2\,\text{V}$ where period-1 motion will bifurcate into subharmonic motion, and a critical value of $U_x = 28.3\,\text{V}$ where chaotic motion occurs. Thus, the DSPM drive can spontaneously produce chaotic motion by tuning the input voltage, which actually can be easily realized by using a transformer. Figure 7.30 shows the corresponding normalized maximum Lyapunov exponent. It should be noted that when $U_x > 40.0\,\text{V}$, the corresponding normalized maximum Lyapunov exponent is zero, but clusters of points can be observed from the bifurcation diagram at this region. This indicates that there is a quasiperiodic motion in this region.

Firstly, the behaviors of the DSPM drive under different ψ_{PM} values are simulated with a fixed value of $U_x = 35.4\,\text{V}$. Figure 7.31 shows its trajectory and Poincaré map when $\psi_{PM} = 0.11$ Wb. The corresponding dynamics exhibit random-like but bounded behavior, which is a typical characteristic of chaos. Then, when $\psi_{PM} = 0.086$ Wb, the system exhibits a higher order subharmonic motion, as shown in Figure 7.32. Consequently, when $\psi_{PM} = 0.075$ Wb, it becomes a second-order subharmonic motion as shown in Figure 7.33. Finally, when $\psi_{PM} = 0.065$ Wb, the system operates with normal period-1 motion, as shown in Figure 7.34.

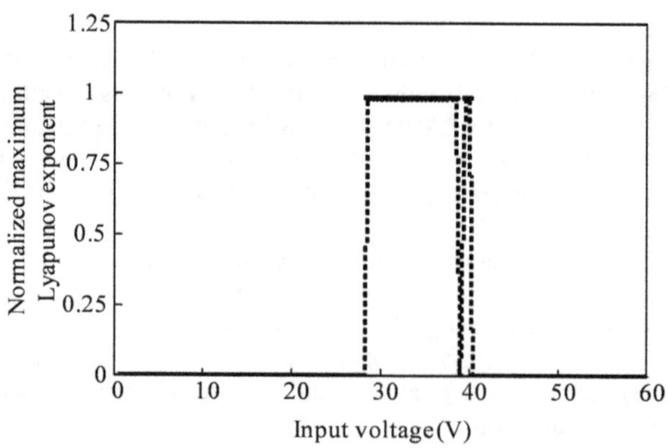

Figure 7.30 Normalized maximum Lyapunov exponent against input voltage of DSPM drive

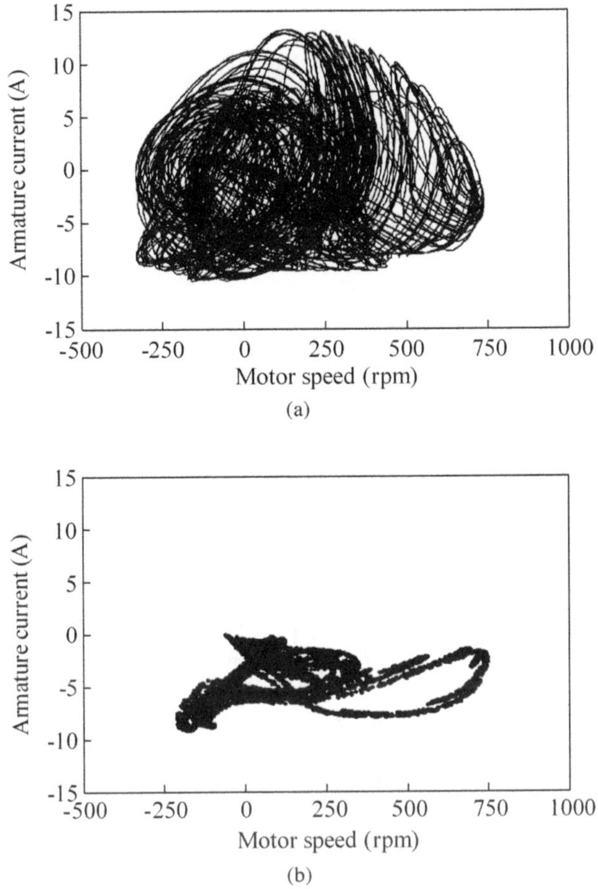

Figure 7.31 Behaviors of DSPM drive under $U_x = 35.4$ V and $\psi_{PM} = 0.11$ Wb. (a) Trajectory. (b) Poincaré map

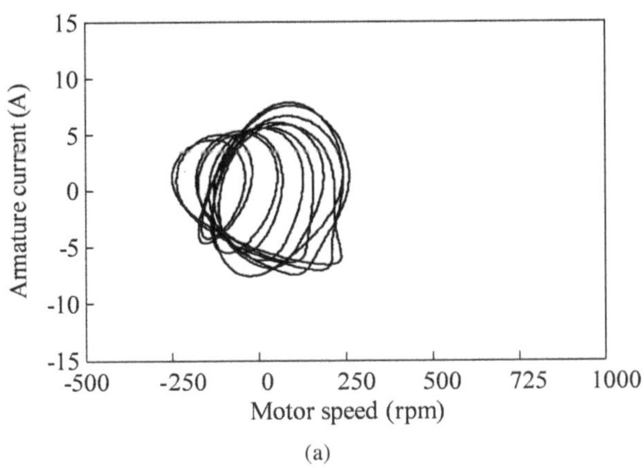

Figure 7.32 Behaviors of DSPM drive under $U_x = 35.4$ V and $\psi_{PM} = 0.086$ Wb. (a) Trajectory. (b) Poincaré map

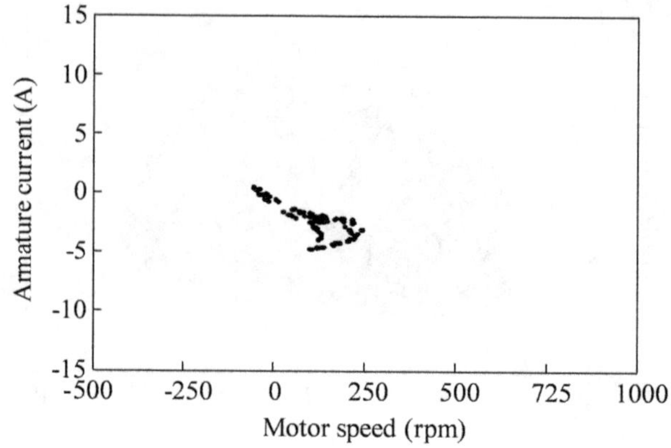

(b)

Figure 7.32 *(Continued)*

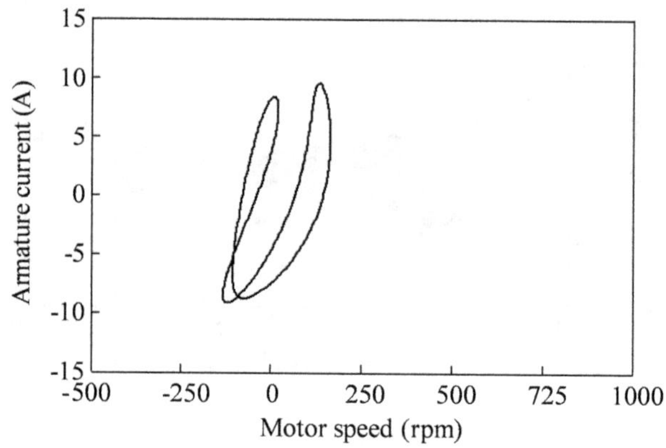

(a)

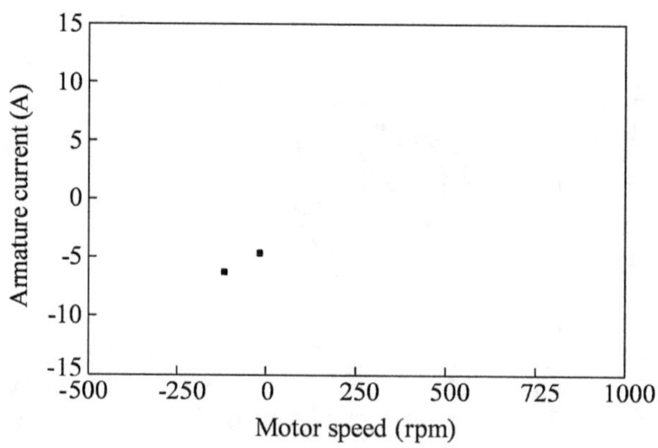

(b)

Figure 7.33 Behaviors of DSPM drive under $U_x = 35.4$ V and $\psi_{PM} = 0.075$ Wb. (a) Trajectory. (b) Poincaré map

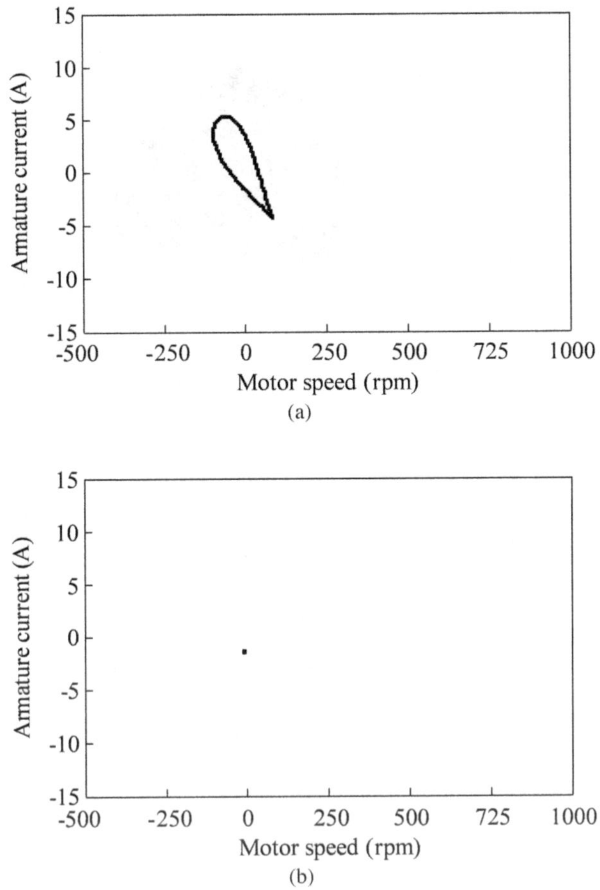

Figure 7.34 Behaviors of DSPM drive under $U_x = 35.4$ V and $\psi_{PM} = 0.065$ Wb. (a) Trajectory. (b) Poincaré map

Secondly, the behaviors of the DSPM drive under different U_x values are simulated with a fixed value of $\psi_{PM} = 0.1045$ Wb. Figure 7.35 shows its trajectory and the Poincaré map when $U_x = 50$ V. It can be seen that the trajectory distributes densely, while the Poincaré map possesses a closed orbit. This indicates that the DSPM drive exhibits a quasiperiodic motion. Then, when $U_x = 35.4$ V, it exhibits a chaotic motion, as shown in Figure 7.36. Consequently, when $U_x = 25$ V, second-order subharmonic motion occurs, as shown in Figure 7.37. Finally, when $U_x = 10$ V, the system operates with a normal period-1 motion as shown in Figure 7.38.

7.2.2 *Shaded-Pole Induction Drive System*

The shaded-pole induction motor is one of the most popular single-phase induction motors for domestic electric appliances. It offers the definite advantages of simple structure and low cost, as well as being highly rugged and reliable (Veinott and Martin, 1987). Its unique feature is the use of an auxiliary single-turn winding, which differs from the distributed windings of other single-phase induction motors, to produce the desired starting torque.

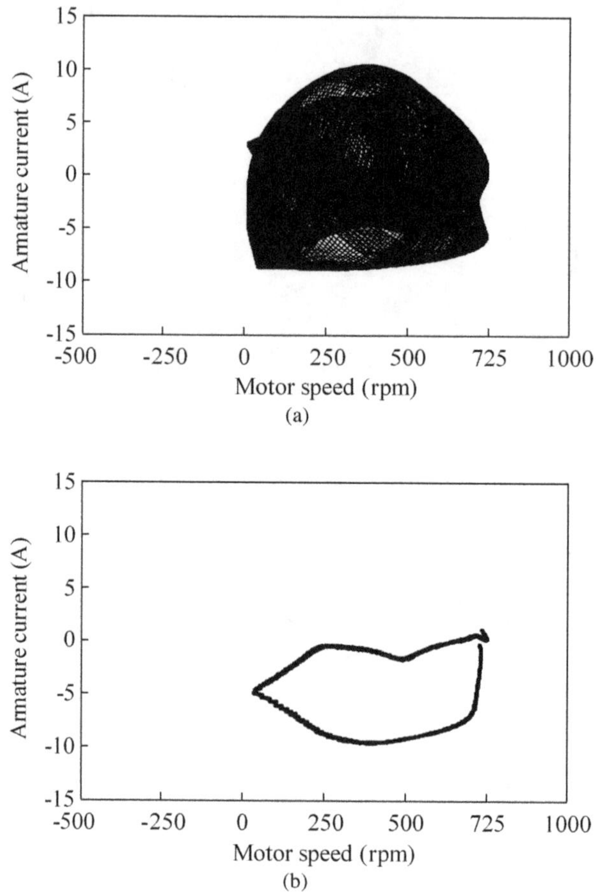

Figure 7.35 Behaviors of DSPM drive under $U_x = 50$ V and $\psi_{PM} = 0.1045$ Wb. (a) Trajectory. (b) Poincaré map

As shown in Figure 7.39(a), the shaded-pole induction motor has a squirrel-cage rotor and a salient-pole stator. It does not need a starting device or switch. Wound on the stator are the main and auxiliary windings. The auxiliary winding is a short-circuited copper band located in a corner of each pole, the so-called shaded-pole winding. This shaded-pole winding has no electrical connection with the input voltage supply. When the main winding is connected to the input voltage supply, the flux through the shaded portion of the pole lags the flux through the unshaded portion owing to the fact that the induced current in the auxiliary winding tends to oppose or delay the generated flux. As a result, the stator flux has two components, one of which lags the other. This leads to the development of a kind of rotating field, thus inducing the rotor current and hence the rotor flux. The interaction of the stator and rotor fluxes produces the starting torque whose direction is from the unshaded portion to the shaded portion of the pole.

Figure 7.39(b) shows the stationary α–β axis model of a shaded-pole induction motor. When the motor is connected to the input voltage supply, its main winding voltage has the form of $v_m = V_m \sin(2\pi ft + \theta)$, where V_m is the voltage amplitude, f is the supply frequency, and θ is the initial phase angle. The system dynamical behavior can then be expressed as (Desai and Mathew, 1971;

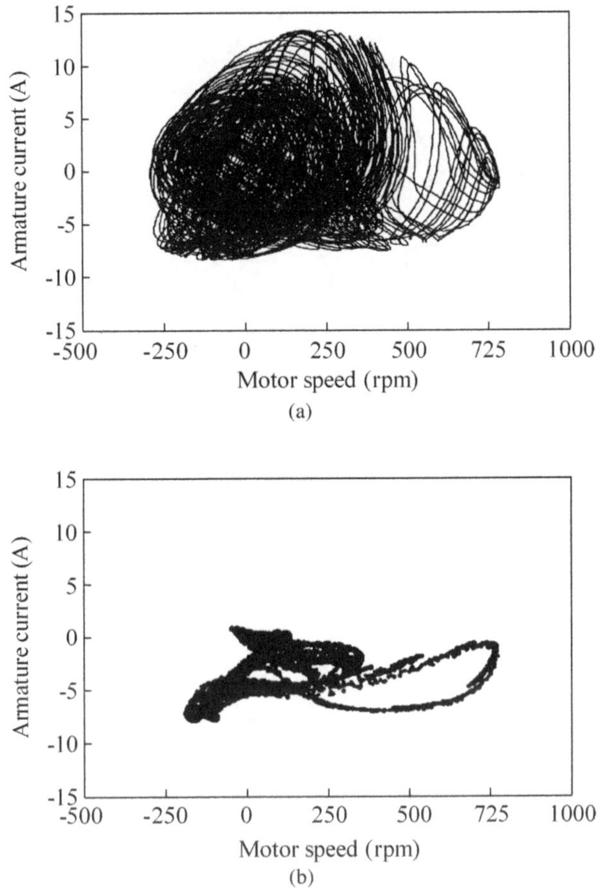

Figure 7.36 Behaviors of DSPM drive under $U_x = 35.4$ V and $\psi_{PM} = 0.1045$ Wb. (a) Trajectory. (b) Poincaré map

Osheiba, Ahmed, and Rahman, 1991):

$$\begin{bmatrix} v_m \\ 0 \\ 0 \\ 0 \end{bmatrix} = \mathbf{L} \begin{bmatrix} \mathrm{d}i_m/\mathrm{d}t \\ \mathrm{d}i_a/\mathrm{d}t \\ \mathrm{d}i_\alpha/\mathrm{d}t \\ \mathrm{d}i_\beta/\mathrm{d}t \end{bmatrix} + \mathbf{R} \begin{bmatrix} i_m \\ i_a \\ i_\alpha \\ i_\beta \end{bmatrix} \tag{7.27}$$

$$\mathbf{L} = \begin{bmatrix} L_{mm} & L_{ma} & L_{m\alpha} & 0 \\ L_{ma} & L_{aa} & L_{ar}\cos\delta & -L_{ar}\sin\delta \\ L_{m\alpha} & L_{ar}\cos\delta & L_{\alpha\alpha} & 0 \\ 0 & -L_{ar}\sin\delta & 0 & L_{\beta\beta} \end{bmatrix} \tag{7.28}$$

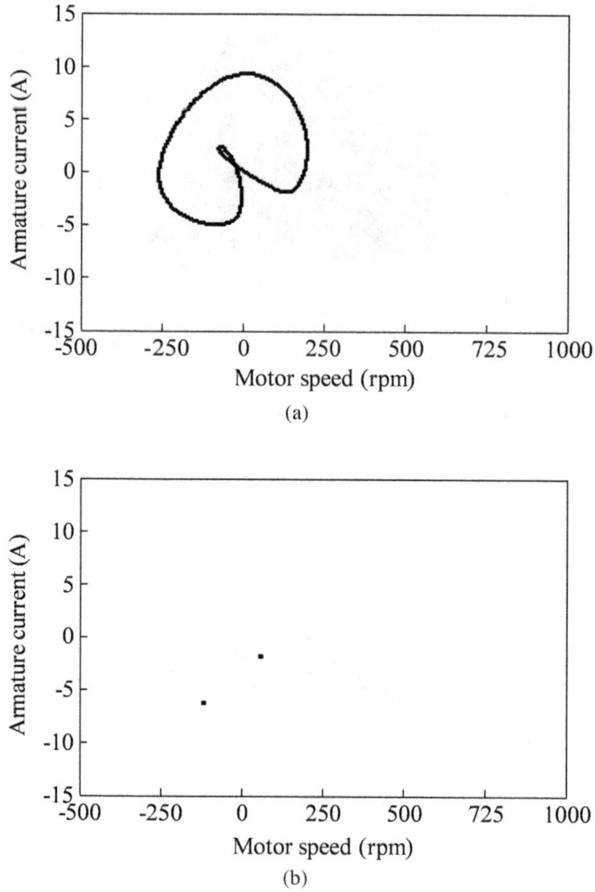

Figure 7.37 Behaviors of DSPM drive under $U_x = 25$ V and $\psi_{PM} = 0.1045$ Wb. (a) Trajectory. (b) Poincaré map

$$\mathbf{R} = \begin{bmatrix} R_m & 0 & 0 & 0 \\ 0 & R_a & 0 & 0 \\ 0 & n_p\omega_r L_{ar}\cos\delta & R_r & -n_p\omega_r L_{\beta\beta} \\ n_p\omega_r L_{m\alpha} & -n_p\omega_r L_{ar}\sin\delta & n_p\omega_r L_{\alpha\alpha} & R_r \end{bmatrix} \tag{7.29}$$

$$T_e = n_p[L_{m\alpha} i_m i_\beta + L_{ar} i_a i_\alpha \sin\delta + L_{ar} i_a i_\beta \cos\delta + (L_{\alpha\alpha} - L_{\beta\beta}) i_\alpha i_\beta] \tag{7.30}$$

$$J\frac{\mathrm{d}\omega_r}{\mathrm{d}t} = T_e - B\omega_r - T_L \tag{7.31}$$

where $\mathbf{R}$ is the resistance matrix and $\mathbf{L}$ is the inductance matrix; i_m and i_a are the currents of the main winding and auxiliary winding, respectively; i_α and i_β are the rotor currents with respect to the

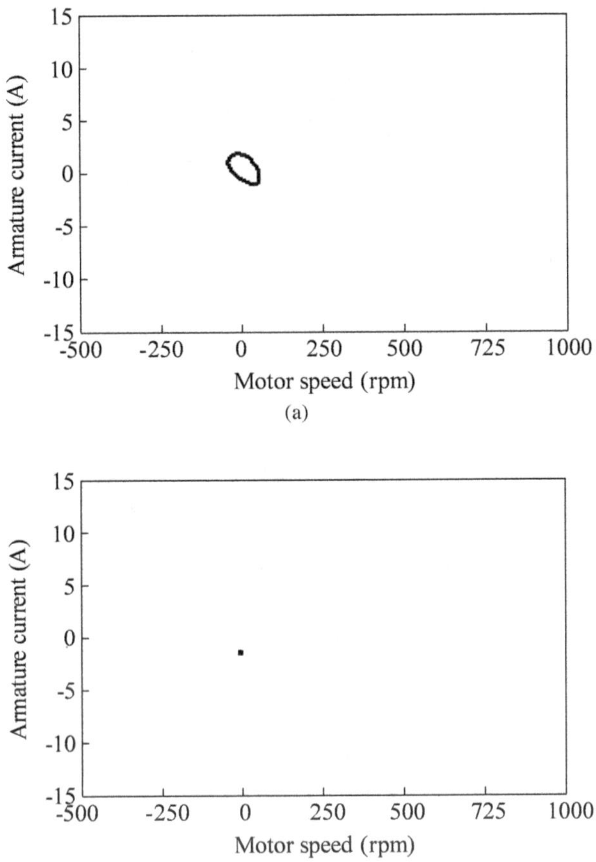

Figure 7.38 Behaviors of DSPM drive under $U_x = 10$ V and $\psi_{PM} = 0.1045$ Wb. (a) Trajectory. (b) Poincaré map

α–axis and β–axis, respectively; R_m, R_a, and R_r are the resistances of the main winding, auxiliary winding, and rotor winding, respectively; L_{mm}, L_{aa}, $L_{\alpha\alpha}$, and $L_{\beta\beta}$ are the self-inductances of the main winding, auxiliary winding, α–axis rotor winding, and β–axis rotor winding, respectively; $L_{m\alpha}$ is the mutual inductance between the main winding and the α–axis rotor winding; L_{ma} is the mutual inductance between the main winding and the auxiliary winding; L_{ar} is the mutual inductance between the auxiliary winding and the rotor winding when the rotor is aligned with the auxiliary winding axis; ω_r is the motor speed; n_p is the number of pole pairs; δ is the angle between the main winding and the auxiliary winding; J is the rotor inertia; B is the viscous coefficient; and T_e and T_L are the electromagnetic torque and load torque, respectively. The model described by (7.27)–(7.31) is a periodically driven nonautonomous system, which is apt to subharmonic oscillations and chaos (Moon and Holmes, 1979).

A practical shaded-pole induction motor with its parameters listed in Table 7.3 is used for exemplification. Based on the model described by (7.27)–(7.31) and the Poincaré mapping, speed bifurcation diagrams with respect to different system parameters can be obtained. Since the motor

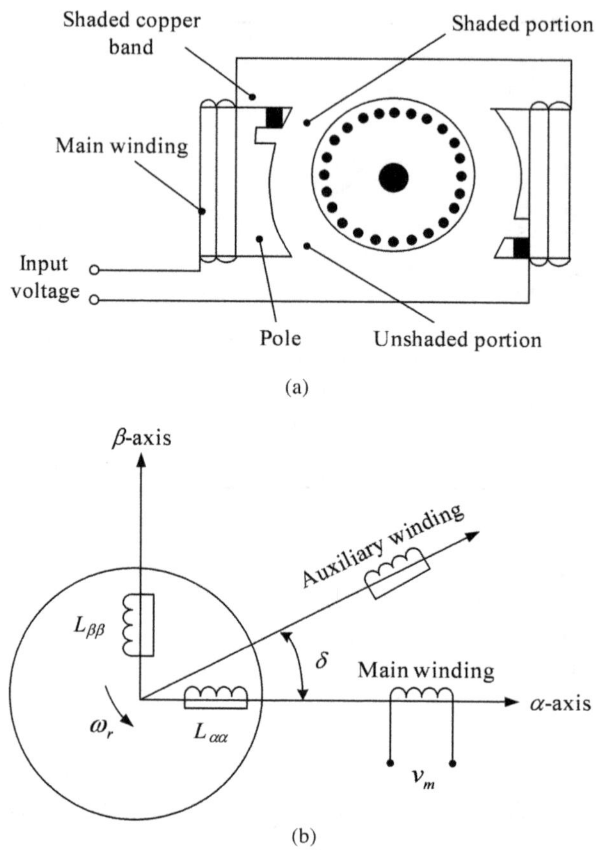

Figure 7.39 Shaded-pole induction motor. (a) Configuration. (b) Model

Table 7.3 Key parameters of shaded-pole induction motor

Number of pole pairs n_p	1
Self inductance of main winding L_{mm}	411.1 mH
Self inductance of auxiliary winding L_{aa}	0.01102 mH
Self inductance of α-axis rotor winding $L_{\alpha\alpha}$	410.5 mH
Self inductance of β-axis rotor winding $L_{\beta\beta}$	376.7 mH
Mutual inductance between main and α-axis rotor windings $L_{m\alpha}$	352.8 mH
Mutual inductance between main and auxiliary windings L_{ma}	1.5 mH
Mutual inductance between auxiliary and rotor windings when rotor aligns with auxiliary axis L_{ar}	1.7 mH
Resistance of main winding R_m	5.630 Ω
Resistance of auxiliary winding R_a	0.212 Ω
Resistance of rotor winding R_r	25.0 Ω
Angle between main and auxiliary windings δ	28°
Rotor inertia J	2.130×10^{-5} kgm^2
Viscous coefficient B	1.47×10^{-4} Nm/rad/s

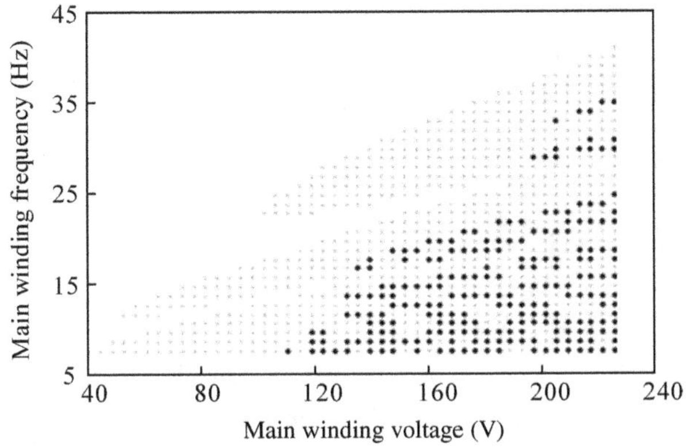

Figure 7.40 Categories of steady-state solutions in bifurcation parameter plane

parameters, such as the pole number, inductances, and resistances are essentially fixed after production, the frequency and amplitude of the applied voltage are selected as the bifurcation parameters. The steady-state behavior against the bifurcation parameter plane is depicted in Figure 7.40 in which the steady-state solutions of the period-1, period-k ($k > 1$), and the others are represented by the plain white, small dot, and large dot, respectively. The corresponding bifurcation diagrams are depicted in Figure 7.41. It can be observed that the drive behavior varies greatly with the change of these operating parameters. Periodic oscillation, quasiperiodic, and chaos can all be found.

As shown in Figure 7.41(a), the drive works as a normal shaded-pole induction motor in the frequency range of 40–60 Hz. Namely, it operates at constant speed with a small amplitude of period-1 oscillation. With a decrease in frequency, a period-2 oscillation occurs. When the frequency is further decreased, complex behaviors (such as quasiperiodic and chaos) arise. Despite of different patterns, similar findings can be observed from the bifurcation diagrams in Figures 7.41(b) and 7.41(c). Therefore, different chaotic boundaries and even types of chaos can result by varying either the frequency or the amplitude of the applied voltage.

Figures 7.42 and 7.43 show the measured speed and main winding current waveforms, respectively, at various periodic-speed operations, namely period-1, period-2, and period-3. It can be found that the period-1 waveform is the normal speed waveform of a conventional shaded-pole induction drive, whereas the period-2 and period-3 waveforms correspond to its abnormal subharmonic operations. Their speed ripples actually reflect the well-known phenomenon of torque pulsation caused by the auxiliary winding.

As can be seen in the speed bifurcation diagrams, the drive exhibits chaotic behavior at certain ranges of the bifurcation parameters. Figures 7.44 and 7.45 show the measured speed and main winding current waveforms, respectively, at various chaotic speed operations. It can be found that the chaotic speed waveforms offer a well-known property of chaos, namely random-like but bounded oscillations. Also, these waveforms are aperiodic and very sensitive to the initial conditions. Although, the main winding current waveforms seem to be more regular than the speed waveforms, they are still random-like and aperiodic.

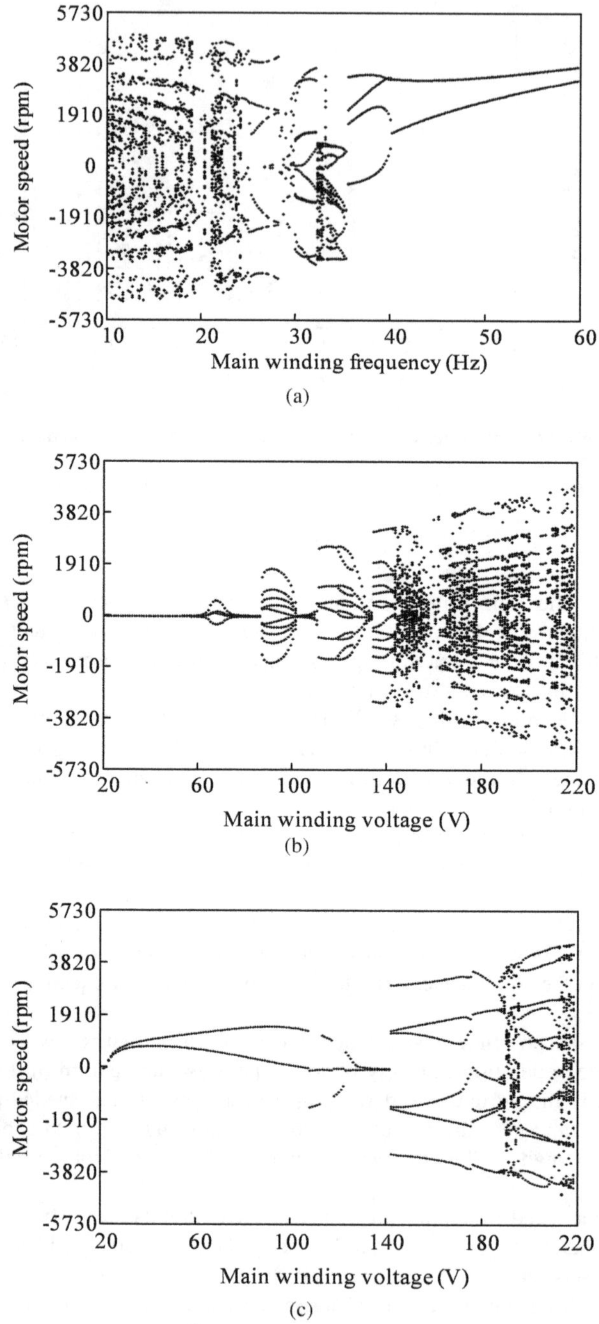

Figure 7.41 Speed bifurcation diagrams. (a) Varying frequency with $V_m = 220\,\text{V}$. (b) Varying amplitude with $f = 10\,\text{Hz}$. (c) Varying amplitude with $f = 22\,\text{Hz}$

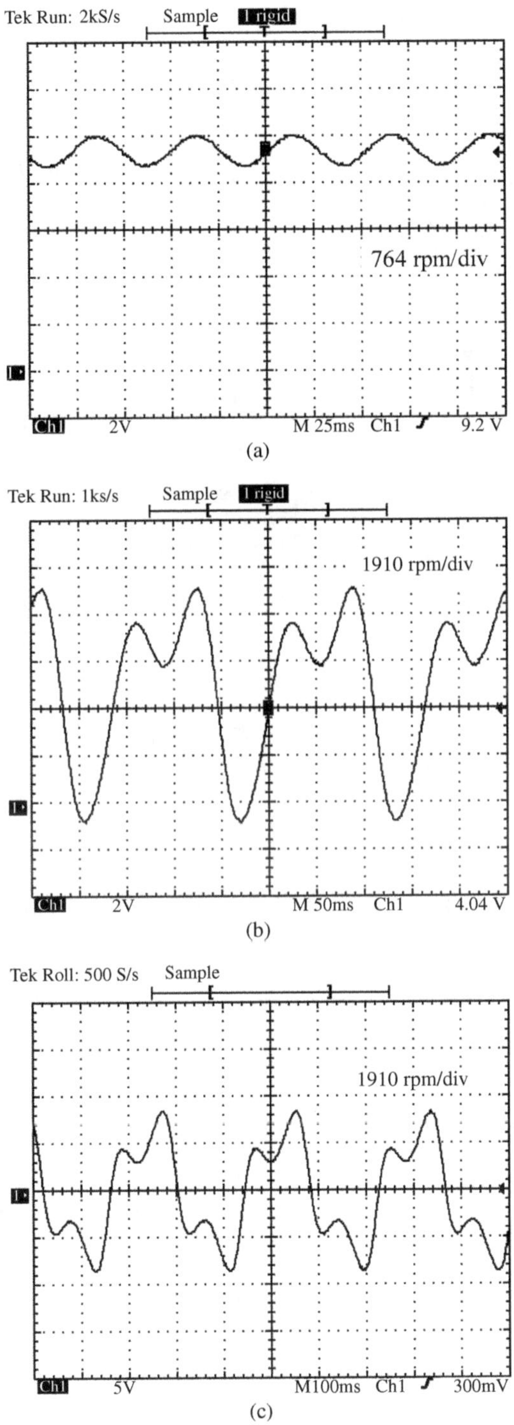

Figure 7.42 Measured periodic speed waveforms. (a) $V_m = 220\,\text{V}, f = 60\,\text{Hz}$. (b) $V_m = 220\,\text{V}, f = 38\,\text{Hz}$. (c) $V_m = 160\,\text{V}, f = 22\,\text{Hz}$

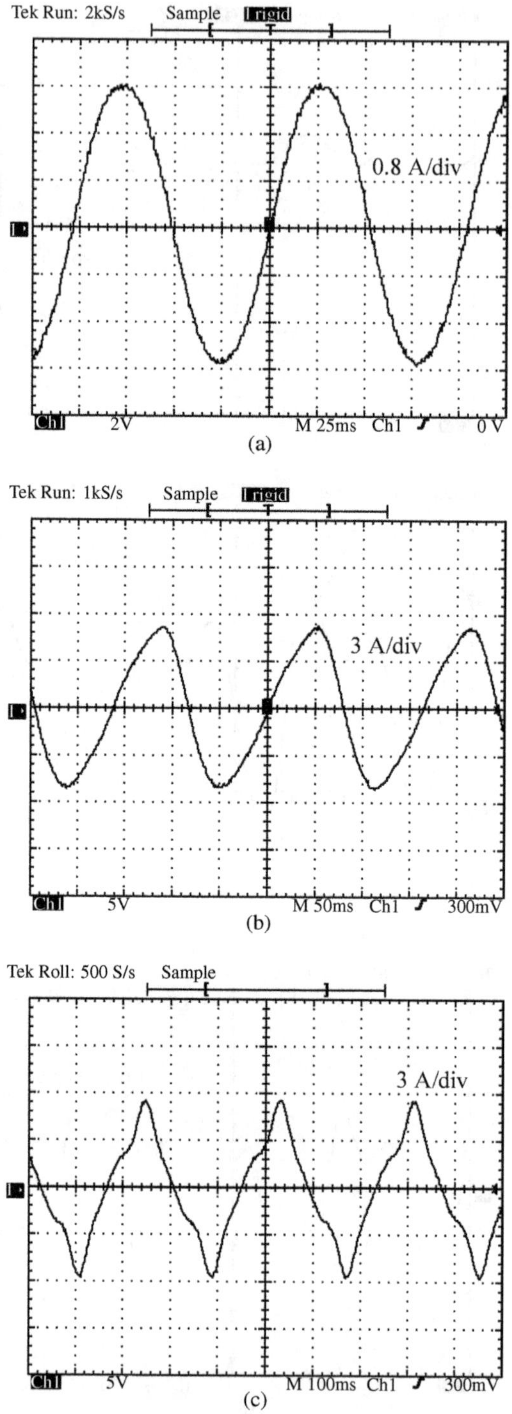

Figure 7.43 Measured periodic main winding current waveforms. (a) $V_m = 220\,\text{V}$, $f = 60\,\text{Hz}$. (b) $V_m = 220\,\text{V}$, $f = 38\,\text{Hz}$. (c) $V_m = 160\,\text{V}$, $f = 22\,\text{Hz}$

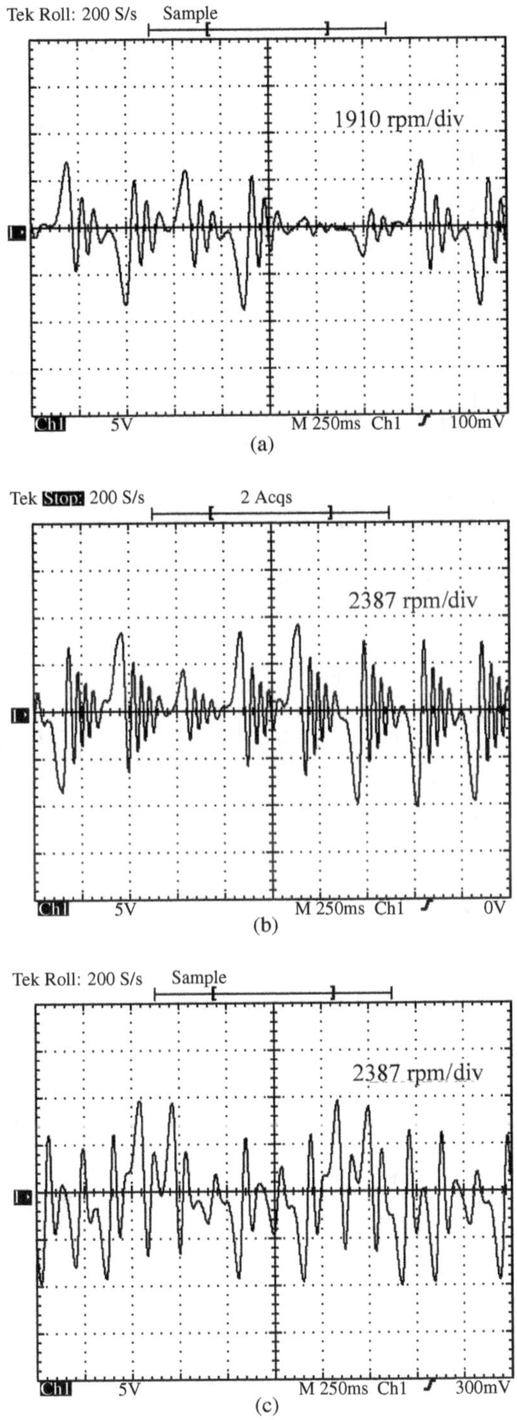

Figure 7.44 Measured chaotic speed waveforms. (a) $V_m = 150$ V, $f = 10$ Hz. (b) $V_m = 220$ V, $f = 10$ Hz. (c) $V_m = 220$ V, $f = 18$ Hz

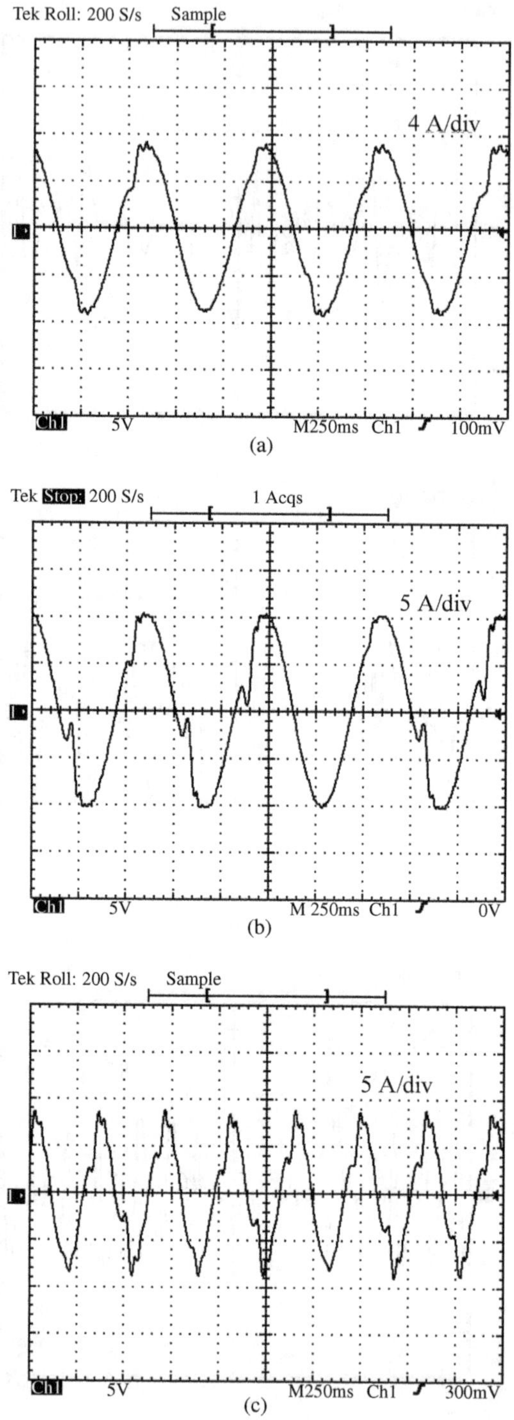

Figure 7.45 Measured chaotic main winding current waveforms. (a) $V_m = 150\,\text{V}, f = 10\,\text{Hz}$. (b) $V_m = 220\,\text{V}, f = 10\,\text{Hz}$. (c) $V_m = 220\,\text{V}, f = 18\,\text{Hz}$

References

Cheng, M., Chau, K.T., and Chan, C.C. (2001) Design and analysis of a new doubly salient permanent magnet motor. *IEEE Transactions on Magnetics*, **37**, 3012–3020.

Desai, B.G. and Mathew, M.A. (1971) *Transient analysis* of shaded pole motor. *IEEE Transactions on Power Apparatus and Systems*, **PAS-90**, 484–494.

Gao, Y. and Chau, K.T. (2002) Chaotification of permanent-magnet synchronous motor drives using time-delay feedback. Proceedings of IEEE Industrial Electronics Conference, pp. 762–766.

Ge, Z.M., Cheng, J.W., and Chen, Y.S. (2004) Chaos anticontrol and synchronization of three time scales brushless DC motor system. *Chaos, Solitons and Fractals*, **22**, 1165–1182.

Liu, C., Chau, K.T., Jiang, J.Z., and Niu, S. (2008) Comparison of stator-permanent-magnet brushless machines. *IEEE Transactions on Magnetics*, **44**, 4405–4408.

Moon, F.C. and Holmes, P.J. (1979) A magnetoelastic strange attractor. *Journal of Sound Vibration*, **65**, 275–296.

Osheiba, A.M., Ahmed, K.A., and Rahman, M.A. (1991) Performance prediction of shaded pole induction motors. *IEEE Transactions on Industry Applications*, **27**, 876–882.

Strogatz, S.H. (1994) *Nonlinear Dynamics and Chaos*, Addison-Wesley, Reading, MA.

Veinott, C.G. and Martin, J.E. (1987) *Fractional and Subfractional Horsepower Electric Motors*, McGraw-Hill, New York.

Wang, X.F., Chen, G., and Yu, X. (2000) Anticontrol of chaos in continuous-time systems via time-delay feedback. *Chaos*, **10**, 771–779.

Ye, S. and Chau, K.T. (2007) Chaoization of DC motors for industrial mixing. *IEEE Transactions on Industrial Electronics*, **54**, 2024–2032.

Part Four

Application of Chaos in Electric Drive Systems

8

Application of Chaos Stabilization

The stabilization of chaos in electric drive systems has been investigated in Chapter 6. This can regulate the chaotic motion in such a way that any harmful vibration can be avoided. It has also been discovered that chaotic behaviors exist in many applications. In particular, some electromechanical systems exhibit chaotic behaviors due to their special configurations or operating characteristics. Such chaotic behaviors not only deflect the system performance but also harm the safety of human beings. Because of the high controllability and flexibility of electric drives, they offer the ability to stabilize those undesirable chaotic behaviors.

In this chapter, the stabilization of chaos in three different applications – namely the wiper system in automobiles, the centrifugal governor system for internal combustion engines, and the rate gyro system in space vehicles – are investigated. The corresponding modeling, analysis, and stabilization are discussed.

8.1 Chaos Stabilization in Automotive Wiper Systems

Various harmful vibrations have been identified in automotive wiper systems (Grenouillat and Leblanc, 2002). As there is a trend for cars to become increasingly quieter, these vibrations not only decrease the wiping efficiency, but also degrade the driving comfort. Also, the disturbance creates a safety hazard. The vibrations are classified into three main groups: squeal noise with a frequency above 1 kHz, reversal noise with a frequency of approximately 500 Hz, and chattering with a frequency below 100 Hz (Goto, Takahashi, and Oya, 2001). Chattering is mainly caused by the stick-slip motion of the rubber blades which leave a stripped pattern on the windshield. There exists a critical wiping speed beyond which chattering will disappear. Actually, chattering is a common phenomenon of mechatronic systems with stick-slip friction (Owen and Croft, 2003). Also, chaotic behaviors are detected under certain wiping speeds in the chattering region (Suzuki and Yasuda, 1998).

Many control methods, including feedforward schemes and feedback schemes, have been put forward to suppress vibrations in various mechatronic systems (Park *et al.*, 2006). For automotive wiper systems, an attempt has been made to reduce chattering by adjusting the attack angle between the blades and the windshield (Grenouillat and Leblanc, 2002). However, the control of attack angles is impractical and may not be implemented. Recently, a linear state feedback control method has also been proposed to stabilize chattering (Chang and Lin, 2004), but this control method requires the online measurement of the angular deflection of the wiper arms, which again is impractical to realize.

Electric motors are considered to be the heart of many mechatronic systems (Straete *et al.*, 1998). Rather than using mechanical means which involve the design of wipers and feedback of wiper motion, the use of electrical means can directly stabilize the driving device – that is, the electric motor – in such a way

Chaos in Electric Drive Systems: Analysis, Control and Application, First Edition. K.T. Chau and Zheng Wang.
© 2011 John Wiley & Sons (Asia) Pte Ltd. Published 2011 by John Wiley & Sons (Asia) Pte Ltd.

that the chaotic chattering in wipers can be suppressed. The key is to apply a proper control method to regulate the electric motor based on an electrical parameter feedback rather than a mechanical parameter feedback. Hence, it offers two distinct advantages: high practicality and high effectiveness. The former can enable practical implementation based on a reasonable cost, while the latter can ensure effective stabilization in various conditions.

In order to achieve high practicality, the feedback parameters should be easily measurable. Thus, electrical parameters, such as the voltage and current of the wiper motor, are preferred to mechanical parameters, such as the angular deflection and angular speed of the wiper arms. Since the armature current is directly proportional to the generated torque of the wiper motor – which is actually a permanent magnet DC (PMDC) motor – it is selected as the measurable feedback parameter to stabilize chaotic vibration in the wiper system.

8.1.1 Modeling

An automotive wiper system is composed of three main parts: an electric motor, a mechanical linkage assembly, and two wipers (one on the driver's side and one on the passenger's side). Each wiper consists of an arm and a rubber blade. As shown in Figure 8.1, the electric motor provides the torque for the mechanical linkage which, in turn, generates the desired motion for the wiper arms and blades. It has been identified that the mechanical linkage between the two wipers can be described by stiffness and damping, and the frictional force on the windshield can be approximated by a cubic polynomial (Suzuki and Yasuda, 1998). Also, the electric motor in this wiper system can be represented by a second-order dynamical equation (Hsu and Ling, 1990). Moreover, the linkage between the wipers and the motor can be described by stiffness (Lévine, 2004). Based on these three works, the dynamical model of the whole wiper system can be formulated as:

$$I_i(\ddot{\theta}_i - n\dot{\omega}) = \begin{cases} -R_i - D_i - M_i(\dot{\xi}_i) & (\dot{\xi}_i \neq 0) \\ -R_i - D_i - M_i(\dot{\xi}_i) & (\dot{\xi}_i = 0, |R_i| \geq N_i l_i \mu_0) \\ 0 & (\dot{\xi}_i = 0, |R_i| < N_i l_i \mu_0) \end{cases} \tag{8.1}$$

$$J_m\dot{\omega} = K_T i_a - B\omega - nR_M \tag{8.2}$$

$$L_a\dot{i}_a = V_{in} - K_E\omega - R_a i_a \tag{8.3}$$

$$\dot{\xi}_i = (\dot{\theta}_i - n\omega)l_i \tag{8.4}$$

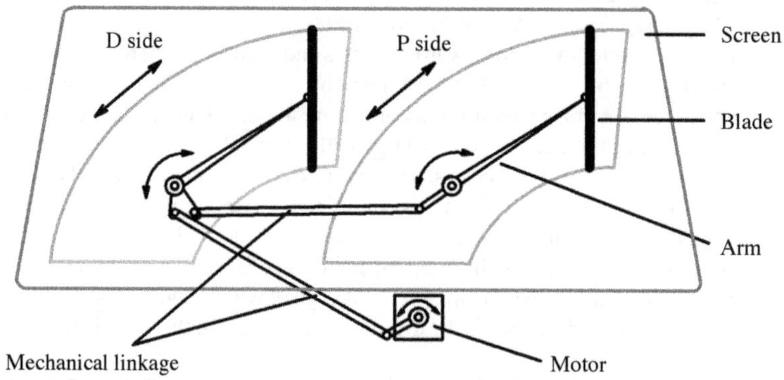

Figure 8.1 Structure of an automotive wiper system

where $i = D, P$ indicate the driver's side and the passenger's side respectively, I_i are the moments of inertia of the wiper arms, l_i are the lengths of the wiper arms, θ_i are the angular deflections of the wiper arms with respect to their positions when no deflections occur, n is the speed reduction ratio between the mechanical linkage and the motor, ω is the motor speed, $\dot{\xi}_i$ are the relative velocities of the wiper blades with respect to the screen; with respect to the motor, V_{in} is the input voltage, i_a is the armature current, K_E is the back EMF constant, K_T is the torque constant, B is the viscous damping, R_a is the armature resistance, L_a is the armature inductance, and J_m is the moment of inertia. The corresponding equivalent-model is shown in Figure 8.2.

By defining R_i as the torques produced by the elastic forces, D_i as the torques produced by the damping forces, R_M as the restoring torque produced by the motor, and M_i as the torques produced by the frictional forces between the wiper blades and screen, they can be expressed as:

$$R_D = k_D\theta_D - k_{DP}\theta_P \tag{8.5}$$

$$R_P = k_P\theta_P - k_{DP}\theta_D \tag{8.6}$$

$$D_D = c_D\dot{\theta}_D - c_{DP}\dot{\theta}_P \tag{8.7}$$

$$D_P = c_P\dot{\theta}_P - c_{DP}\dot{\theta}_D \tag{8.8}$$

$$R_M = k_{DM}\theta_D + k_{PM}\theta_P \tag{8.9}$$

$$M_i(\dot{\xi}_i) = N_i l_i \mu(\dot{\xi}_i) \tag{8.10}$$

$$\mu(\dot{\xi}_i) = \mu_0 \operatorname{sgn}(\dot{\xi}_i) + \mu_1\dot{\xi}_i + \mu_2(\dot{\xi}_i)^3 \tag{8.11}$$

where k_D and k_P are the self-stiffness of the wipers, which are represented by

$$K_D(K_P + K_M)/(K_D + K_P + K_M) \text{ and } K_P(K_P + K_M)/(K_D + K_P + K_M),$$

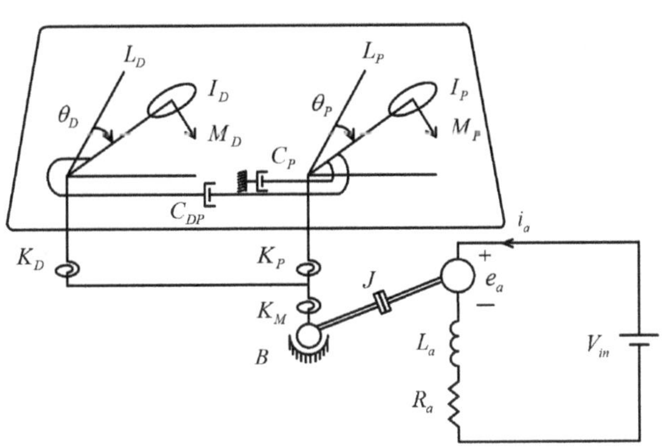

Figure 8.2 Equivalent model of an automotive wiper system

respectively; k_{DP}, k_{DM}, and k_{PM} are the mutual stiffness among the wipers and the motor which are described as

$$K_D K_P/(K_D + K_P + K_M), K_D K_M/(K_D + K_P + K_M) \text{ and } K_P K_M/(K_D + K_P + K_M),$$

respectively; c_D and c_P are the self-damping of the wipers, which are represented as C_{DP} and $C_{DP} + C_P$, respectively; c_{DP} is the mutual damping between the wipers, which equals C_{DP}; K_D, K_P, and K_M are the stiffness coefficients on the driver's side, the passenger's side and the motor's side, respectively; C_P and C_{DP} are the damping coefficients on the passenger's side and the mutual damping coefficients between the driver's side and the passenger's side, respectively; N_i are the forces pressed on the screen by the blades; μ is the coefficient of dry friction between the wiper blades and the screen; and μ_0, μ_1, and μ_2 are constants.

8.1.2 *Analysis*

As listed in Table 8.1, the parameters of a practical automotive wiper system are adopted for exemplification. Since the electric motor is a PMDC motor, the wiping behaviors under different V_{in} are analyzed. By using the mathematical model described by (8.1)–(8.11), the bifurcation diagram of θ_D versus V_{in} is shown in Figure 8.3. The corresponding points represent the locally maximum and minimum values of θ_D at each V_{in}. It can be seen that the bifurcation diagram is divided into five regions. In region V, where $V_{in} > 14.5$ V, θ_D is a fixed value so that chattering does not occur. On the other hand, when $V_{in} \leq 14.5$ V, θ_D oscillates between the maximum and minimum values, indicating that chattering occurs.

As shown in Figure 8.3, the chattering region can be further divided into regions I to IV. For each region, a typical V_{in} is chosen to observe the motion. Namely, Figure 8.4 plots the trajectories of $\dot{\theta}_D$ versus θ_D under $V_{in} = 4$ V, $V_{in} = 8$ V, $V_{in} = 10.8$ V, and $V_{in} = 12$ V. It can be seen that the trajectories display period-1 vibration under $V_{in} = 8$ V in region II and under $V_{in} = 12$ V in region IV; the trajectory shows subharmonic motion under $V_{in} = 10.8$ V in region III; whereas the trajectory exhibits an irregular but bounded behavior

Table 8.1 Parameters of an automotive wiper system

Parameter	Value
Inertia of the wiper arm at driver's side I_D	1.91×10^{-2} kgm^2
Inertia of the wiper arm at passenger's side I_P	1.65×10^{-2} kgm^2
Length of the wiper arm at driver's side l_D	4.70×10^{-1} m
Length of the wiper arm at passenger's side l_P	4.50×10^{-1} m
Force pressed by blade at driver's side N_D	7.35 N
Force pressed by blade at passenger's side N_P	5.98 N
Dry friction coefficient component μ_0	1.18
Dry friction coefficient component μ_1	-9.84×10^{-1}
Dry friction coefficient component μ_2	4.74×10^{-1}
Stiffness coefficient on driver's side K_D	7.20×10^{2} Nm/rad
Stiffness coefficient on passenger's side K_P	7.51×10^{2} Nm/rad
Stiffness coefficient on motor's side K_M	3.53×10^{2} Nm/rad
Damping coefficient on passenger's side C_P	1.00×10^{-2} Nm/rad/s
Mutual damping coefficient C_{DP}	1.00×10^{-2} Nm/rad/s
Speed reduction ratio n	1.59×10^{-2}
Torque constant of motor K_T	1.36×10^{-1} Nm/A
Back EMF constant of motor K_E	1.36×10^{-1} V/rad/s
Viscous damping of motor B	1.91×10^{-5} Nm/rad/s
Inertia of motor J_m	2.30×10^{-5} kgm^2
Armature resistance of motor R_a	9.00×10^{-1} Ω
Armature inductance of motor L_a	3.00×10^{-3} H

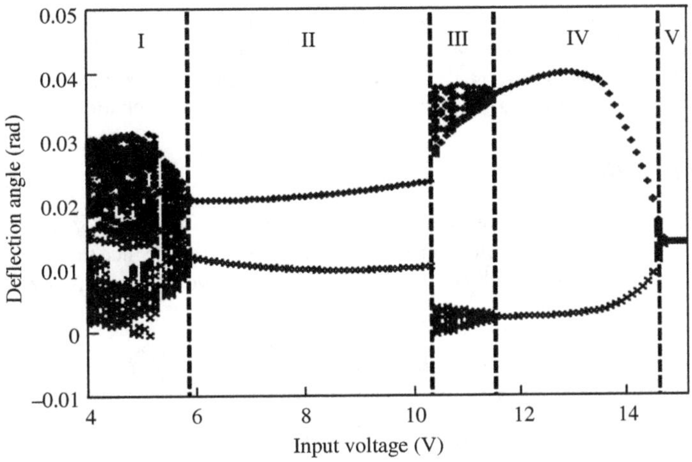

Figure 8.3 Bifurcation diagram of deflection angle versus input voltage of a wiper system

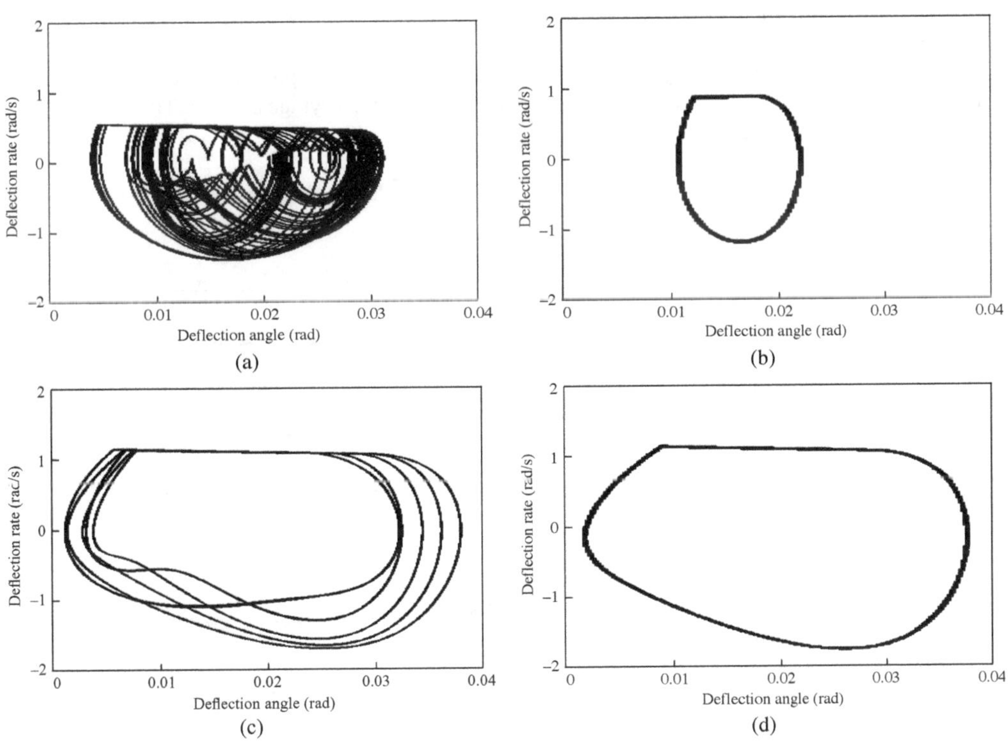

Figure 8.4 Trajectories of deflection rate versus deflection angle of a wiper system under different input voltages. (a) 4 V. (b) 8 V. (c) 10.8 V. (d) 12 V

under $V_{in} = 4$ V in region I. Although an irregular but bounded behavior is an important property of chaos, it is not mathematically sufficient to confirm that the chattering in region I is chaotic. Therefore, the maximum Lyapunov exponent λ_{max} needs to be calculated. When $V_{in} = 4$ V, λ_{max} is computed to be 0.346. A positive value of λ_{max} mathematically confirms that there is chaotic chattering in region I.

8.1.3 Stabilization

A time-delay feedback control enables the stabilization of chaos without a prior analytical knowledge of the system dynamics or the desired reference signals (Chen *et al.*, 2000). Since the current of the motor can readily be measured, the delayed feedback of current is used to derive V_{in}^* for the motor. The corresponding control strategy is given by:

$$\dot{i}_a = \frac{1}{L_a}(V_{in} - K_E\omega - R_a i_a) + \frac{K}{L_a}[i_a(t-\tau) - i_a(t)] \tag{8.12}$$

$$V_{in}^* = V_{in} + K\,[i_a(t-\tau) - i_a(t)] \tag{8.13}$$

where τ is the time delay and K is the feedback gain. It has been proved that chaos can be stabilized into a newly produced periodic orbit even if τ does not equal the period of the embedded unstable periodic orbit (UPO) and the control signal does not become zero (Franceschini, Bose and Schöll, 1999). For the sake of simplicity, τ is chosen to be 0.033 s, which is the period of wiper motion under $V_{in} = 8$ V and is also less than the period of wiper motion under $V_{in} = 12$ V. Consequently, the bifurcation diagram of locally maximum and minimum values of θ_D against K under $V_{in} = 12$ V is plotted as shown in Figure 8.5. It can be seen that chattering can be suppressed within the range of $K \in [1.5\text{ V/A}, 10.2\text{ V/A}]$. However, such range of K is relatively narrow and is insufficient to stabilize chattering over the whole range of V_{in}.

The extended time-delay autosynchronization (ETDAS) control incorporates the advantages of time-delay feedback control while offering a wide operating range of K (Pyragas, 1995). The corresponding control strategy is given by:

$$\dot{i}_a = \frac{1}{L_a}(V_{in} - K_E\omega - R_a i_a) + \frac{K}{L_a}\left[(1-R)\sum_{q=1}^{\infty} R^{q-1} i_a(t-q\tau) - i_a(t)\right] \tag{8.14}$$

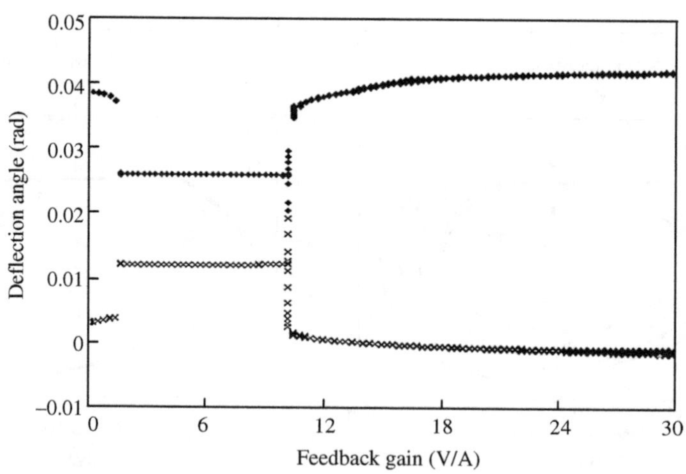

Figure 8.5 Bifurcation diagram of deflection angle versus feedback gain of a wiper system using a time-delay feedback control

$$V_{in}^* = V_{in} + K\left[(1-R)\sum_{q=1}^{\infty} R^{q-1} i_a(t-q\tau) - i_a(t)\right] \tag{8.15}$$

where $R \in [0, 1)$ is a regressive parameter. By defining $S(t) = \sum_{q=1}^{\infty} R^{q-1} i_a(t-q\tau+\tau)$ and $\sigma = q-1$, this yields:

$$\begin{aligned} S(t) &= \sum_{\sigma=0}^{\infty} R^{\sigma} i_a(t-\sigma\tau) = \sum_{\sigma=1}^{\infty} R^{\sigma} i_a(t-\sigma\tau) + i_a(t) \\ &= R\sum_{\sigma=1}^{\infty} R^{\sigma-1} i_a(t-\sigma\tau) + i_a(t) \end{aligned} \tag{8.16}$$

After taking $\gamma = \sigma-1$, (8.16) can be rewritten as:

$$S(t) = R\sum_{\gamma=0}^{\infty} R^{\gamma} i_a(t-\gamma\tau-\tau) + i_a(t) = RS(t-\tau) + i_a(t) \tag{8.17}$$

Thus, (8.14) and (8.15) can be rewritten as:

$$V_{in}^* = V_{in} + K[(1-R)S(t-\tau) - i_a(t)] \tag{8.18}$$

$$\dot{i}_a = \frac{1}{L_a}(V_{in} - K_E\omega - R_a i_a) + \frac{K}{L_a}[(1-R)S(t-\tau) - i_a(t)] \tag{8.19}$$

The block diagram of an ETDAS control is shown in Figure 8.6. The key to realize (8.19) is to determine proper values of R, τ, and K. Firstly, the value of R is chosen. To examine the sensitivity of feedback perturbation, a transfer function $G(\omega) = \Delta V(\omega)/i_a(\omega)$ is introduced where $\Delta V(\omega)$ and $i_a(\omega)$ are the Fourier transformation of the feedback perturbation $K[(1-R)S(t-\tau) - i_a(t)]$ and $i_a(t)$, respectively. So, for the ETDAS control (Pyragas, 1995), $G(\omega)$ is given by:

$$G(\omega) = K\frac{e^{j\omega\tau}-1}{1-e^{j\omega\tau}R} \tag{8.20}$$

When $R \to 1$, this gives $G(\omega) \approx -K$ for all frequencies except those narrow windows around $\omega = 2n\pi/\tau$. So, by increasing R, the feedback perturbation becomes more sensitive and offers a wider range of K for stabilization. Therefore, R is chosen to be 0.95. A corresponding bifurcation diagram of θ_D with respect to K under $V_{in} = 12$ V is plotted, as shown in Figure 8.7. It can be seen that this method offers a wide operating range of $K \in [3.3$ V/A, 19.2 V/A]. The bifurcation diagrams of θ_D with respect to K/L_a at four representative values of V_{in} are plotted, as shown in Figure 8.8. It can be seen that when K/L_a is chosen to be 2167 V/AH, a period-1 motion can be attained throughout all regions. Since L_a is equal to 0.003 H, K is chosen to be 6.5 V/A.

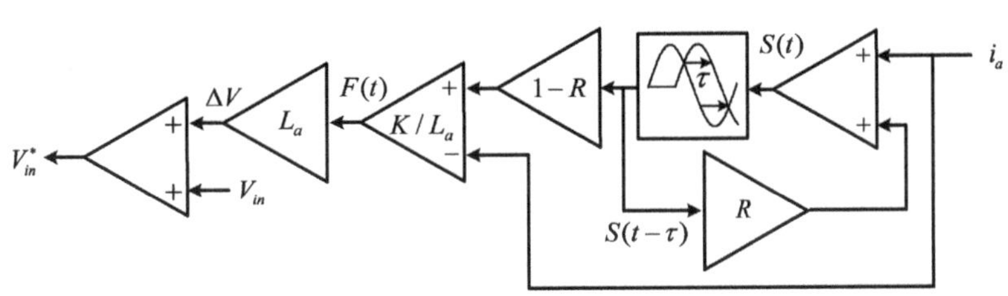

Figure 8.6 Block diagram of the ETDAS control of a wiper system

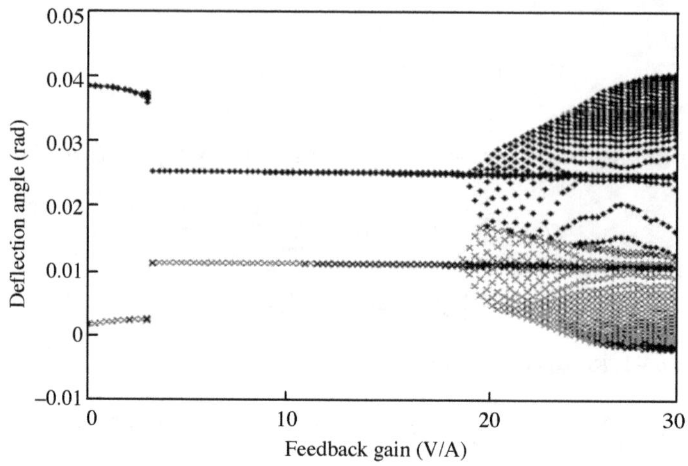

Figure 8.7 Bifurcation diagram of deflection angle versus feedback gain of a wiper system using ETDAS control

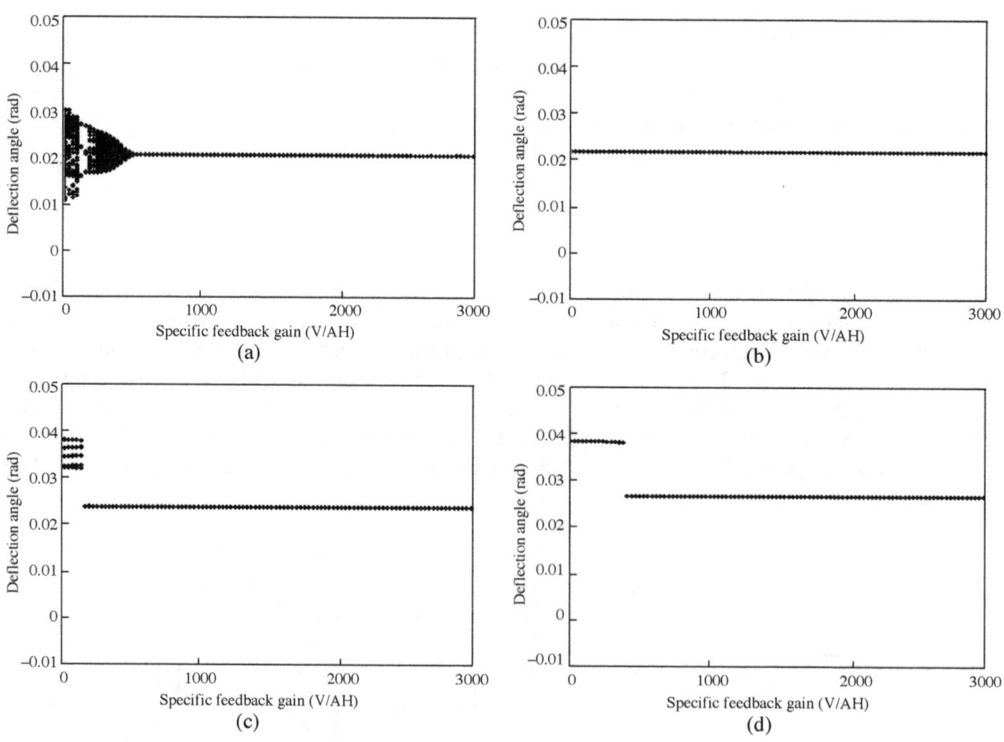

Figure 8.8 Bifurcation diagrams of deflection angle versus specific feedback gain of a wiper system using ETDAS control under different input voltages. (a) 4 V. (b) 8 V. (c) 10.8 V. (d) 12 V

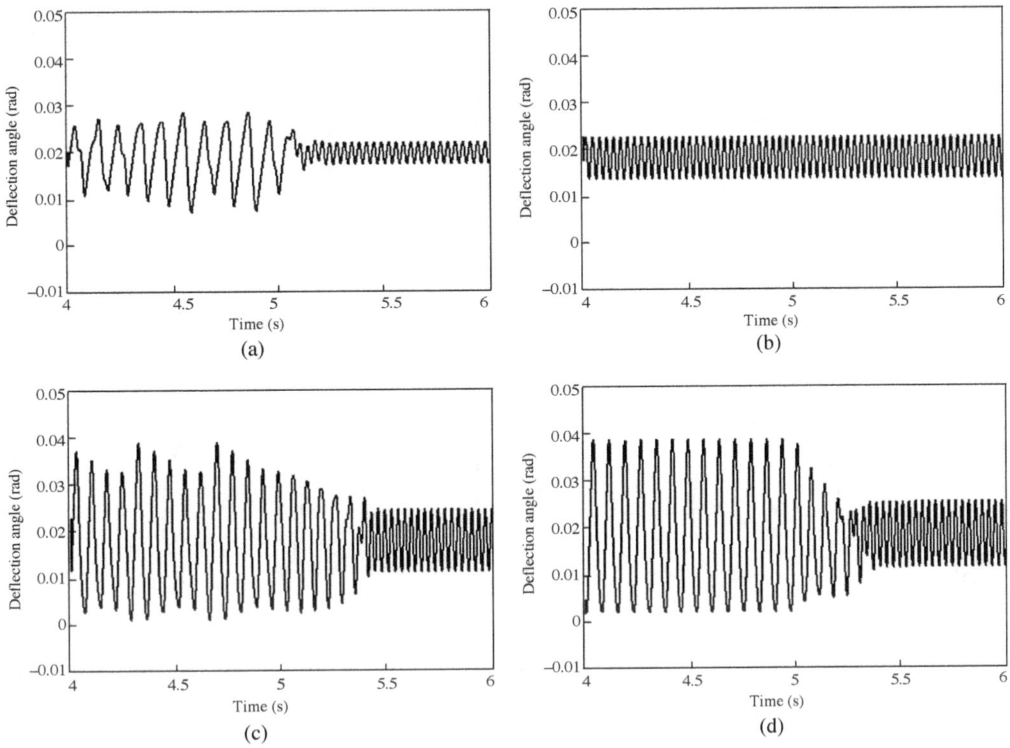

Figure 8.9 Waveforms of deflection angle of a wiper system using ETDAS control under different input voltages. (a) 4 V. (b) 8 V. (c) 10.8 V. (d) 12 V

After substituting $R = 0.95$, $\tau = 0.033$ s, and $K = 6.5$ V/A into (8.18) and (8.19), the time-domain waveforms of θ_D under four typical values of V_{in} are shown in Figure 8.9. It should be noted that the ETDAS control is applied at the instant $t = 5$ s, and all stabilization processes can be successfully completed within 0.4 s. Figure 8.10 shows the corresponding trajectories $\dot{\theta}_D$ versus θ_D. It can be seen that the behavior of the wiper can be stabilized into period-1 motion under all conditions. So, with respect to V_{in} using the ETDAS control with $K = 6.5$ V/A and $\tau = 0.033$ s, the bifurcation diagram of θ_D is plotted in Figure 8.11. Compared with the bifurcation diagram without control, as shown in Figure 8.3, it can be seen that chattering can be stabilized over the whole range of V_{in}.

Experiments are developed to examine the stabilization performance of the ETDAS control. The experimental setup of an automotive wiper assembly is taken from a commercial automobile. It should be pointed out that the modeling of frictional forces between the wiper blades and the screen is based on a flat windshield, which is actually an assumption that is commonly used in the available literature. However, the experimental example is an actual automotive windshield which has a curved surface. The control method is digitally implemented by a TMS320F240 DSP microcontroller which is equipped with an A/D conversion and PWM module. The sampling rate is 10 kHz, and the motor current is sampled once in every switching interval. Since the electrical time constant is much larger than the sampling interval, the motor current can be constructed using this sampling rate. The microcontroller generates proper switching pulses for a switched mode power supply which, in turn, provides a controllable input voltage of the motor. The analog accelerometer ADXL311 is mounted on the wiper to record its acceleration for the purpose of display.

Figure 8.12(a) shows the measured acceleration of the wiper without control under $V_{in} = 12$ V. It can be observed that there exists chattering vibration on the wiper which has also been predicted in Section 8.1.2.

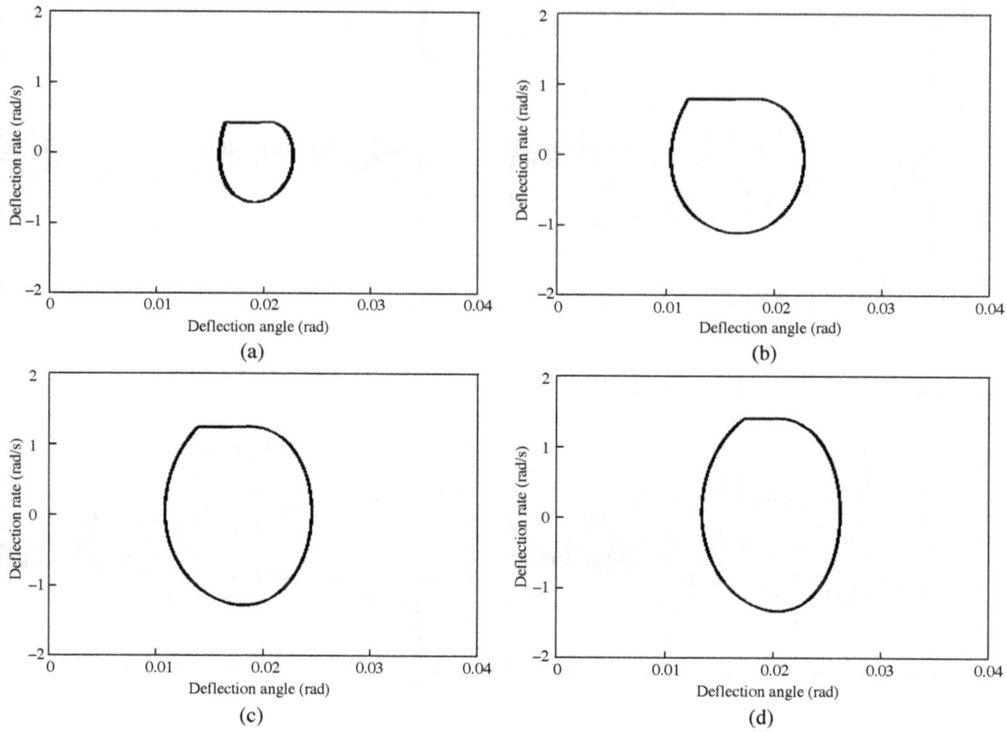

Figure 8.10 Trajectories of deflection rate versus deflection angle of a wiper system using ETDAS control under different input voltages. (a) 4 V. (b) 8 V. (c) 10.8 V. (d) 12 V

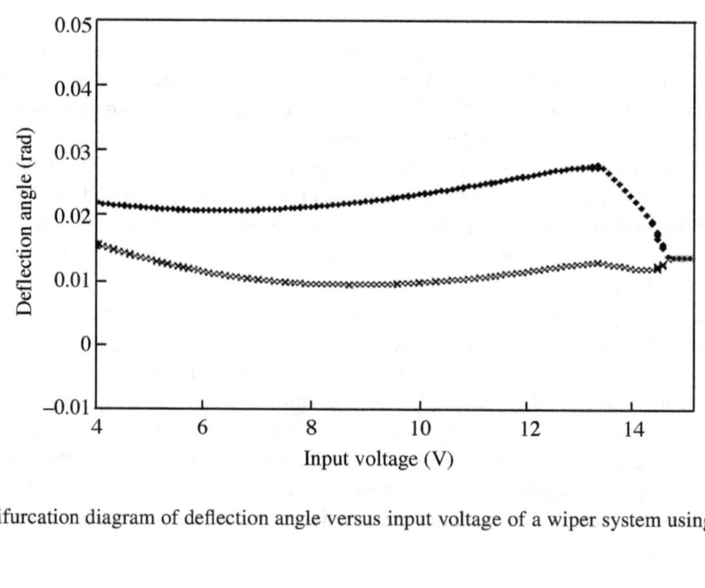

Figure 8.11 Bifurcation diagram of deflection angle versus input voltage of a wiper system using ETDAS control

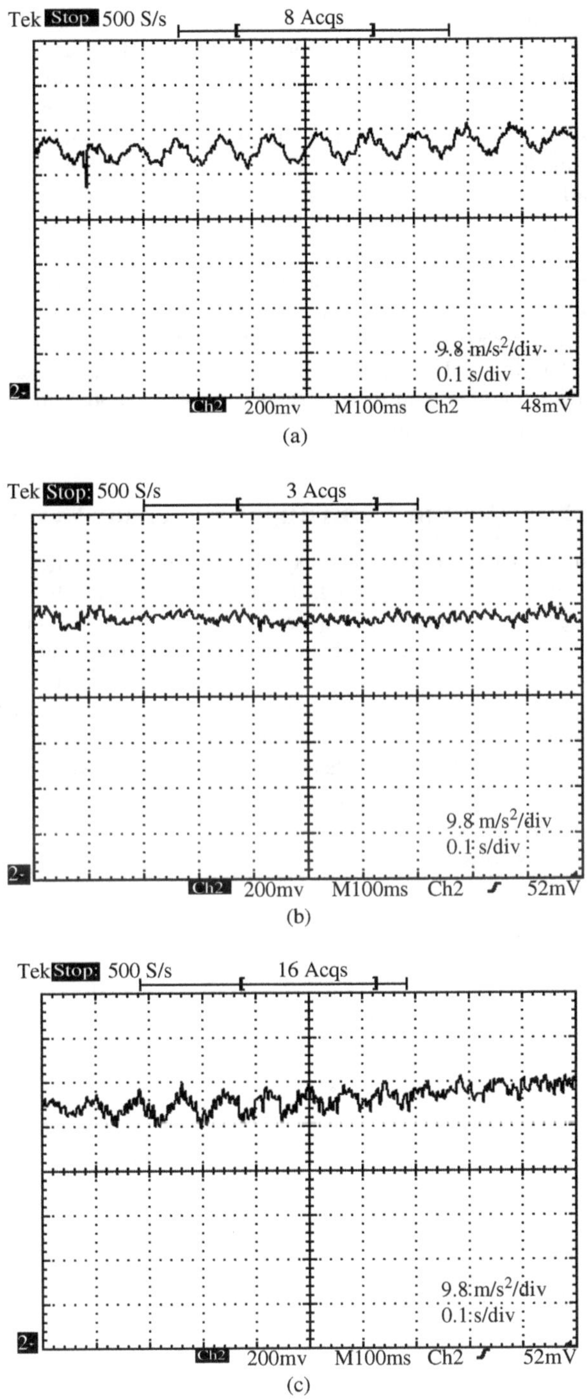

Figure 8.12 Measured acceleration of the wiper arm. (a) Without control. (b) With ETDAS control. (c) With time-delay feedback control

As previously mentioned, this is due to the stick-slip motion of the wipers on the surface of the windshield. Then, the ETDAS control is implemented using a DSP microcontroller to suppress this chattering vibration. In the experimentation, the wet condition of the windshield surface is kept to be almost the same for fair comparison. It should be noted that it is not really possible to match the parameters of the experimental wiper system with those for the aforementioned analysis. So, given $\tau = 0.05$ s, which is smaller than the period of chattering, different values of K are selected for the ETDAS control until the optimal performance for the suppression of chattering is achieved. Thus, Figure 8.12(b) shows the measured acceleration of the wiper using the ETDAS control with $K = 4$ V/A under $V_{in} = 12$ V. It can be observed that the chattering amplitude is significantly suppressed. Also, the time-delay feedback control is implemented for comparison. Again, different values of K and τ are selected until optimal performance is achieved. Figure 8.12(c) shows the performance using a time-delay feedback control with $K = 4$ V/A and $\tau = 0.05$ s under $V_{in} = 12$ V. It can be observed that the time-delay feedback control cannot provide the same performance as the ETDAS control, owing to the fact that its operating range of K is not wide enough to suppress the chattering of this experimental system.

8.2 Chaos Stabilization in Centrifugal Governor Systems

The centrifugal governor functions to control the speed of the engine automatically and prevent the engine from damage caused by a sudden change of load torque. Figure 8.13 shows the configuration of a hexagonal centrifugal governor coupled to an engine (Ge and Lee, 2003). The engine drives the rotational axis of the governor. There are four rods which are joined to the hinges at the two ends of the axis. The upper and lower rods are attached to two fly-balls, and a linear spring is attached to the sleeve. If the speed of the engine drops below the desired speed, the centrifugal force acting on the fly-balls will decrease, the fuel injection control valve will open wider, and, as more fuel will be supplied, the speed of the engine will increase until equilibrium is reached. On the other hand, if the engine speed increases, the fuel supply will be reduced and hence the speed will be reduced accordingly.

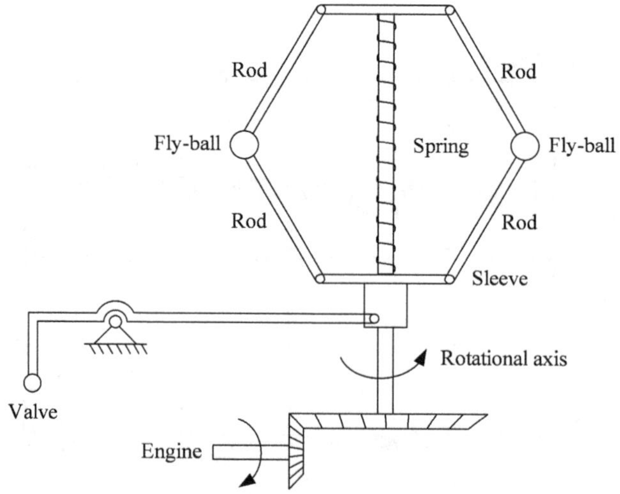

Figure 8.13 Configuration of a centrifugal governor system

It has been identified that chaotic behaviors exist in the centrifugal governor system under a harmonic load torque (Ge and Lee, 2003). Various control methods have been proposed to stabilize this chaotic motion, such as the addition of constant motor torque, the addition of periodic force, the use of a periodic impulse input, a delayed feedback control, an adaptive control, a bang-bang control, an external force control, and an optimal control.

8.2.1 Modeling

The dynamical model of the centrifugal governor system can be obtained by calculating its kinetic and potential energies as follows (Ge and Lee, 2003):

$$\begin{aligned} KE &= 2\times\left\{\frac{1}{2}m\left[(r+l\sin\phi)^2\eta^2+l^2\phi^2\right]\right\} \\ &= m\eta^2(r+l\sin\phi)^2+ml^2\dot{\phi}^2 \end{aligned} \tag{8.21}$$

$$PE = 2kl^2(1-\cos\phi)^2+2mgl(1-\cos\phi) \tag{8.22}$$

where KE is the kinetic energy, PE is the potential energy, m is the mass of each fly-ball, k is the stiffness of the spring, η is the speed of the rotational axis, l is the length of each rod, r is the distance between the rotational axis and the hinge, and ϕ is the angle between the rotational axis and the rod. Thus, the Lagrange equation is given by:

$$\begin{aligned} L &= KE-PE \\ &= m\eta^2(r+l\sin\phi)^2+ml^2\dot{\phi}^2-2kl^2(1-\cos\phi)^2-2mgl(1-\cos\phi) \end{aligned} \tag{8.23}$$

Consequently, the dynamical equation of the governor can be derived as:

$$\begin{aligned} &2[ml^2\ddot{\phi}-mrl\eta^2\cos\phi-(2k+m\eta^2)l^2\sin\phi\cos\phi \\ &+(2kl+mg)l\sin\phi]+c\dot{\phi}=0 \end{aligned} \tag{8.24}$$

where c is the damping coefficient.

For the rotational machine, the net torque is the difference between the engine torque and the load torque. The dynamics of the rotational machine can therefore be expressed as:

$$J\dot{\omega} = \gamma\cos\phi-\beta \tag{8.25}$$

where ω is the speed of the engine, J is the inertia of the rotational machine, $\gamma > 0$ is a proportional constant, and β is an equivalent torque of the load. By defining the time scale $\varsigma = \alpha t$, the dynamical system represented by (8.24) and (8.25) can be rewritten as:

$$\frac{\mathrm{d}\phi}{\mathrm{d}\varsigma} = \varphi \tag{8.26}$$

$$\frac{\mathrm{d}\varphi}{\mathrm{d}\varsigma} = D\omega^2\cos\phi+(E+P\omega^2)\sin\phi\cos\phi-\sin\phi-B\varphi \tag{8.27}$$

$$\frac{\mathrm{d}\omega}{\mathrm{d}\varsigma} = Q\cos\phi-F \tag{8.28}$$

where

$$Q=\frac{\gamma}{J\alpha},\ F=\frac{\beta}{J\alpha},\ D=\frac{n^2mr}{2kl+mg},\ E=\frac{2kl}{2kl+mg},$$

$$P=\frac{n^2ml}{2kl+mg},\ B=\frac{c}{2ml^2\eta},\ \alpha=\sqrt{\frac{2kl+mg}{ml}},$$

and n is the gear ratio between the rotational axis and the engine shaft. Hence, the dynamics of this centrifugal governor system is a three-dimensional autonomous system.

8.2.2 Analysis

When the load torque is harmonic, it can be represented by a constant term and a harmonic term $F+A\sin\kappa t$, where F, A, and κ are constants. Thus, the dynamical equations (8.26)–(8.28) can be expressed as:

$$\frac{d\phi}{d\varsigma}=\varphi \tag{8.29}$$

$$\frac{d\varphi}{d\varsigma}=D\omega^2\cos\phi+(E+P\omega^2)\sin\phi\cos\phi-\sin\phi-B\varphi \tag{8.30}$$

$$\frac{d\omega}{d\varsigma}=Q\cos\phi-F-A\sin\kappa t \tag{8.31}$$

Given the parameters $D=0.08$, $E=0.8$, $P=0.04$, $F=1.942$, $A=0.6$, $B=0.4$, $\kappa=1$, and $\omega=1$, the bifurcation diagram of the deflection angle ϕ with respect to the control parameter Q is plotted as shown in Figure 8.14. It can be observed that the centrifugal governor exhibits periodic motions when Q is small. If Q increases beyond the threshold, chaotic motion will occur in the system. With $Q=2$, the trajectory of φ versus ϕ is plotted as shown in Figure 8.15. As can be seen, the system exhibits a period-1 motion. On the other hand, the trajectory depicted in Figure 8.16 exhibits chaotic motion under $Q=2.6$. These results are in good agreement with the bifurcation diagram shown in Figure 8.14.

8.2.3 Stabilization

Contrary to the previous stabilization methods (Ge and Lee, 2003), the ETDAS control is used to stabilize any chaotic motion in this centrifugal governor system. The principle of the ETDAS control for this system can be represented as:

$$S(t)=R\sum_{\gamma=0}^{\infty}R^{\gamma}\omega(t-\gamma\tau-\tau)+\omega(t)=RS(t-\tau)+\omega(t) \tag{8.32}$$

$$T^*=K\left[(1-R)S(t-\tau)-\omega(t)\right] \tag{8.33}$$

$$\frac{d\omega}{d\varsigma}=Q\cos\phi-F-A\sin\kappa t+K\left[(1-R)S(t-\tau)-\omega(t)\right] \tag{8.34}$$

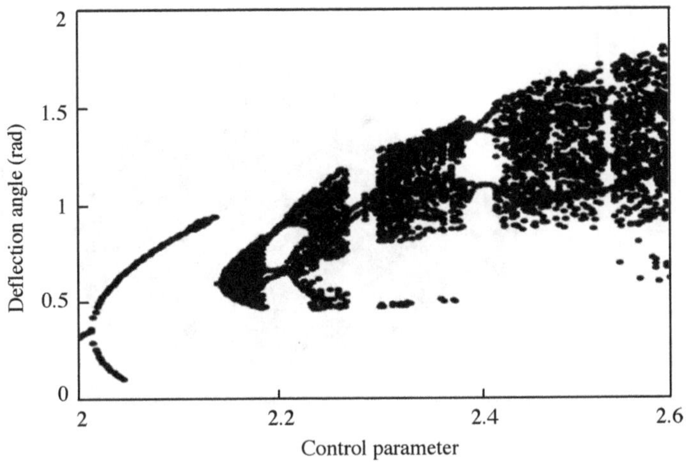

Figure 8.14 Bifurcation diagram of deflection angle versus control parameter of a governor system

where R, τ, and K are selectable parameters for ETDAS control, and T^* is the additional torque imposed on the engine to stabilize the chaotic behaviors in the centrifugal governor system. This additional torque can be easily generated by using an electric drive that is directly coupled to the shaft of the engine. By tuning the feedback gain K in (8.34), a bifurcation diagram of the system under $Q = 2.6$ is obtained, as depicted in Figure 8.17. It can be observed that when K is increased beyond the low threshold, the chaotic motion of the system can be stabilized into period-1 motion. Hence, by setting $K = 50$, the trajectory of φ versus ϕ with the ETDAS control is as plotted in Figure 8.18(a). Also, the transient waveform of ϕ is plotted in Figure 8.18(b). It can be found that any chaotic motion can be stabilized effectively and promptly by the ETDAS control.

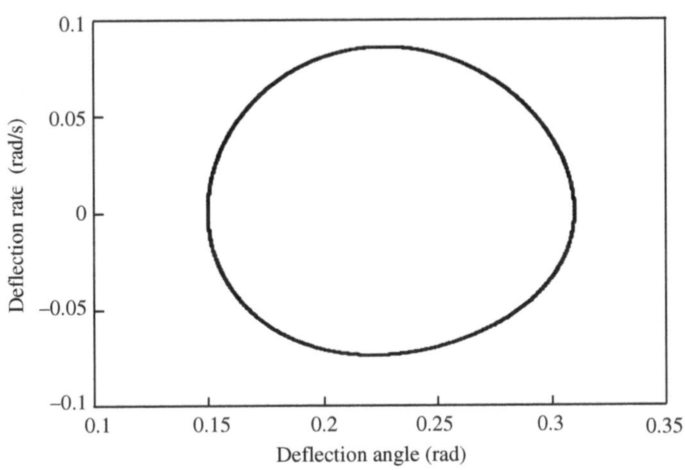

Figure 8.15 Trajectory of deflection rate versus deflection angle of a governor system under $Q = 2$

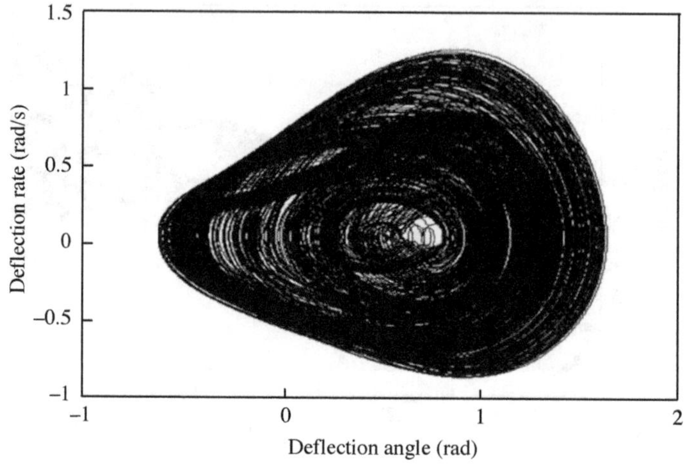

Figure 8.16 Trajectory of deflection rate versus deflection angle of a governor system under $Q = 2.6$

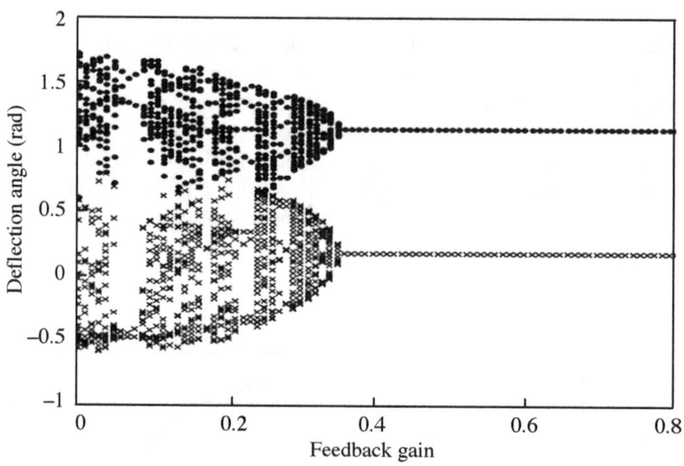

Figure 8.17 Bifurcation diagram of deflection angle versus feedback gain of a governor system using EDTAS control under $Q = 2.6$

8.3 Chaos Stabilization in Rate Gyro Systems

The gyroscope is widely used in the navigation and control system of space vehicles. It functions to measure the angular velocity in spinning space vehicles. Thus, the stability of motion of the gyro is critical for the accurate measurement of angular velocity. Recently, it has been identified that chaotic motion occurs in a rate gyro with feedback control mounted on a space vehicle that is spinning with an uncertain angular velocity (Ge and Chen, 1998).

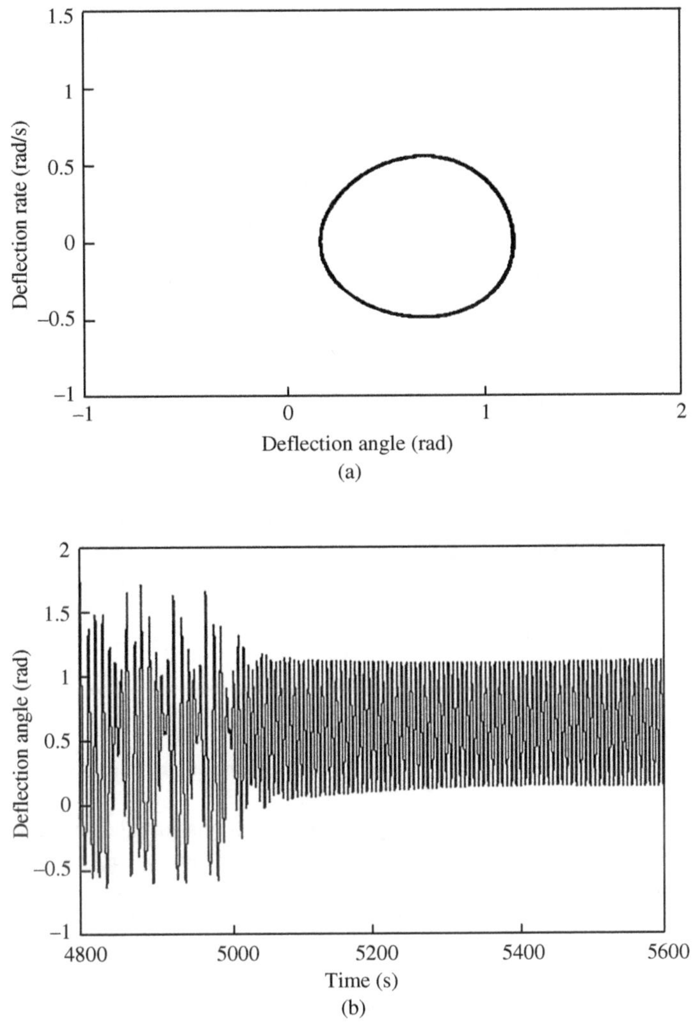

Figure 8.18 Performance of the governor system using EDTAS control under $Q = 2.6$. (a) Trajectory. (b) Waveform

8.3.1 Modeling

The configuration of a single-rate gyro system with feedback control mounted on a space vehicle is shown in Figure 8.19. The gimbal can turn about the output X-axis with a deflection angle θ. The corresponding motion is damped by the damping torque $C_d\dot{\theta}$, where C_d is the damping coefficient. By using the Lagrange equation, the dynamical equation of a rate gyro system with feedback control can be derived as (Chen, 2004):

$$\begin{aligned} T_c = & (A + A_g)\,\ddot{\theta} + C_d\dot{\theta} + C_R(\omega_Y \cos\theta + \omega_Z \sin\theta) \\ & + (A + B_g - C_g)\,(\omega_Y \cos\theta + \omega_Z \sin\theta)(\omega_Y \sin\theta - \omega_Z \cos\theta) \\ & + (A + A_g)\dot{\omega}_X \end{aligned} \tag{8.35}$$

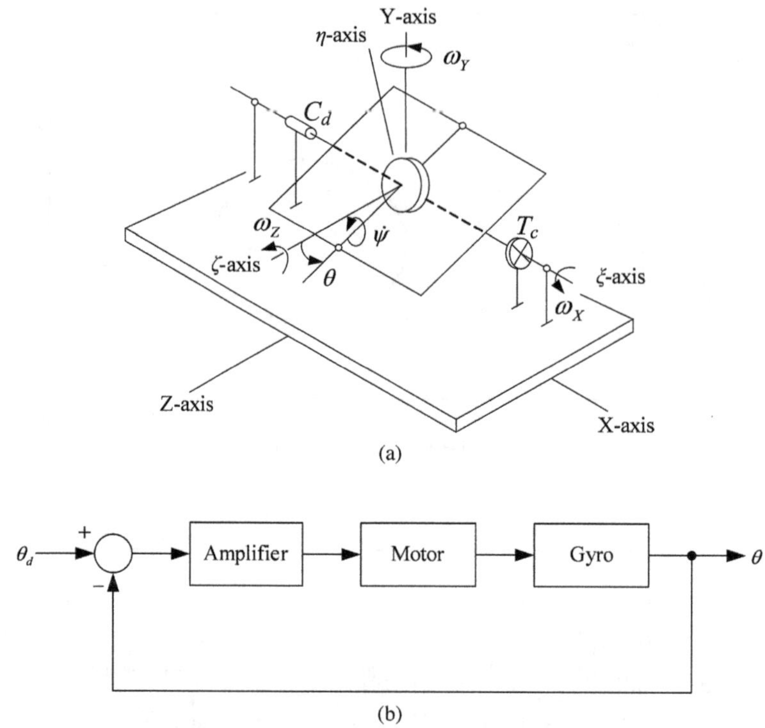

Figure 8.19 Single-rate gyro system. (a) Configuration. (b) Control diagram

where T_c is the controlled motor torque along the output X-axis to balance the gyroscopic torque; $C_R = C(\dot{\psi} - \omega_Y \sin\theta + \omega_Z \cos\theta)$ is a constant; ω_X, ω_Y, and ω_Z are the angular velocity components of the platform along the output X-axis, input Y-axis, and normal Z-axis, respectively; $A, B = A, C$ and A_g, B_g, C_g are the inertias of rotor and gimbals about the gimbal axes ξ, η, ζ, respectively.

Normally, the electric motor is a permanent magnet DC (PMDC) motor. The corresponding dynamical model is given by:

$$T_c = K_T I \tag{8.36}$$

$$L\dot{I} + RI = K_a(\theta_d - \theta) - K_E\dot{\theta} \tag{8.37}$$

where I, R, L, K_E, and K_T are the armature current, armature resistance, armature inductance, back EMF constant, and torque constant of the PMDC motor, respectively, and the corresponding feed-in voltage is proportional to the difference between the desired deflection angle θ_d and the actual deflection angle θ, with the amplifier gain K_a.

The desired motion of the gyro θ_d needs to be fixed at the origin, namely $\theta_d = 0$. In particular, when the space vehicle undergoes an uncertain angular velocity of $\omega_Z(t)$ about the spinning Z-axis, an acceleration of $\dot{\omega}_X(t)$ about the output X-axis, and a zero $\omega_Y(t)$ about the input Y-axis, the dynamical equation of the rate gyro system with feedback control can be rewritten as:

$$\begin{cases} \dot{x} = y \\ \dot{y} = -K_1 y + K_2 z - K_3 \omega_Z(t) \sin x + 1/2 K_4 \omega_Z^2(t) \sin 2x - \dot{\omega}_X \\ \dot{z} = -K_5 z - K_6 x - K_7 y \end{cases} \tag{8.38}$$

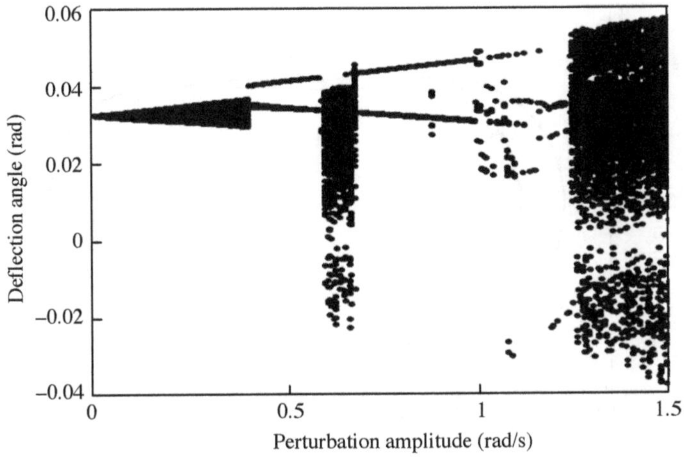

Figure 8.20 Bifurcation diagram of deflection angle versus perturbation amplitude of a gyro system

where $x = \theta$, $y = \dot{\theta}$, $z = I$, $K_1 = C_d/(A + A_g)$, $K_2 = K_T/(A + A_g)$, $K_3 = C_R/(A + A_g)$, $K_4 = (A + B_g - C_g)/(A + A_g)$, $K_5 = R/L$, $K_6 = K_a/L$, and $K_7 = K_E/L$.

8.3.2 Analysis

As $\dot{\omega}_X(t)$ is time-varying and its value is small, it can be assumed to be zero. Also, $\omega_Z(t) = \omega_{Z0} + Z_m \sin \omega t$ is used to represent the uncertain angular velocity with which the rate gyro in the space vehicle spins about its Z-axis. Thus, the dynamical equation of the system can be expressed as:

$$\begin{cases} \dot{x} &= y \\ \dot{y} &= -K_1 y + K_2 z - K_3(\omega_{Z0} + Z_m \sin \omega t) \sin x \\ & \quad + 1/2K_4(\omega_{Z0} + Z_m \sin \omega t)^2 \sin 2x \\ \dot{z} &= -K_5 z - K_6 x - K_7 y \end{cases} \tag{8.39}$$

Given the parameters $K_1 = 1$, $K_2 = 10$, $K_3 = 2000$, $K_4 = 1$, $K_5 = 25$, $K_6 = 250$, $K_7 = 1$, and $\omega_{Z0} = 2000$, the bifurcation diagram of θ versus the perturbation amplitude Z_m is plotted as shown in Figure 8.20. It can be observed that chaotic behaviors occur in some regions of Z_m. By setting $Z_m = 0.5$ rad/s and $Z_m = 1.31$ rad/s, the trajectories of $\dot{\theta}$ versus θ of the system are plotted, as shown in Figures 8.21 and 8.22, respectively. This exhibits periodic motion under $Z_m = 0.5$ rad/s, but chaotic motion under $Z_m = 1.31$ rad/s.

8.3.3 Stabilization

Similar to the stabilization of the automotive wiper system and centrifugal governor system, the ETDAS control is applied to stabilize the chaotic motion in this rate gyro system. The principle of the ETDAS for this system can be expressed as:

$$S(t) = R \sum_{\gamma=0}^{\infty} R^{\gamma} y(t - \gamma\tau - \tau) + y(t) = RS(t - \tau) + y(t) \tag{8.40}$$

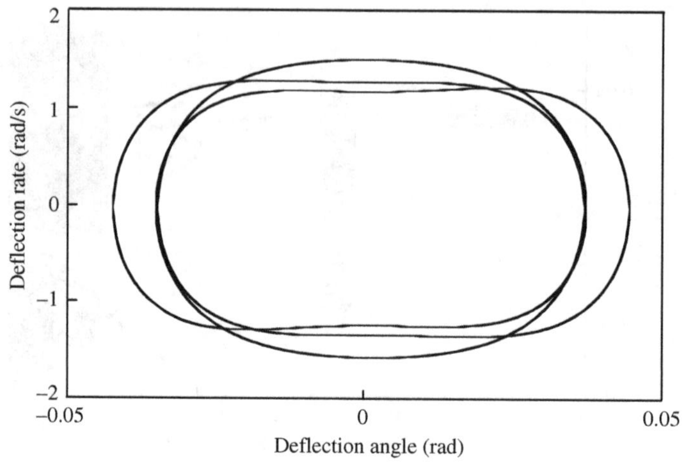

Figure 8.21 Trajectory of deflection rate versus deflection angle of a gyro system under $Z_m = 0.5$ rad/s

$$V_{in}^* = K\left[(1-R)S(t-\tau)-y(t)\right] \tag{8.41}$$

$$\dot{z} = -K_5 z - K_6 x - K_7 y + K\left[(1-R)S(t-\tau)-y(t)\right] \tag{8.42}$$

where R, τ. and K are tunable parameters for the ETDAS control, and V_{in}^* is the feed-in voltage of the electric motor.

By tuning the control parameter K, a bifurcation diagram of the system with EDTAS control under $Z_m = 1.31$ rad/s is depicted in Figure 8.23. It can be observed that the chaotic motion of the system can be stabilized into periodic motion within some regions of K. Namely, when K is chosen to be 50, the chaotic motion under $Z_m = 1.31$ rad/s can be effectively stabilized into a period-2 motion, as shown in Figure 8.24.

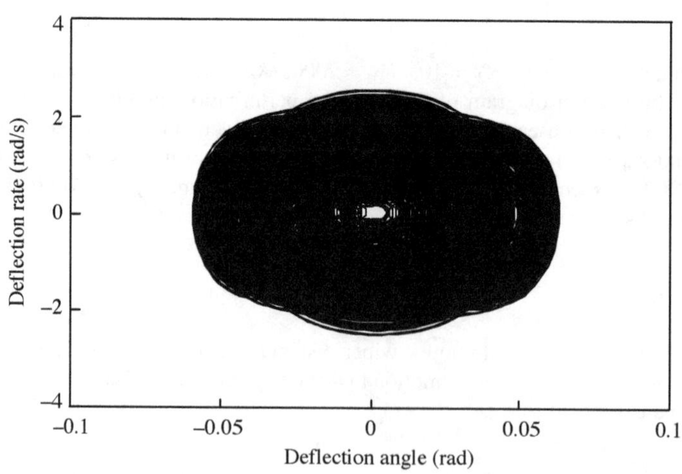

Figure 8.22 Trajectory of deflection rate versus deflection angle of a gyro system under $Z_m = 1.31$ rad/s

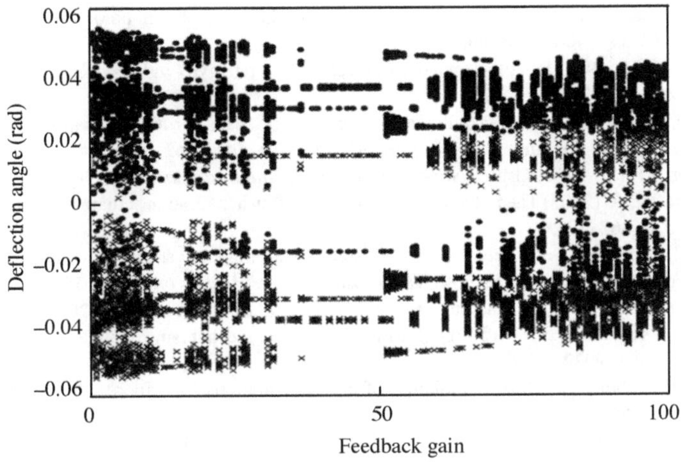

Figure 8.23 Bifurcation diagram of deflection angle versus feedback gain of a gyro system using ETDAS control under $Z_m = 1.31$ rad/s

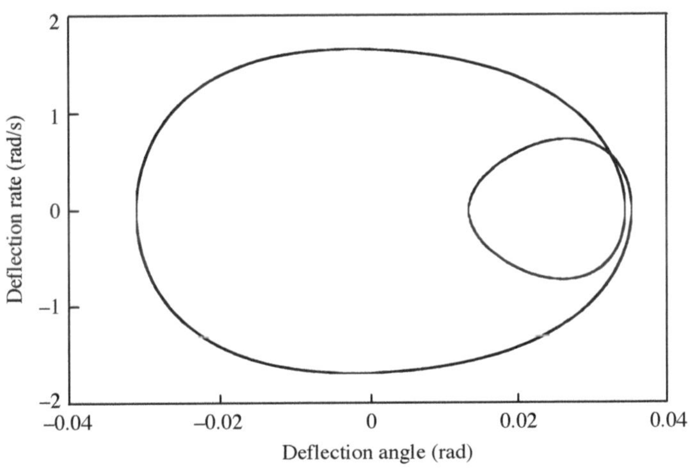

Figure 8.24 Trajectory of deflection rate versus deflection angle of a gyro system using ETDAS control under $Z_m = 1.31$ rad/s

References

Chang, S.C. and Lin, H.P. (2004) Chaos attitude motion and chaos control in an automotive wiper system. *International Journal of Solids and Structures*, **41**, 3491–3504.

Chen, H.H. (2004) Stability and chaotic dynamics of a rate gyro with feedback control under uncertain vehicle spin and acceleration. *Journal of Sound and Vibrations*, **273**, 949–968.

Chen, J.H., Chau, K.T., Siu, S.M., and Chan, C.C. (2000) Experimental stabilization of chaos in a voltage-mode DC drive system. *IEEE Transactions on Circuits and Systems - I: Fundamental Theory and Applications*, **47**, 1093–1095.

Franceschini, G., Bose, S., and Schöll, E. (1999) Control of chaotic spatiotemporal spiking by time-delay autosynchronization. *Physical Review E*, **60**, 5426–5434.

Ge, Z.M. and Chen, H.H. (1998) Double degeneracy and chaos in a rate gyro with feedback control. *Journal of Sound and Vibration*, **209**, 753–769.

Ge, Z.M. and Lee, C. (2003) Non-linear dynamics and control of chaos for a rotational machine with a hexagonal centrifugal governor with a spring. *Journal of Sounds and Vibrations*, **262**, 845–864.

Hsu, B.S. and Ling, S.F. (1990) Windshield wiper system design. *International Journal of Vehicle Design*, **11**, 63–78.

Owen, W.S. and Croft, E.A. (2003) The reduction of stick-slip friction in hydraulic actuators. *IEEE/ASME Transactions on Mechatronics*, **8**, 362–371.

Park, J., Chang, P.H., Park, H.S., and Lee, E. (2006) Design of learning input shaping technique for residual vibration suppression in an industrial robot. *IEEE/ASME Transactions on Mechatronics*, **11**, 55–65.

Pyragas, K. (1995) Control of chaos via extended delay feedback. *Physics Letters A*, **206**, 323–330.

Goto, S., Takahashi, H., and Oya, T. (2001) Clarification of the mechanism of wiper blade rubber squeal noise generation. *JSAE Review*, **22**, 57–62.

Grenouillat, R. and Leblanc, C. (2002) Simulation of chatter vibrations for wiper systems. *Society of Automotive Engineers Papers*, **2002-01-1239**, 1–8.

Lévine, J. (2004) On the synchronization of a pair of independent windshield wipers. *IEEE Transactions on Control System Technology*, **12**, 787–795.

Straete, H.J.V., Degezelle, P., Schutter, J.D., and Belmans, R.J.M. (1998) Servo motor selection criterion for mechatronic applications. *IEEE/ASME Transactions on Mechatronics*, **3**, 43–50.

Suzuki, R. and Yasuda, K. (1998) Analysis of chatter vibration in an automotive wiper assembly. *JSME International Journal Series C*, **41**, 616–620.

9

Application of Chaotic Modulation

With the advent of pulse width modulation (PWM), switching power converters have received a great attraction for application in electric drive systems, because of the advantages of flexible power control, compact size, and high efficiency. In general, the PWM DC–DC converter (usually called the DC chopper) functions to control the applied voltage or current for DC drive systems, whereas the PWM DC–AC converter (usually called the AC inverter) is to simultaneously control the applied voltage or current and frequency for AC drive systems. However, because of the nature of switching, those advantages are counterbalanced by the generation of harmonics, electromagnetic interference (EMI), and acoustic noise. Increasingly, there is a trend for pushing up the switching frequency of modern electric drive systems, hence reducing their volume and weight. This trend inevitably contributes to an increasing level of EMI. It also degrades the electromagnetic compatibility (EMC) of electronic devices.

Conventionally, the aforementioned EMI problems are alleviated by filtering the output or shielding the setup. In recent years, attention has been focused on using signal processing rather than filtering or shielding – namely, developing various PWM schemes for inverter-fed AC drive systems.

In this chapter, chaos is applied to PWM schemes for inverter-fed AC drive systems, hence reducing the corresponding audible noise and mechanical vibration. The open-loop and closed-loop control of chaotic PWM inverter drive systems are discussed.

9.1 Overview of PWM Schemes

PWM schemes have been the subject of intensive research during the last few decades. A large variety of methods, different in concept and performance, have been developed and described. As there are many power electronics books and survey papers comprehensively discussing various PWM schemes (Bowes and Clements, 1982; Holtz, 1992), this section aims to give a brief overview of various PWM schemes.

Basically, PWM schemes can be categorized as voltage control and current control. The voltage-controlled schemes generally operate in an open-loop feed-forward fashion, while the current-controlled schemes generally operate in a closed-loop feedback fashion.

9.1.1 *Voltage-Controlled PWM Schemes*

Over the years, numerous voltage-controlled PWM schemes have been developed. In this section, five representative schemes that are widely accepted for inverter-fed AC drive systems are selected for discussion: namely, the sinusoidal PWM, regular PWM, optimal PWM, delta PWM, and random PWM.

Chaos in Electric Drive Systems: Analysis, Control and Application, First Edition. K.T. Chau and Zheng Wang.
© 2011 John Wiley & Sons (Asia) Pte Ltd. Published 2011 by John Wiley & Sons (Asia) Pte Ltd.

9.1.1.1 Sinusoidal PWM

The sinusoidal PWM is the earliest PWM scheme and was developed for analog implementation (Mokrytzki, 1967; Bowes, 1975). Basically, a triangular carrier wave (known as a sampling signal) is compared directly with a sinusoidal modulating wave (known as a reference signal) to determine the switching instants and, hence, the resultant pulse widths. Since the switching edges of the pulse are determined by the instantaneous intersection of the two waves, the resultant pulse widths are proportional to the amplitudes of the modulating wave at the switching instants. This causes the centers of the pulses in the PWM waveform to be unequally spaced, and presents difficulty in analytically expressing the pulse widths of the PWM waveform. Actually, such pulse widths can only be expressed by using a transcendental equation, and in terms of a series of Bessel functions. So, it is not feasible to calculate the pulse widths directly in real time.

This sinusoidal PWM takes the advantages of real-time generation using low-cost analog hardware, linear amplification, and constant average switching frequency. However, it is ill-suited for digital implementation, which is actually the trend of development for modern electric drives.

9.1.1.2 Regular PWM

The regular PWM is recognized to have a definite advantage over the sinusoidal PWM when implemented using digital or microprocessor techniques (Bowes, 1975; Bowes and Mount, 1981). Basically, the amplitude of the modulating wave at the sample instant is stored by a sampling circuit (which is also operated at the carrier frequency), and is held constant until the next sample is taken. This produces a sample-and-hold version of the modulating wave. Then, the intersection of this "staircase" modulating wave and the triangular carrier wave determine the switching instants, and hence the pulse widths. Since the staircase modulating wave has constant amplitude when each sample is taken, the pulse widths are proportional to the amplitudes of the modulating wave at regularly spaced sampling intervals.

It is an important feature of regular PWM that the sampling positions and sampled values can be defined unambiguously – that is, the pulses produced are well defined both in width and position, which is not the case for the sinusoidal PWM. Because of this feature, it is possible to derive a simple trigonometric function to calculate the pulse widths of regular PWM. Hence, the real-time generation of a regular PWM can readily be performed by using a low-cost microprocessor or microcontroller.

Rather than modulating the pulse edge symmetrically, the regular PWM can be further extended to modulate the pulse edges by different amounts. Namely, the leading and trailing edges of each pulse are determined by using two different samples of modulating wave. Although the pulse widths of this asymmetric regular PWM waveform can still be expressed as a simple trigonometric function, the required number of calculations is almost double that for the original regular PWM waveform. This will significantly increase the computational time for PWM generation, and thus reduce the maximum allowable inverter output frequency. Nevertheless, since more information about a modulating wave is associated with an asymmetric regular PWM waveform, its harmonic spectrum is superior to that produced using the symmetric one. Because of the sample-and-hold process, the performance of a regular PWM is inevitably inferior to that of a sinusoidal PWM, especially at low pulse numbers.

9.1.1.3 Optimal PWM

The optimal PWM offers the definite advantage of the capability of optimization (Buja and Indri, 1977). With the advent of microprocessor technology, the implementation of this scheme is becoming feasible. Contrary to both sinusoidal PWM and regular PWM, which are generated on the basis of well-defined modulation processes, the optimal PWM is first defined by a general PWM waveform in terms of a set of switching angles which are then determined using numerical methods.

A typical optimal PWM waveform has an odd pulse number N and is of quarter-wave symmetry. Thus, only odd harmonics exist. Optimization is generally classified as harmonic elimination and an objective function. For harmonic elimination, it aims to eliminate a well-defined number of lower order harmonics from the spectrum. Hence, it can eliminate all torque harmonics having six times the fundamental frequency at $N=5$ and so on (Patel and Hoft, 1973). A well-accepted objective function is the minimization of the harmonic current distortion or the peak current at very low pulse numbers. The objective function that defines the optimization problem generally exhibits a large number of local minimums, which makes the numerical calculation of the optimal switching angles extremely time consuming, even with modern computers. These angles can subsequently be preprogrammed into the microprocessor memory and used to generate the optimal PWM waveform in real-time.

9.1.1.4 Delta PWM

The delta PWM is particularly attractive for low-cost inverter-fed AC drive systems in which variable-voltage variable-frequency (VVVF) control is adopted (Ziogas, 1981). Basically, it utilizes a sinusoidal modulating wave and a delta-shaped carrier wave which is forced to oscillate within a predefined window extending equally above and below the sinusoidal modulating wave. The minimum window width and the maximum carrier slope determine the maximum switching frequency of the resultant PWM waveform. This forced oscillation ensures that the fundamental components of the carrier wave and the modulating wave have the same amplitude, and the dominant harmonics of the carrier wave and the resultant PWM waveform cluster close to the carrier frequency. Since the ratio of the fundamental component to the frequency of the resultant PWM waveform is directly proportional to the amplitude of the modulating wave, this ratio is constant until the output frequency reaches its critical value – that is, the base frequency at which the PWM waveform becomes a square wave. After the base frequency, the voltage amplitude remains constant.

The delta PWM offers three advantageous intrinsic features. Firstly, it provides inherent constant volts per hertz control for output frequencies below the base frequency, and constant voltage above the base frequency. Also, the transition between these two modes of operation is inherently smooth. It should be noted that all other PWM schemes have to rely on additional control circuitry or algorithms to perform VVVF control. Secondly, as the corresponding dominant harmonic frequencies are close to the carrier frequency, low-order harmonics are attenuated. Thirdly, it can be implemented by using a very simple and low-cost circuit, which involves only three operational amplifiers. Nevertheless, the delta PWM suffers from some drawbacks, such as a reduction of the fundamental output voltage and the possible generation of subharmonics (Rahman, Quacioe, and Chowdhury, 1987).

9.1.1.5 Random PWM

The random PWM is particularly attractive to suppress the annoying acoustic noise in AC drive systems, which is caused by the interaction of the fundamental and harmonic flux densities in the motor (Habetler and Divan, 1991; Blaabjerg *et al.*, 1996). Basically, by randomly modulating the triangular carrier wave in sinusoidal PWM, the spectral tones around the switching frequency are spread out with a subsequent reduction in peak values, hence eliminating the annoying whine. Although the total level of the acoustic noise emitted by the motor remains constant, the acoustic noise is more pleasing to the ear since the noise is now random. This random PWM can maintain the advantages of sinusoidal PWM, including the capability of real-time generation, linear amplification, and constant average switching frequency.

The adoption of a very high bandwidth white noise to modulate the triangular carrier wave can result in a very wide spectrum for the PWM waveform, including substantial content at low frequencies. These low-order harmonics are difficult to filter, and will cause low-frequency currents flowing in the motor. In order to eliminate these low-order harmonics, the noise source needs to be a pink noise in which the power spectral density is inversely proportional to the frequency, and each octave carries an equal amount of noise power. Apart from controlling the bandwidth and amplitude of the noise, it may also be desired to vary other characteristics of the noise, such as adding pre-emphasis. However, the use of such techniques to reduce the low-order harmonics definitely complicates the implementation of random PWM.

The major shortcoming of random PWM is that it is difficult to generate a true random signal. Also, the corresponding switching frequency may clash with the system's natural frequency, which inevitably increases the possibility of creating mechanical resonance (Lo *et al.*, 2000).

9.1.2 Current-Controlled PWM Schemes

Although voltage-controlled PWM schemes have been widely accepted for industrial AC drive systems, current control is becoming more and more attractive because of the following advantages (Kazmierkowski and Malesani, 1998):

- direct control of instantaneous current and hence the developed torque;
- inherent peak current protection;
- inherent overload rejection;
- good dynamic response;
- compensation for load parameter variations;
- compensation for device voltage drops and converter dead times;
- compensation for voltage changes.

In recent years, many current-controlled PWM schemes have been developed. In this section, two of the most representative schemes are selected for discussion – namely, hysteresis-band PWM and space vector PWM, which are well accepted for high-performance AC drive systems.

9.1.2.1 Hysteresis-Band PWM

The hysteresis-band PWM is the most popular current-controlled PWM scheme for inverter-fed AC drive systems because of its simple hardware implementation, fast response, and inherent peak current limiting capability (Plunkett, 1979). Basically, the motor phase currents are measured and compared with the respective sinusoidal command currents so that the resulting errors are fed into the hysteresis-band current controllers which, in turn, drive the PWM inverter. Consequently, the motor phase currents are forced to swing between the upper and lower hysteresis bands and hence track with the sinusoidal command currents.

The hysteresis bands are normally fixed and are the same for all phases. However, the use of fixed hysteresis bands has the drawback that the modulation frequency varies within a band. As a result, the phase current contains rich harmonics, which cause additional machine heating. The difficulty of vector conversion of those harmonic-rich feedback currents also causes control problems of the drive system. Moreover, since the fixed hysteresis band has to be designed on the worst-case basis, the drive system generally operates with nonoptimal phase current ripples (Bose, 1990). Therefore, adaptive control may be incorporated into this hysteresis-band PWM in which the band is modulated as a function of system parameters to maintain an almost constant modulation frequency.

9.1.2.2 Space Vector PWM

The space vector PWM is well accepted for high-performance AC drive systems because it can offer low harmonic current content in steady state and fast current response in transient state (Nabae, Ogasawara, and Akagi, 1986). Traditionally, these two requirements contradict one another. This space vector PWM utilizes the current deviation vector to satisfy both requirements. For steady-state operation, the switching mode with the smallest current-deviation derivative is chosen to suppress the current harmonic content and hence the torque ripple and acoustic noise. For transient operation, the switching mode with the largest current deviation derivative is chosen to produce a high-speed current response.

Basically, the inverter output voltage is expressed as eight possible voltage vectors, based on the on–off state of the six switching devices. After detecting the back EMF vector, the current-deviation derivative can be deduced, which is the most important control variable to command the harmonic current content or current response. However, this space vector PWM may suffer from inaccurate estimation of the back EMF vector due to the derivative of high-frequency current components.

In recent years, many improved versions of space vector PWM have been developed, such as the use of three-level hysteresis comparators to select proper voltage vectors via switching electrically programmable read-only memory (EPROM) table (Kazmierkowski, Dzieniakowski, and Sulkowski, 1991), or the incorporation of an asymmetrical modulating function and randomly varied pulse rate to improve the power spectrum and reduce switching losses (Trzynadlowski, Kirlin, and Legowski, 1997).

9.2 Noise and Vibration

In general, the total conducted EMI is caused by two mechanisms: the common-mode (CM) noise is related to the capacitive coupling of voltages with the line impedance stabilizing network (LISN), and the differential-mode (DM) noise is related to the voltage difference among phases. It has been verified that the high slew rate ($\mathrm{d}v/\mathrm{d}t$) of the PWM inverter voltages are mainly responsible for the conducted EMI in electric drives. Figure 9.1 illustrates the conducted EMI of an induction drive system fed from a voltage source PWM inverter. The CM current flows between the phases and the ground, and the corresponding excitation source is given by $V_{com} = (V_A + V_B + V_C)/3 + V_d/2$ (Ran *et al.*, 1998b). On the other hand, the DM current flows between different phases, and the corresponding excitation source is given by $V_{dif} = V_i - V_j$ $(i, j = A, B, C)$ (Ran *et al.*, 1998a).

The mechanical vibration of the induction drive system is mainly due to the electromagnetic force that occurred at the stator inner surface of the induction motor (Stemmler and Eilinger, 1994). The corresponding electromagnetic force density $D_f(t, \varphi)$, which is both time t and position φ dependent,

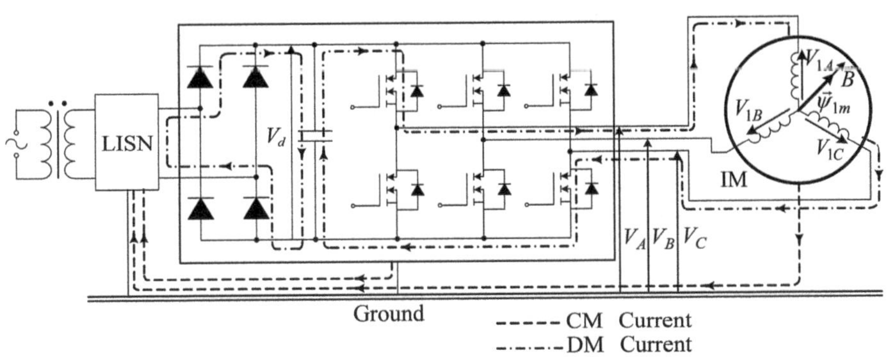

Figure 9.1 Conducted EMI in induction drive system

can be expressed as:

$$D_f(t,\varphi)=\frac{B(t,\varphi)^2}{2\mu_0}=\sum_{m=0}^{\infty}D_m\cos(\omega_m t-m\varphi+\varphi_{m0}) \tag{9.1}$$

$$B(t,\varphi)=\mathrm{Re}\left(\vec{B}\,e^{-j\varphi}\right) \tag{9.2}$$

where

$$\vec{B}=\frac{\vec{\psi}_m}{wdl} \tag{9.3}$$

$$\vec{\psi}_m=\frac{\vec{V}_1(1-\sigma_s)}{j\omega_{s1}}+\frac{\vec{V}_n(1-\sigma_s)\sigma_r}{j\omega_{sn}(\sigma_s+\sigma_r-\sigma_s\sigma_r)} \tag{9.4}$$

$$\begin{aligned}\vec{V}&=\vec{V}_1+\vec{V}_n\\&=\frac{2}{3}\left(V_A+V_B e^{j\frac{2\pi}{3}}+V_C e^{-j\frac{2\pi}{3}}\right)\end{aligned} \tag{9.5}$$

where $\vec{V}_1$ and $\vec{V}_n$ are respectively the fundamental and nth harmonic components of the inverter output voltage vector; σ_s and σ_r are respectively the leakage factor of the stator and rotor; ω_{s1} and ω_{sn} are respectively the fundamental and nth harmonic components of the angular speed of the inverter output voltage vector; w, d and l are respectively the stator number of turns of phase winding, inner diameter and active length; $\vec{\psi}_m$, $\vec{B}$ and $B(t,\varphi)$ are respectively the magnetic flux linkage vector, magnetic flux density vector and magnetic flux density at t and φ; and D_m, ω_m, and φ_{m0} are respectively the magnitude, angular frequency, and initial phase angle of the electromagnetic force of each vibration mode m. From (9.1)–(9.5), it can be observed that $D_f(t,\varphi)$ is governed by the inverter output voltages V_A, V_B, and V_C. If the spectrum of $D_f(t,\varphi)$ overlaps with the natural frequencies of the induction motor, mechanical resonance will occur, causing annoying audible noise and even disastrous mechanical damage. Thus, a proper PWM scheme is highly desirable to avoid the mechanical resonance in the induction drive system.

To design a PWM strategy to avoid the mechanical resonance, the natural frequencies of the induction motor should be predicted in advance. In general, the natural frequencies of the stator of the induction motor can be determined by the motor parameters (Tímár, 1989):

$$f_{0r}=\frac{83750}{R_{ay}\sqrt{\Delta}}\quad(m=0) \tag{9.6}$$

$$f_{1r}=\sqrt{\frac{3Ed^4}{32000\pi^2[l_r(4R_r^2-d^2)+l_b d^2]l_t^3}}\quad(m=1) \tag{9.7}$$

$$f_{1m}=f_{0r}\sqrt{\frac{B}{2A}\left(1\pm\sqrt{1-\frac{4AC}{B^2}}\right)}\quad(m\geq 2) \tag{9.8}$$

where

$$A = 1 + \frac{h_y^2}{12R_{ay}^2}\left[m^2\left(3 + \frac{\Delta_m}{\Delta}\right) - 3\right] \tag{9.9}$$

$$B = \frac{h_y^2}{12R_{ay}^2}(m^2 - 1)\left[m^2\left(4 + \frac{\Delta_m}{\Delta}\right) + 3\right] + (m^2 + 1) \tag{9.10}$$

$$C = \frac{h_y^2}{12R_{ay}^2} m^2 (m^2 - 1)^2 \tag{9.11}$$

$$\Delta = 1 + \frac{G_{t1} + G_{w1}}{G_{y1}} \tag{9.12}$$

$$\Delta_m = 1 + \frac{S_n l_t b_t}{2\pi R_{ay} h_y^3}(4l_t^2 + 6l_t h_y + 3h_y^2) \tag{9.13}$$

where R_{ay} is the stator yoke mean radius, h_y is the stator yoke height, E is the modulus of elasticity for iron, R_r is the outside radius of rotor core, l_r is the length of rotor core, l_b is the bearing distance, d is the shaft diameter, G_{t1} is the stator tooth weight, G_{w1} is the stator winding weight, G_{y1} is the stator yoke weight, l_t is the stator tooth length, b_t is the mean width of stator tooth, and S_n is the stator slot number.

9.3 Chaotic PWM

Among the many available PWM schemes, random PWM is becoming more and more attractive. Random PWM can effectively spread the discrete spectral power over a continuous spectrum and significantly suppress the maximum acoustic noise that occurs at the switching frequency. However, it generally ignores the consideration of low-order harmonic frequencies and the system natural frequency, thus introducing low-order noises and increasing the possibility of creating mechanical resonance.

Since chaos possesses a random-like but bounded feature, chaotic PWM is becoming an alternative to (or even supersedes) random PWM for application to electric drive systems. Chaotic PWM is useful for both DC and AC motor drives. Among them, the application to the induction drive system is more desirable and challenging, since it has long been affected by the EMI problem and acoustic noise. Recently, it has been identified that the use of chaotic PWM to replace sinusoidal PWM can reduce the EMI in drive systems (Bellini *et al.*, 2001; Balestra *et al.*, 2004). This chaotic PWM scheme employs a Bernoulli shift map to chaoize an amplitude-modulated signal which then modulates the carrier frequency of sinusoidal PWM. The corresponding harmonic reduction can not only reduce the size of power filters, but also suppress the acoustic noise in PWM drive systems.

Compared with random PWM, the available chaotic PWM offers the following advantages:

- Some chaotic signals can provide better spectral performance than some random signals, namely better range resolution than the Gaussian distributed signal and equal to the range resolution of the uniformly distributed signal (Ashtari *et al.*, 2003).
- Chaotic PWM exhibits a 9–14 dB improvement over random PWM which randomizes the pulse positions and/or widths (Balestra *et al.*, 2004).
- Chaotic PWM performs no worse than random PWM in terms of harmonic spectra, whereas the implementation of chaotic generators is much easier than that of random generators. Notice that chaos

can be governed by a simple equation, while a chaotic generator can be easily implemented by a few tens of transistors (Delgado-Restituto and Rodíguez-Vázquez, 2002).

- In the case of fast modulations, chaotic PWM can outperform random PWM in terms of harmonic spectra (Callegari, Rovatti, and Setti, 2002).
- The truly random sources are rigorously unrealizable. All the viable random techniques only generate pseudorandom sequences. Sophisticated pseudorandom sources enable a roughly continuous spectrum, but it will be more costly (Callegari, Rovatti, and Setti, 2003). Chaos has an inherent continuous spectrum which is very similar to that of true randomness. The correlation between the chaotic series will be decreased with the elapse of time. Consequently, there is no performance loss for the chaotic series to take the place of the truly random sources.

Although the available chaotic PWM offers a better advantage of easier implementation than random PWM, which needs a truly random source, it is inflexible to tune the spectral power distribution and is limited to those chaotic maps satisfying some specific characteristics, namely the mixing rate and probability density function. Also, the corresponding switching frequency may clash with the natural frequency of mechanical resonance.

In this section, two newly developed chaotic modulation schemes (Wang and Chau, 2007), namely chaotically amplitude-modulated frequency modulation (CAFM) and chaotically frequency-modulated frequency modulation (CFFM) are introduced. They are then used to modulate the switching frequency of both sinusoidal PWM and space vector PWM (Wang, Chau, and Liu, 2007). Consequently, they are applied to both the open-loop control and the closed-loop control of induction drive systems (Wang, Chau, and Cheng, 2008).

The two chaotic modulation schemes, namely CAFM and CFFM, are derived from the standard sinusoidal frequency modulation. Figure 9.2 shows the corresponding schematic diagram in which v_{pwm} is the PWM signal for switching the power inverter, v_r is the reference signal for power flow control, and v_c is the carrier signal operated at a frequency much higher than the bandwidth of v_r. They are related by:

$$v_{pwm} = \operatorname{sgn}(v_r - v_c) \tag{9.14}$$

For induction drive systems, v_r is generally expressed as:

$$v_r = M \sin 2\pi f t \tag{9.15}$$

where M is the modulation index as defined by $\sqrt{2}V/V_d$, V is the desired output voltage, V_d is the input DC voltage, and f is the desired output frequency. These two parameters are generally controlled in such a way

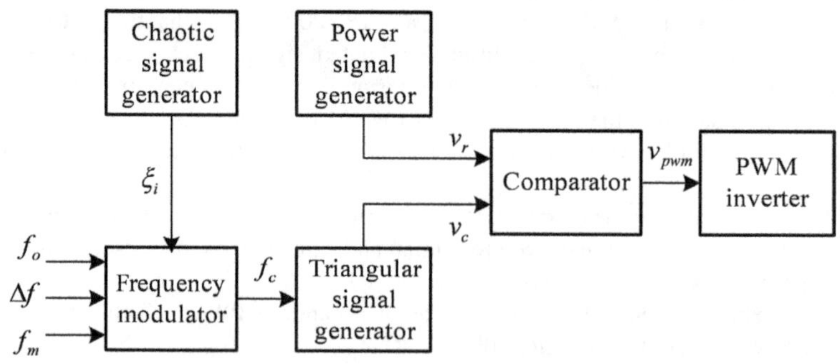

Figure 9.2 Block diagram of chaotic PWM

that the ratio V/f is kept constant at speeds below the rated speed for so-called constant-torque operation, or V is kept at the rated voltage while f is increased beyond the rated frequency at speeds above the rated speed for so-called constant-power operation. On the other hand, v_c is a jittered triangular wave which is expressed as:

$$v_c = \arcsin(\cos(2\pi f_c t)) \tag{9.16}$$

where f_c is the carrier frequency which is expressed as:

$$f_c = f_o + \Delta f \sin(2\pi f_m t) \tag{9.17}$$

where f_o is the nominal carrier frequency, Δf is the carrier deviation frequency, and f_m is the carrier modulation frequency. So, the spectral power of v_c is given by:

$$S = 2\left[\sum_{n=1}^{\infty} F_n(X+Y)\right]^2 \tag{9.18}$$

$$X = J_0(n\beta)\delta(f - nf_o) \tag{9.19}$$

$$Y = \sum_{k=1}^{\infty} J_k(n\beta)[\delta(f - nf_o - kf_m) + (-1)^k \delta(f - nf_o + kf_m)] \tag{9.20}$$

where F_n is the nth harmonic coefficient of the unmodulated carrier signal, $\beta = \Delta f / f_m$ is the frequency modulation index, $J_k(\cdot)$ is the kth-order Bessel function, and $\delta(\cdot)$ is the impulse function. So, the original spectral power around nf_o is distributed around the discrete terms at $(nf_o \pm kf_m)$. According to Carson's rule, 98% of the spectral power lies within the frequency range $[nf_o - (n\beta + 1)f_m,\ nf_o + (n\beta + 1)f_m]$. If $\beta \gg 1$, this frequency range will be $[nf_o - n\Delta f,\ nf_o + n\Delta f]$.

9.3.1 *Chaotic Sinusoidal PWM*

Based on the use of the standard sinusoidal frequency modulation scheme to modulate the switching frequency of sinusoidal PWM, the inverter output voltages can be expressed as:

$$\begin{aligned} V_{A,B,C} &= \frac{V_d}{2} + \frac{V_d}{2} A_m \cos(2\pi f_r t + \varphi_{A,B,C}) \\ &\quad + \sum_{n=1}^{\infty}\sum_{k=0}^{\infty}\sum_{h=-\infty}^{\infty} M_{n,k,h} J_k(n\beta) J_k\left(n\frac{\pi}{2}A_m\right) N \end{aligned} \tag{9.21}$$

$$\begin{aligned} N &= \cos\left[2\pi(nf_o + kf_m + hf_r)t + h\varphi_{A,B,C} - \frac{k\pi}{2}\right] \\ &\quad + \cos\left[2\pi(nf_o - kf_m + hf_r)t + h\varphi_{A,B,C} - \frac{k\pi}{2}\right] \end{aligned} \tag{9.22}$$

$$M_{n,h,k} = \frac{V_d K H}{2n\pi} \sin\left[(n+k)\frac{\pi}{2}\right], \quad K = \begin{cases} 2 & (k=0) \\ 0 & (k \neq 0) \end{cases}, \quad H = \begin{cases} 2 & (h=0) \\ 0 & (h \neq 0) \end{cases} \tag{9.23}$$

where $A_m = A_r/A_c$ is the amplitude modulation index, A_r is the amplitude of the sinusoidal reference signal, A_c is the amplitude of the carrier signal, f_r is the frequency of the sinusoidal reference signal, and $\varphi_{A,B,C} = 0°, 120°, 240°$ are, respectively, the initial phase angles of the phase A, B, C sinusoidal reference signals. From (9.21)–(9.23), it can be found that the spectral power of the output voltage $V_{A,B,C}$ clusters around the frequencies at $(hf_r + nf_o \pm kf_m)$. Also, based on Carson's rule, the majority of spectral power lies within the frequency range $[nf_o - (n\beta + 1)f_m - (n\pi A_m f_r/2 + f_r),\ nf_o + (n\beta + 1)f_m + (n\pi A_m f_r/2 + f_r)]$. If $\beta \gg 1$ and $f_o \gg f_r$, this frequency range will become $[nf_o - n\Delta f,\ nf_o + n\Delta f]$.

In order to even out the spectral power around the frequencies at $(hf_r + nf_o \pm kf_m)$ and hence to further reduce the EMI of the drive system, a chaotic sequence $\{\xi_i\} \in 0, 1)$ is utilized to chaoize the sinusoidal frequency modulator. The first method is to substitute Δf in (9.17) by $\xi_i \Delta f$. Hence, the amplitude of the frequency modulator can be chaoized: the so-called CAFM. The second method is to substitute f_m in (9.17) by $\xi_i f_m$. Hence, the frequency of the frequency modulator is chaoized: the so-called CFFM. While the chaotic sequence exhibits the nature of continuous spectral power, the power spectrum of the inverter output voltage can be smoothed out. Since $\xi_i \leq 1$, the power spectrum of the inverter output voltage can be maintained in the frequency range $[nf_o - n\Delta f,\ nf_o + n\Delta f]$ when $\beta \gg 1$ and $f_o \gg f_r$. This property can not only effectively reduce the EMI, but also avoid overlapping with the natural frequency of mechanical vibration.

There are many ways to generate the desired chaotic signal. In general, these chaos generators can be classified into two categories (Delgado-Restituto and Rodíguez-Vázquez, 2002):

- Discrete-time chaos generators (chaotic discrete maps), such as the Logistic map, tent map, Bernoulli map, and Henon map.
- Continuous-time chaos generators (chaotic oscillators), such as the double-scroll-like oscillator, Colpitts oscillator, Chua's oscillator, and Lorenz system.

The use of a Logistic map is the most commonly adopted method to generate the chaotic sequence $\{\xi_i\}$ for chaotic PWM inverter-fed induction drive systems. It is governed by the map $\xi_{i+1} = A\xi_i \in (1 - \xi_i)$, where $A \in [0, 4]$. Figure 9.3(a) depicts the bifurcation diagram of ξ_i versus A of the Logistic map. The corresponding Lyapunov exponent versus A is shown in Figure 9.3(b). It can be seen that when $A \in [0, 1)$, $\{\xi_i\}$ exhibits a zero value with $\xi_i = 0$. This denotes that the chaotic frequency modulator does not modulate the frequency of the carrier signal, which is equivalent to the traditional sinusoidal PWM. When $A \in [1, 3)$, $\{\xi_i\}$ takes a fixed value with $\xi_i > 0$. This denotes that the chaotic frequency modulator works as a standard sinusoidal frequency modulator. When A is further increased, $\{\xi_i\}$ begins to bifurcate with multiple values. When $A > 3.57$, $\{\xi_i\}$ exhibits infinite values. Also, the boundary of such infinite values changes with the value of A. As the corresponding Lyapunov exponents are positive, this confirms that it is a chaotic series. Thus, by properly tuning the value of A, this chaotic frequency modulator can offer various power spectra.

The use of a Bernoulli map is another commonly adopted method to generate the chaotic sequence for chaotic PWM inverter-fed induction drive systems. It is attractive because of its simplicity for the evaluation of its rate of mixing (Setti *et al.*, 2002) and the implementation of its integrated circuit (Delgado-Restituto and Rodíguez-Vázquez, 2002). Figure 9.4 shows a typical four-way Bernoulli shift map, which is expressed as $B_4(\cdot) \in (0, 1)$: $\xi_{i+1} = 4\xi_i \bmod (0.25)$.

The use of a Chua circuit is another method to generate the chaotic sequence for chaotic PWM inverter-fed induction drive systems (Cui *et al.*, 2006). Since the sequence is directly obtained from sampling the Chua circuit, it inherits the stochastic nature of the Chua circuit, hence exhibiting the ideal nature of autocorrelation and cross-correlation. Figure 9.5 shows a typical Chua circuit in

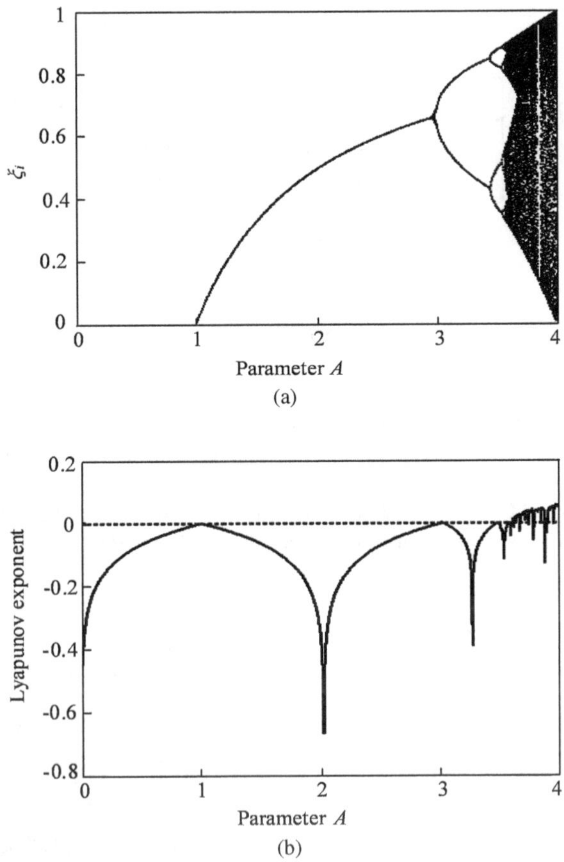

Figure 9.3 Logistic map. (a) Bifurcation diagram. (b) Lyapunov exponent

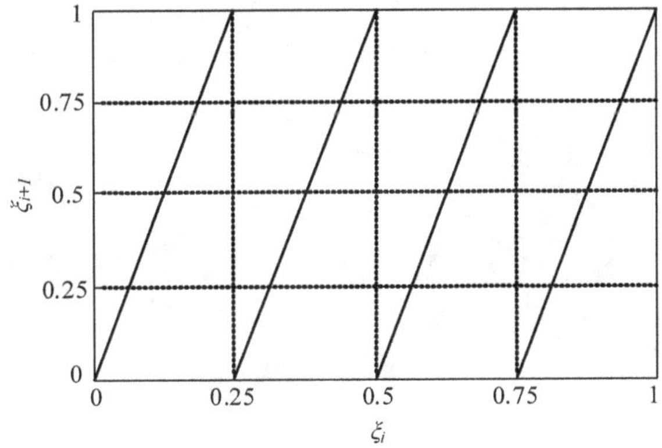

Figure 9.4 Four-way Bernoulli shift map

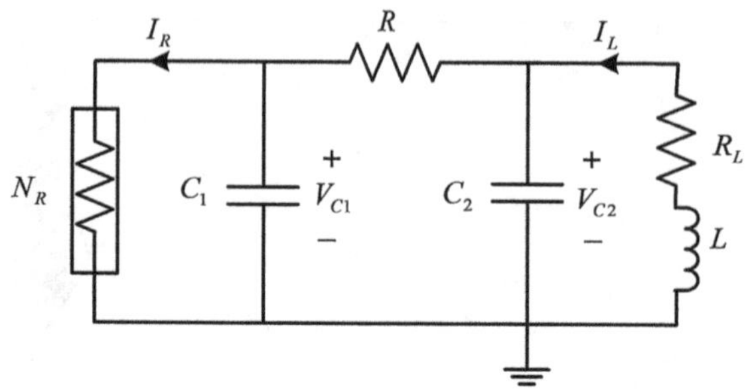

Figure 9.5 Chua circuit

which the nonlinear resistor N_R is chosen to have a piecewise linear V-I characteristic. Its major drawback is that the digital implementation of a Chua circuit is much more difficult than that of a Logistic map or Bernoulli map.

By applying CAFM and CFFM to modulate the switching frequency of the sinusoidal PWM (SPWM), CAFM-SPWM and CFFM-SPWM can be obtained. Figures 9.6 and 9.7 show the power spectra of the

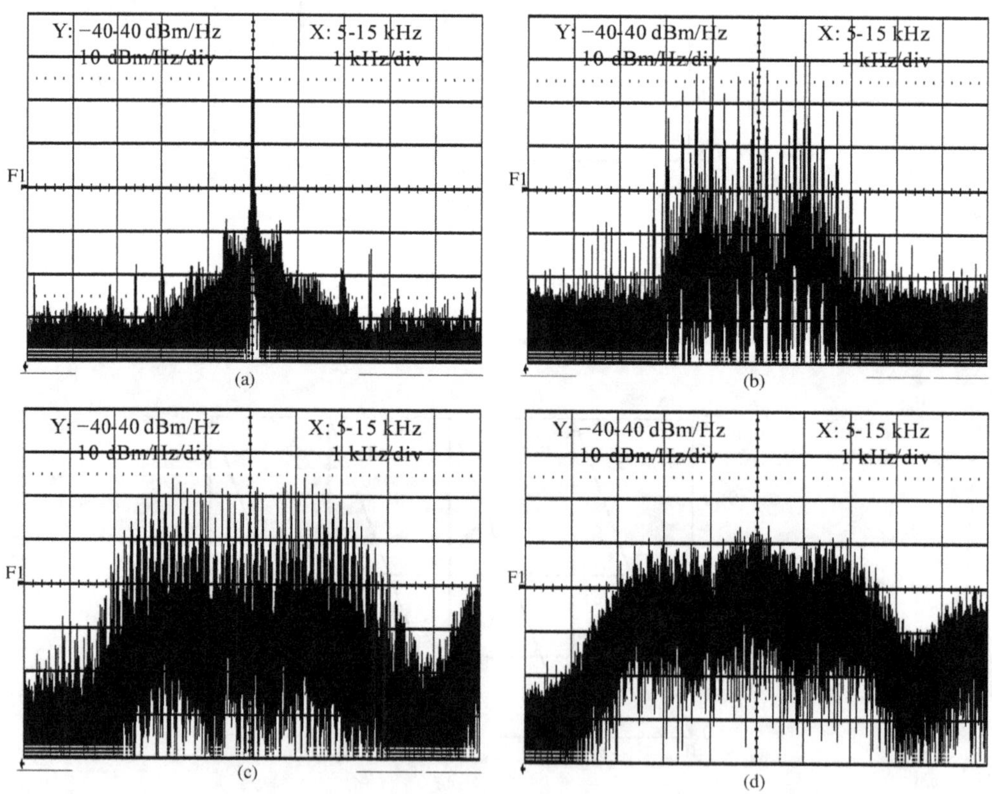

Figure 9.6 CAFM-SPWM power spectra. (a) $A = 0$. (b) $A = 2$. (c) $A = 3.2$. (d) $A = 4$

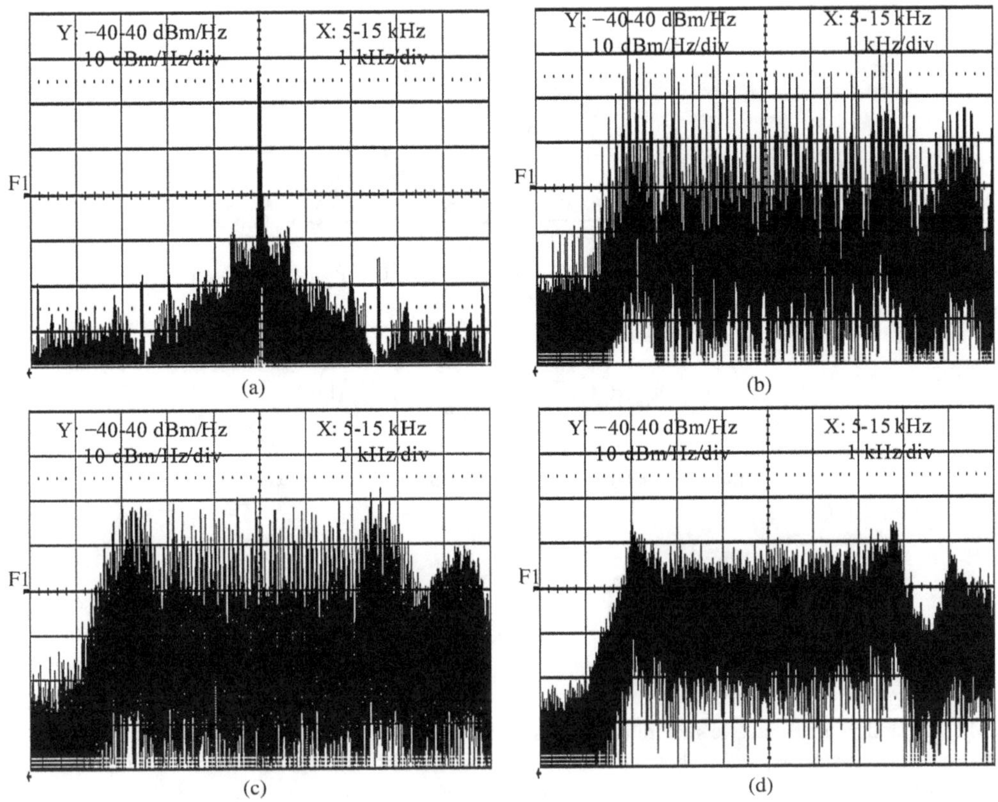

Figure 9.7 CFFM-SPWM power spectra. (a) $A = 0$. (b) $A = 2$. (c) $A = 3.2$. (d) $A = 4$

PWM output voltage when using CAFM-SPWM and CFFM-SPWM with various values of A, respectively, in which $f_r = 10\,\text{Hz}, f_o = 10\,\text{kHz}, \Delta f = 3\,\text{kHz}$ and $f_m = 0.3\,\text{kHz}$ are adopted. It can be observed that when $A = 0$, the power spectra have significant spectral spikes, which cause undesirable peaky EMI. When $A = 2$, the magnitude of such spectral spikes are alleviated. When $A = 3.2$, such spectral spikes can be further reduced. When $A = 4$, the power spectra are significantly smoothed out so that the EMI is effectively suppressed.

9.3.2 *Chaotic Space Vector PWM*

Compared with sinusoidal PWM, space vector PWM offers the definite advantages of lower harmonic distortion, lower switching loss, and better utilization of DC link voltage. Figure 9.8 shows the diagram of the inverter output voltage space vectors. There are eight switching states of the three upper-leg power switches of the inverter: $S_0 = (0,0,0)$, $S_1 = (1,0,0)$, $S_2 = (1,1,0)$, $S_3 = (0,1,0)$, $S_4 = (0,1,1)$, $S_5 = (0,0,1)$, $S_6 = (1,0,1)$, $S_7 = (1,1,1)$. The three lower-leg power switches have inverted switching states. $S_1 \sim S_6$ correspond to the basic voltage vector $\vec{V}_{0^\circ} \sim \vec{V}_{300^\circ}$. In each switching period, the sequence

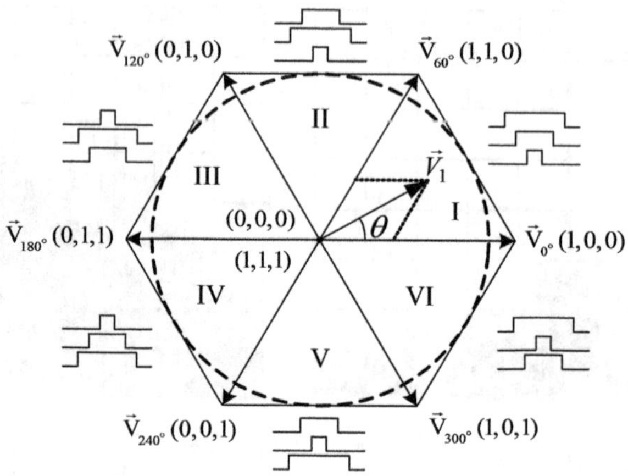

Figure 9.8 Diagram of inverter output voltage space vectors

of the switching states are:

$$S_0(T_0/4) \rightarrow S_A(T_1/2) \rightarrow S_B(T_2/2) \rightarrow S_7(T_0/2) \rightarrow S_B(T_2/2) \rightarrow S_A(T_1/2) \rightarrow S_0(T_0/4),$$

where S_A and S_B are the active switching states. The active switching states correspond to the two adjacent basic voltage vectors $\vec{V}_x$ and $\vec{V}_y$ between which $\vec{V}_1$ locates. For the sectors I to VI, they are represented by $(\vec{V}_x = \vec{V}_{0^\circ},\ \vec{V}_y = \vec{V}_{60^\circ})$, $(\vec{V}_x = \vec{V}_{60^\circ},\ \vec{V}_y = \vec{V}_{120^\circ})$, $(\vec{V}_x = \vec{V}_{120^\circ},\ \vec{V}_y = \vec{V}_{180^\circ})$, $(\vec{V}_x = \vec{V}_{180^\circ},\ \vec{V}_y = \vec{V}_{240^\circ})$, $(\vec{V}_x = \vec{V}_{240^\circ},\ \vec{V}_y = \vec{V}_{300^\circ})$, and $(\vec{V}_x = \vec{V}_{300^\circ},\ \vec{V}_y = \vec{V}_{0^\circ})$, respectively. Then, the switching times T_1, T_2 and T_0 can be computed by:

$$T_1 = \frac{2|\vec{V}_1|T_v}{\sqrt{3}|\vec{V}_x|}\sin(60^\circ - \theta) \tag{9.24}$$

$$T_2 = \frac{2|\vec{V}_1|T_v}{\sqrt{3}|\vec{V}_y|}\sin\theta \tag{9.25}$$

$$T_0 = T_v - T_1 - T_2 \tag{9.26}$$

where θ is the phase angle of $\vec{V}_1$, and T_v is the switching period.

By applying CAFM and CFFM to space vector PWM (SVPWM), CAFM-SVPWM and CFFM-SVPWM are obtained. Figure 9.9 shows the flowcharts of generation of the switching period. For CAFM-SVPWM, the modulation frequency is updated at the end of each interval $(1/f_m)$. On the other hand, the modulation frequency of CFFM-SVPWM is updated at the end of each interval $1/(\xi_i f_m)$. In order to avoid waiting too long for updating, the longest limit of each update is with $\xi_i = 0.1$. Namely, if $\xi_i \geq 0.1$, the modulation frequency will be updated at $1/(\xi_i f_m)$; otherwise, if $\xi_i < 0.1$, it will be updated at $1/(0.1 f_m)$. For both CAFM-SVPWM and CFFM-SVPWM, T_v is updated at the end of the jth switching period T_{vj}. As a symmetric regular sampling is used, the sampling period T_{CLK} is kept synchronous with T_{vj}. This synchronous sampling can enable good dynamic performance for induction drive systems.

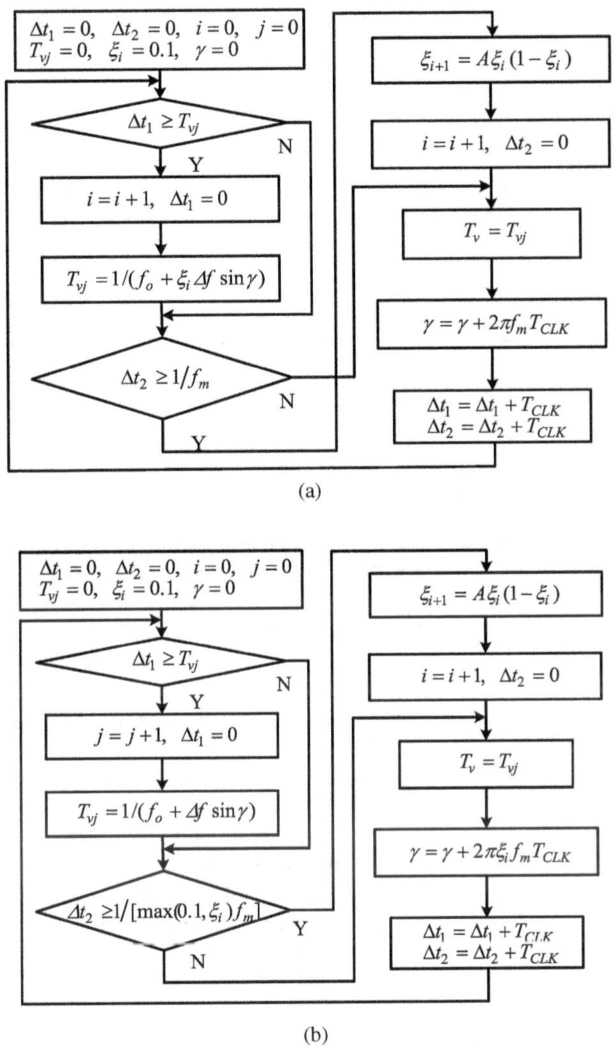

Figure 9.9 Flowcharts of generation of switching period. (a) CAFM-SVPWM. (b) CFFM-SVPWM

9.4 Chaotic PWM Inverter Drive Systems

To investigate the performances of the aforementioned chaotic SVPWM inverters for induction drive systems, a practical 3-phase induction drive system is used for exemplification. The parameters of the induction motor are given in Table 9.1. Throughout the experiment, a power analyzer is used to measure the power spectrum of the PWM output voltage, a current transducer is used to measure the instantaneous stator current, and an encoder is used to measure the instantaneous rotor speed.

Table 9.1 Parameters of induction motor

Rated power	1.5 kW
Rated voltage	220 V
Rated speed	1430 rpm
Poles	4
Stator resistance	3.3 Ω
Stator leakage inductance	43.9 mH
Rotor resistance	3 Ω
Rotor leakage inductance	43.9 mH
Mutual inductance	278 mH
Rotor inertia	6.65×10^{-3} kgm^2
Viscous friction coefficient	5.5×10^{-6} Nm/rad/s

9.4.1 Open-Loop Control Operation

In order to evaluate the performances of CAFM-SVPWM and CFFM-SVPWM, they are compared with the traditional fixed frequency SVPWM (FF-SVPWM) and the recently developed random frequency SVPWM (RF-SVPWM) for the same IGBT based voltage-source PWM inverter and induction motor. For CAFM-SVPWM and CFFM-SVPWM, when selecting $f_r = 10$ Hz, $A = 4$, and $f_o = 10$ kHz, it deduces $\Delta f = 3$ kHz and $f_m = 100$ Hz in order to avoid overlapping with the mechanical natural frequency of around 13.5 kHz. On the other hand, the switching frequency of FF-SVPWM is fixed at 10 kHz, while that of RF-SVPWM is randomly distributed within 10 ± 3 kHz.

The power spectra of the PWM output phase voltage V_A and line voltage V_{AB} using FF-SVPWM, RF-SVPWM, CAFM-SVPWM, and CFFM-SVPWM are compared under the same conditions as shown in Figures 9.10 and 9.11. It can be seen that there exist significant spectral peaks in the spectra of FF-SVPWM. Meanwhile, RF-SVPWM can effectively smooth out the peaky harmonics and provide a flat spectrum. Because of this nature, the possibility of overlapping with the natural frequency of mechanical resonance is significantly increased. On the contrary, both CAFM-SVPWM and CFFM-SVPWM can exhibit an essentially flat spectrum, but associated with a spectral notch around the natural frequency. Thus, these two chaotic SVPWM schemes can reduce the peaky EMI while avoiding the mechanical resonance.

In order to quantitatively assess the effectiveness of the chaotic SVPWM schemes, two important indicators are adopted for comparison. Firstly, since the conducted EMI with frequency exceeding 9 kHz is stringently limited in many countries, the maximum power spectral density (PSD) of V_A and V_{AB} in that frequency range is used as one indicator to compare FF-SVPWM, RF-SVPWM, CAFM-SVPWM, and CFFM-SVPWM. Secondly, since the occurrence of mechanical resonance should be avoided, the spectral power of V_A and V_{AB} within the sideband of 13.4–13.6 kHz around the natural frequency is used as another indicator for comparison. Table 9.2 gives a quantitative comparison of the above SVPWM schemes, confirming that the chaotic SVPWM schemes can effectively reduce the conducted EMI and avoid the mechanical resonance.

To assess whether the use of chaotic SVPWM schemes causes an adverse effect on the open-loop performances of the induction drive system, the steady-state current waveforms, as well as the start-up transient current and speed responses, are recorded as shown in Figures 9.12 and 9.13, respectively. It can be observed that the steady-state current waveforms are very sinusoidal, while the transient responses are very fast, hence confirming that the chaotic SVPWM schemes do not cause any adverse effect on the open-loop performances.

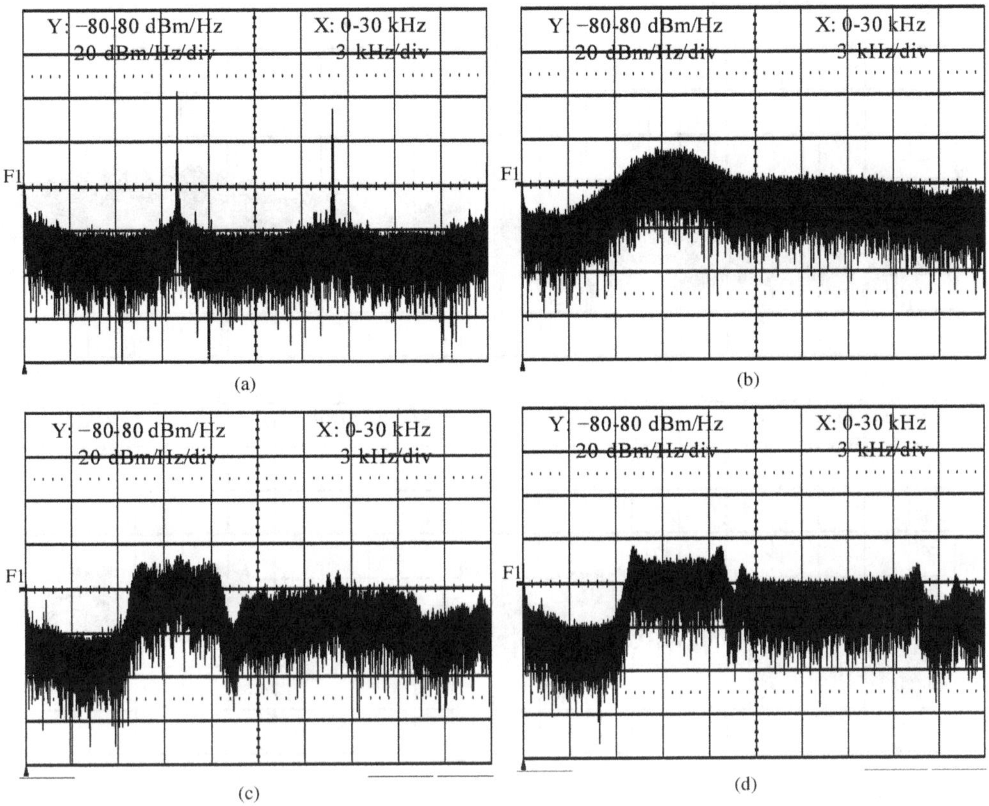

Figure 9.10 Comparison of power spectra of PWM output phase voltage. (a) FF-SVPWM. (b) RF-SVPWM. (c) CAFM-SVPWM. (d) CFFM-SVPWM

9.4.2 Closed-Loop Vector Control Operation

The two chaotic SVPWM schemes are then applied to the vector-controlled induction drive system in order to assess their closed-loop performances. The dynamical equations of the induction drive resulting from rotor field orientation are expressed as:

$$\frac{\mathrm{d}i_d}{\mathrm{d}t} = \frac{R_r L_m^2}{\sigma L_s L_r^2} i_{mr} - \frac{R_s L_r^2 + R_r L_m^2}{\sigma L_s L_r^2} i_d + \omega_r i_q + \frac{u_d}{\sigma L_s} \tag{9.27}$$

$$\frac{\mathrm{d}i_q}{\mathrm{d}t} = -\frac{\omega_n L_m^2}{\sigma L_s L_r} i_{mr} - \frac{R_s L_r^2 + R_r L_m^2}{\sigma L_s L_r^2} i_q - \omega_r i_d + \frac{u_q}{\sigma L_s} \tag{9.28}$$

$$\frac{\mathrm{d}i_{mr}}{\mathrm{d}t} = -\frac{R_r}{L_r} i_{mr} + \frac{R_r}{L_r} i_d \tag{9.29}$$

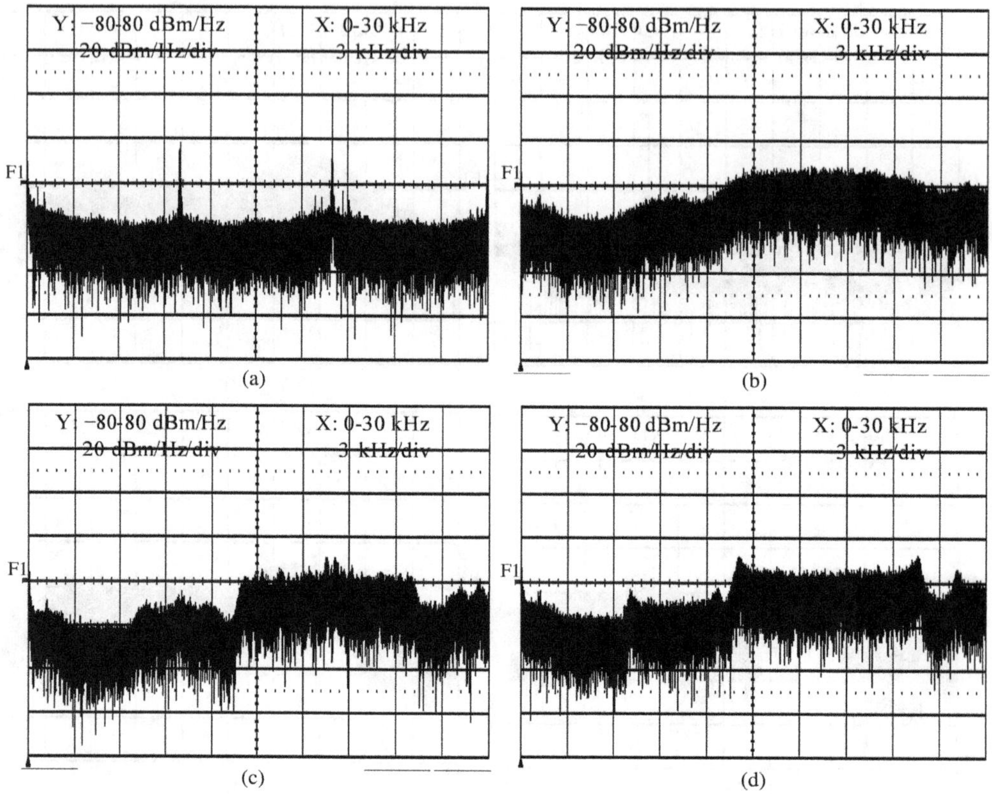

Figure 9.11 Comparison of power spectra of PWM output line voltage. (a) FF-SVPWM. (b) RF-SVPWM. (c) CAFM-SVPWM. (d) CFFM-SVPWM

$$\omega_r = \omega_n + \frac{R_r i_q}{L_r i_{mr}} \tag{9.30}$$

$$\sigma = 1 - \frac{L_m^2}{L_s L_r} \tag{9.31}$$

$$L_s = L_{ls} + L_m \tag{9.32}$$

Table 9.2 Quantitative comparison of various SVPWM schemes

		FF-SVPWM	RF-SVPWM	CAFM-SVPWM	CFFM-SVPWM
Max PSD over 9 kHz (dBm/Hz)	V_A	42.49	17.50	21.03	18.51
	V_{AB}	34.84	5.04	9.41	10.16
Spectral power within 13.4–13.6 kHz (dBm)	V_A	−5.57	21.24	7.19	15.75
	V_{AB}	−5.64	15.57	0.81	2.85

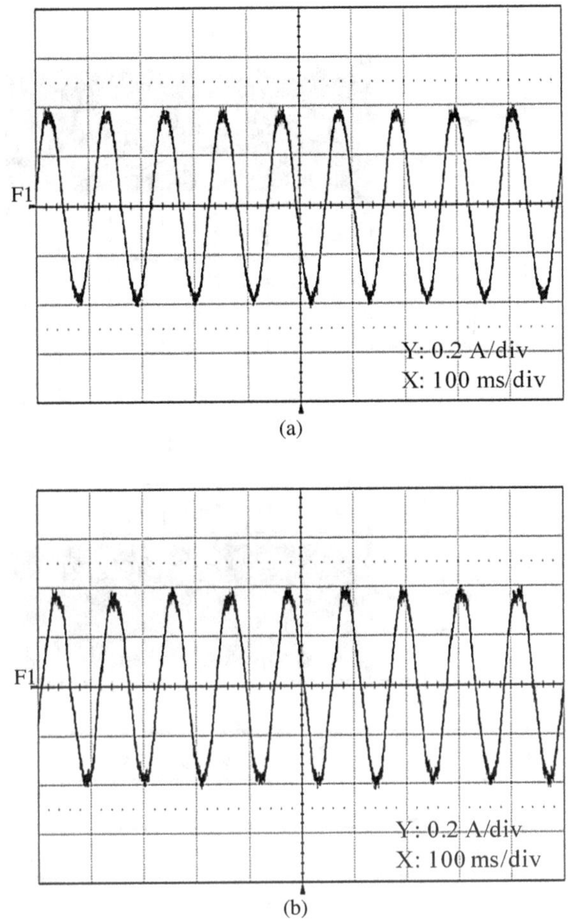

Figure 9.12 Open-loop steady-state current waveforms. (a) CAFM-SVPWM. (b) CFFM-SVPWM

$$L_r = L_{lr} + L_m \tag{9.33}$$

where i_d and i_q are respectively the d-axis and q-axis components of stator current; i_{mr}, ω_r, and ω_n are respectively the rotor magnetizing current, rotor flux speed, and rotor speed; u_d and u_q are respectively the d-axis and q-axis components of stator voltage; R_s and R_r are respectively the stator resistance and rotor resistance; L_s, L_r, L_{ls}, L_{lr}, and L_m are respectively the stator inductance, rotor inductance, stator leakage inductance, rotor leakage inductance, and mutual inductance; and σ is the leakage coefficient. Figure 9.14 shows the corresponding block diagram in which the d-axis and q-axis constitute the reference frame rotating synchronously with the rotor flux, whereas the α-axis and β-axis constitute the stationary reference frame.

There are two closed-loop controllers, namely the inner current loop and the outer speed loop. The proportional-integral (PI) control method is adopted for both the current controller and speed controller. The sampling rate of the current controller varies with the switching period of the SVPWM inverter, while

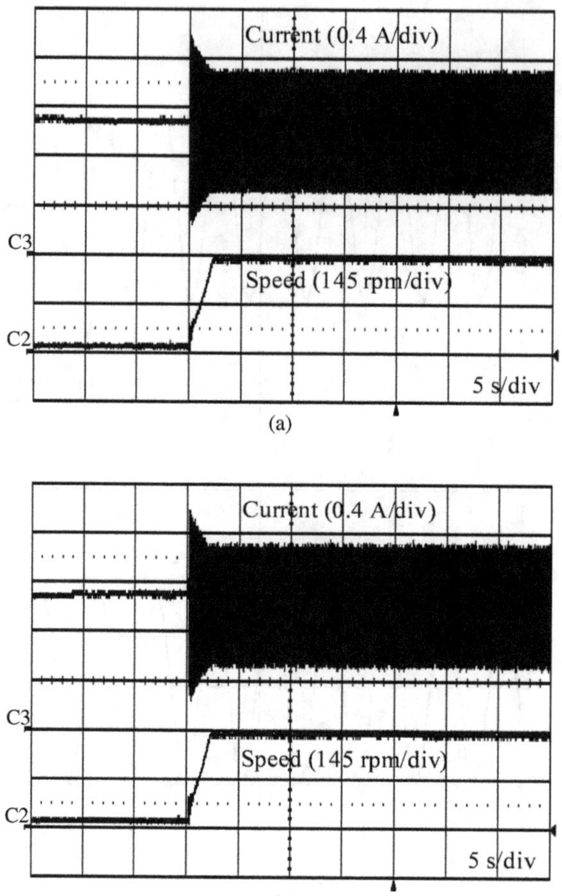

Figure 9.13 Open-loop startup transient responses. (a) CAFM-SVPWM. (b) CFFM-SVPWM

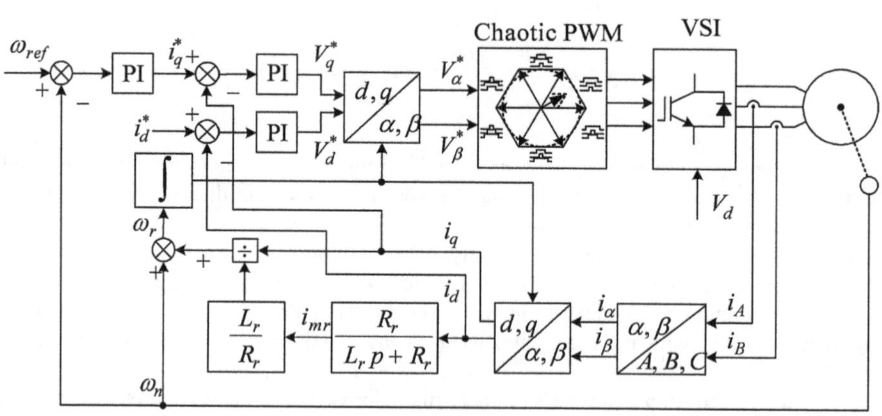

Figure 9.14 Vector control of chaotic SVPWM inverter fed induction drive system

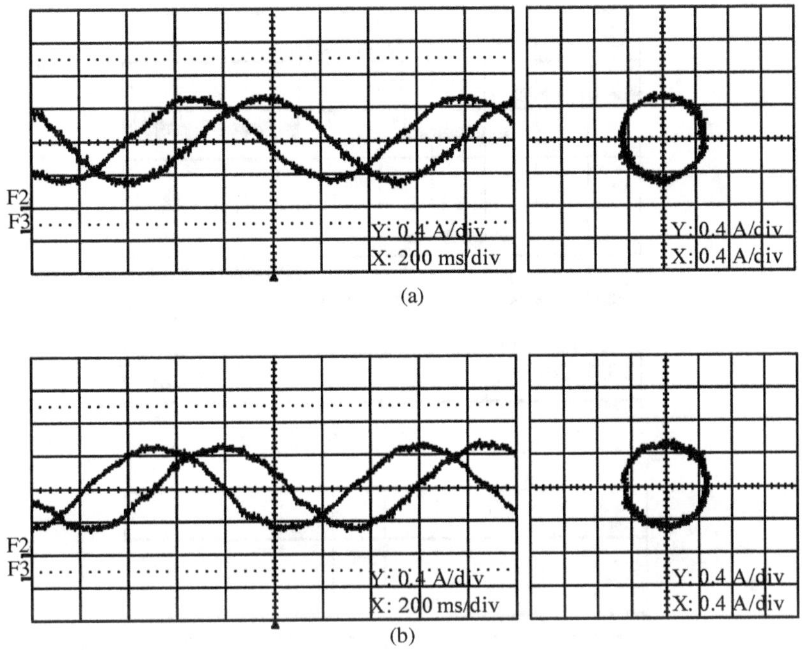

Figure 9.15 Measured α and β components of stator currents. (a) CAFM-SVPWM. (b) CFFM-SVPWM

the sampling rate of the speed controller is kept constant. Due to the variable sampling rate of the current controller, the corresponding PI parameters should be updated at each sampling interval. The criteria are $K_P = K_p$ and $K_I = K_i T_v$ in which K_P and K_I are respectively the discrete proportional parameter and integral parameter, and K_p and K_i are respectively the continuous proportional parameter and integral parameter.

Firstly, the measured power spectra of the PWM output phase voltage V_A and line voltage V_{AB} are the same as the open-loop case, hence confirming that the two chaotic SVPWM schemes can reduce the peaky EMI while avoiding the mechanical resonance.

Secondly, Figure 9.15 shows the measured steady-state waveforms and trajectories of the α-axis and β-axis components of stator current. It can be seen that there is no noticeable distortion in the current waveforms and trajectories.

Thirdly, Figures 9.16–9.18 show the dynamic responses of the two chaotic SVPWM schemes as compared with the traditional FF-SVPWM. Their current responses are based on the d-axis current component step command of 0.5 A → 1 A → 0.5 A → 0.75 A and the q-axis current component step command of 0.58 A → 0.25 A → 1 A → 0.33 A. The speed responses are based on the step command of 28.3 rad/s → 12.4 rad/s → 0 rad/s → 28.3 rad/s. They illustrate that both the current controller and speed controller of the two chaotic SVPWM schemes can track the commands as accurately and as quickly as that of the FF-SVPWM scheme. Hence, it verifies that the chaotic PWM motor drive can not only suppress the conducted EMI while avoiding the mechanical resonance, but also retain the outstanding steady-state and transient performances of vector control.

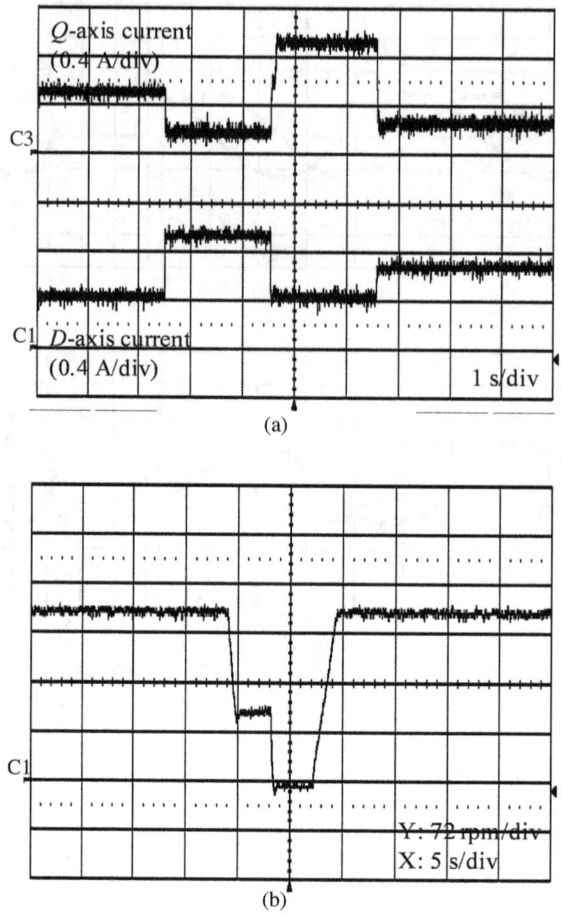

Figure 9.16 Closed-loop transient responses using CAFM-SVPWM. (a) Stator current components. (b) Motor speed

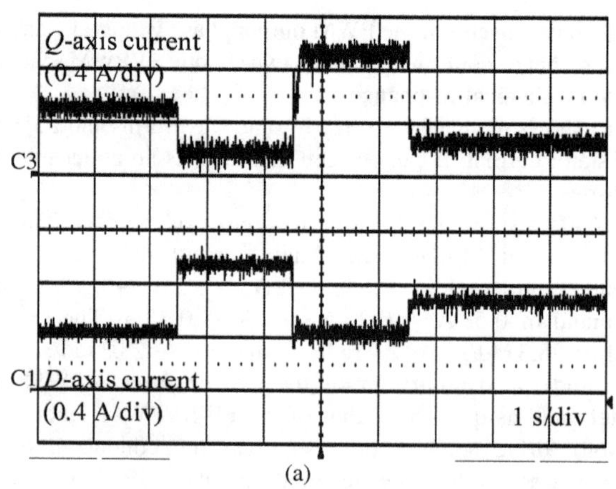

(a)

Figure 9.17 Closed-loop transient responses using CFFM-SVPWM. (a) Stator current components. (b) Motor speed

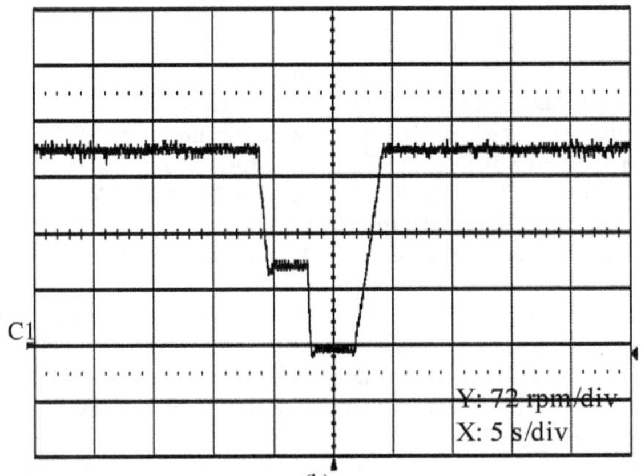

(b)

Figure 9.17 (*Continued*)

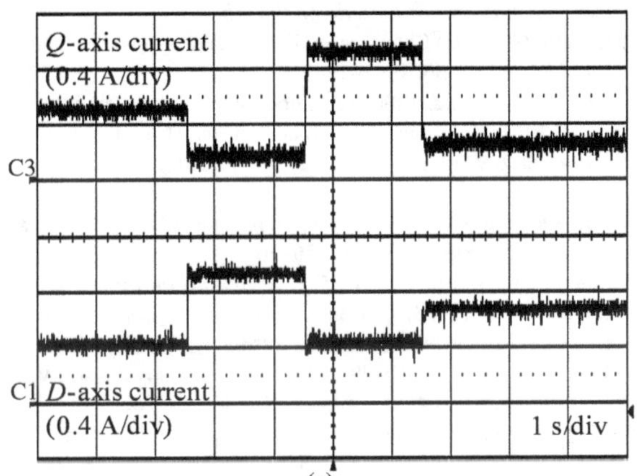

(a)

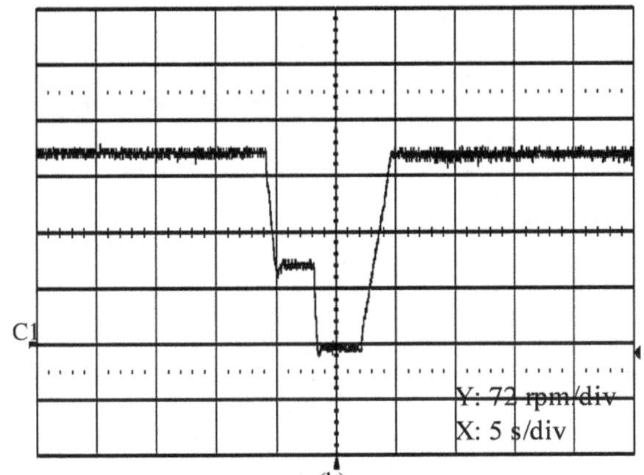

(b)

Figure 9.18 Closed-loop transient responses using FF-SVPWM. (a) Stator current components. (b) Motor speed

References

Ashtari, A., Thomas, G., Kinsner, W., and Flores, B.C. (2003) Sufficient condition for chaotic maps to yield chaotic behavior after FM. *IEEE Transactions on Aerospace and Electronic Systems*, **39**, 202–210.

Balestra, M., Bellini, A., Callegari, S. *et al.* (2004) Chaos-based generation of PWM-like signals for low-EMI induction motor drives: analysis and experimental results. *IEICE Transactions on Electronics*, **E87-C**, 66–75.

Bellini, A., Franceschini, G., Rovatti, R. *et al.* (2001) Generation of low-EMI PWM patterns for induction drives with chaotic maps. Proceedings of IEEE Industrial Electronics Conference, pp. 1527–1532.

Blaabjerg, F., Pedersen, J.K., Oestergaard, L. *et al.* (1996) Optimized and non-optimized random modulation techniques for VSI drives. *EPE Journal*, **6**, 46–53.

Bose, B.K. (1990) An adaptive hysteresis-band current control technique of a voltage-fed PWM inverter for machine drive system. *IEEE Transactions on Industrial Electronics*, **37**, 402–408.

Bowes, S.R. (1975) New sinusoidal pulse width-modulated inverter. *IEE Proceedings*, **122**, 1279–1285.

Bowes, S.R. and Clements, R.R. (1982) Computer-aided design of PWM inverter systems. *IEE Proceedings-B Electric Power Applications*, **129**, 1–17.

Bowes, S.R. and Mount, M.J. (1981) Microprocessor control of PWM inverters. *IEE Proceedings - B, Electric Power Applications*, **128**, 293–305.

Buja, G.S. and Indri, G.B. (1977) Optimal pulsewidth modulation for feeding AC motors. *IEEE Transactions on Industry Applications*, **13**, 38–44.

Callegari, S., Rovatti, R., and Setti, G. (2002) Chaotic modulations can outperform random ones in electromagnetic interference reduction tasks. *IEE Electronics Letters*, **38**, 543–544.

Callegari, S., Rovatti, R., and Setti, G. (2003) Chaos-based FM signals: application and implementation issues. *IEEE Transactions on Circuits and Systems I*, **50**, 1141–1147.

Cui, W., Chau, K.T., Wang, Z., and Jiang, J.Z. (2006) Application of chaotic modulation to ac motors for harmonic suppression. Proceedings of IEEE International Conference on Industrial Technology, pp. 2343–2347.

Delgado-Restituto, M. and Rodíguez-Vázquez, A. (2002) Integrated chaos generators. *IEEE Proceedings*, **90**, 747–767.

Habetler, T.G. and Divan, D.M. (1991) Acoustic noise reduction in sinusoidal PWM drives using a randomly modulated carrier. *IEEE Transactions on Power Electronics*, **6**, 356–363.

Holtz, J. (1992) Pulsewidth modulation – a survey. *IEEE Transactions on Industrial Electronics*, **39**, 410–420.

Kazmierkowski, M.P. and Malesani, L. (1998) Current control techniques for three-phase voltage-source PWM converters: a survey. *IEEE Transactions on Industrial Electronics*, **45**, 691–703.

Kazmierkowski, M.P., Dzieniakowski, M.A., and Sulkowski, W. (1991) Novel space vector based current controllers for PWM-inverters. *IEEE Transactions on Power Electronics*, **6**, 158–166.

Lo, W.C., Chan, C.C., Zhu, Z.Q. *et al.* (2000) Acoustic noise radiated by PWM-controlled induction machine drives. *IEEE Transactions on Industrial Electronics*, **47**, 880–889.

Mokrytzki, B. (1967) Pulse-width-modulated inverters for A.C. motor drives. *IEEE Transactions*, **3**, 493–503.

Nabae, A., Ogasawara, S., and Akagi, H. (1986) A novel control scheme for current-controlled PWM inverters. *IEEE Transactions on Industry Applications*, **22**, 697–701.

Patel, H.S. and Hoft, R.G. (1973) Generalized techniques of harmonic elimination and voltage control in thyristor inverters. *IEEE Transactions on Industry Applications*, **9**, 310–317.

Plunkett, A.B. (1979) A current-controlled PWM transistor inverter drive. Proceedings of IEEE Industry Applications Society Annual Meeting, pp. 785–792.

Rahman, M.A., Quacioe, J.E., and Chowdhury, M.A. (1987) Performance analysis of delta modulated PWM. *IEEE Transactions in Power Electronics*, **2**, 227–233.

Ran, L., Gokani, S., Clare, J. *et al.* (1998a) Conducted electromagnetic emissions in induction motor drive systems – Part I: time domain analysis and identification of dominant modes. *IEEE Transactions on Power Electronics*, **13**, 757–767.

Ran, L., Gokani, S., Clare, J. *et al.* (1998b) Conducted electromagnetic emissions in induction motor drive systems – Part II: frequency domain models. *IEEE Transactions on Power Electronics*, **13**, 768–776.

Setti, G., Mazzini, G., Rovatti, R., and Callegari, S. (2002) Statistical modeling of discrete time chaotic processes: Basic finite dimensional tools and applications. *IEEE Proceedings*, **90**, 662–690.

Stemmler, H. and Eilinger, T. (1994) Spectral analysis of the sinusoidal PWM with variable switching frequency for noise reduction in inverter-fed induction motors. Proceedings of IEEE Power Electronics Specialists Conference, pp. 269–277.

Tímár, P.L. (1989) *Noise and Vibration of Electrical Machines*, Elsevier Science Publishers, Amsterdam.

Trzynadlowski, A.M., Kirlin, R.L., and Legowski, S.F. (1997) Space vector PWM technique with minimum switching losses and a variable pulse rate. *IEEE Transactions on Industrial Electronics*, **44**, 173–181.

Wang, Z. and Chau, K.T. (2007) Design and analysis of a chaotic PWM inverter for electric vehicles. Proceedings of IEEE Industry Applications Society Annual Meeting, pp. 1954–1961.

Wang, Z., Chau, K.T., and Liu, C. (2007) Improvement of electromagnetic compatibility of motor drives using chaotic PWM. *IEEE Transactions on Magnetics*, **43**, 2612–2614.

Wang, Z., Chau, K.T., and Cheng, M. (2008) A chaotic PWM motor drive for electric propulsion. Proceedings of IEEE Vehicle Power and Propulsion Conference, pp. 1–6.

Ziogas, P.D. (1981) The delta modulation techniques in static PWM inverters. *IEEE Transactions on Industry Applications*, **17**, 199–204.

10

Application of Chaotic Motion

There are many possible applications of chaotic motion, including both industrial equipment and domestic appliances. Some industrial equipment (such as compactors, mixers, and grinders) and some domestic appliances (such as clothes-washers, dishwashers, ventilating fans, heaters, and air-conditioners) have successfully employed chaotic motion to significantly improve their performance.

In this chapter, various promising applications of chaotic motion, including compaction, mixing, washing, HVAC, and grinding, are unveiled and elaborated.

10.1 Chaotic Compaction

Compaction is one of the most important operations in civil and geotechnical engineering. In recent years, being fueled by the advancement of mechatronic design and control strategies, there have been tremendous developments in compaction technology. Modern compaction equipment should be not only powerful, economical, and versatile, but also environmental friendly.

A compactor functions to provide a vibratory compressive stress onto soil, granulates, or powders so that the required densities can be attained. This vibratory compressive stress is due to the centrifugal force of a rotating eccentric mass which can be driven by an electric motor or an internal combustion engine. It should be noted that the electric compactor takes the definite advantages of no smelly gas, quiet operation, and free maintenance over its engine-powered counterpart.

The compaction process is to induce the particles to mechanically vibrate in such a way that the internal frictions among particles, and hence the void content, can be reduced. Since the particles are usually of different sizes and shapes, a fixed frequency vibration is only a compromise to reduce the internal frictions. With the use of a variable frequency vibration, various internal frictions can be individually reduced, hence improving the compaction effectiveness. Because of the wideband spectral power density of chaos, it is anticipated that a chaotic vibration can offer a simultaneous reduction of various internal frictions, hence further improving the compaction effectiveness.

The application of chaos to compaction was initiated in a mechanical vibrator for compacting soft soil (Long, 2001). Figure 10.1 shows the configuration of this mechanically chaotic vibrator which consists of three rotating eccentric masses with mass centers C_1, C_2, and C_3. The eccentric mass 1 is driven by an electric motor or an internal combustion engine at a constant rotational speed, while the eccentric masses 2 and 3 are pivoted on O_1 and O_2, respectively. By using this three-mass mechanism for realistic soil compaction, the measured degree of compaction is found to be 12.2 per cent higher than that by the traditional method.

Chaos in Electric Drive Systems: Analysis, Control and Application, First Edition. K.T. Chau and Zheng Wang.
© 2011 John Wiley & Sons (Asia) Pte Ltd. Published 2011 by John Wiley & Sons (Asia) Pte Ltd.

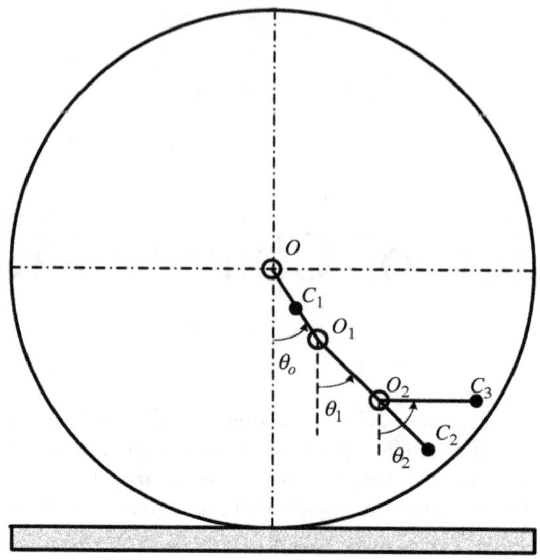

Figure 10.1 Mechanically chaotic vibrator

Instead of using a complicated three-mass mechanism to generate chaotic vibration for compaction, a chaotic drive system can perform the same task more easily. Figure 10.2 shows the structure of an electrically chaotic vibrator, which only needs a single eccentric mass driven by an electric motor. The electric motor can be a permanent magnet DC (PMDC) motor or a single-phase AC

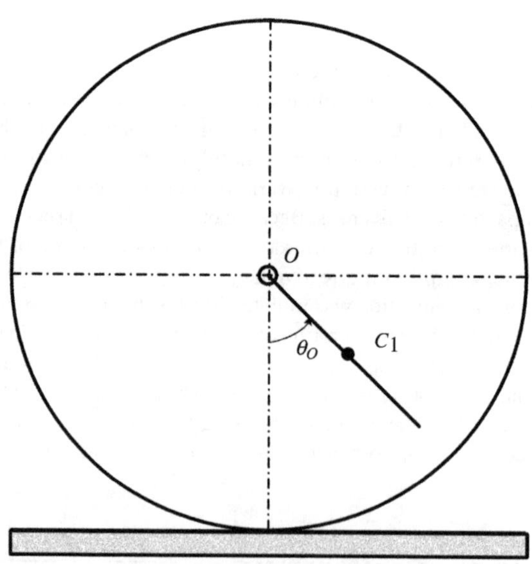

Figure 10.2 Electrically chaotic vibrator

motor for application to a low-power vibratory plate compactor (Chau and Wang, 2005), whereas it may be a 3-phase AC motor for application to a high-power vibratory roller compactor (Chau and Wang, 2006).

Compared with the mechanically chaotic vibratory compactor, the electrically chaotic vibratory compactor offers the following advantages:

- The structure of this electric drive system is very simple and involves only one eccentric rotating mass.
- By using the electric drive, the generated chaotic vibration can be predefined and analytically formulated. Notice that the three-mass mechanism involves a very complicated interaction between three eccentric rotating masses, and the corresponding complex motion can only be determined using numerical simulation.
- The electrically chaotic vibrator can offer various chaotic vibrations that are suitable for compacting different kinds of particles such as soil, granulates, and powders. Notice that the mechanically chaotic vibrator cannot offer such flexibility or selectivity of chaotic vibrations for different kinds of particles.

The purpose of this section is to present an electrically chaotic compactor system. The key is to chaoize a PMDC drive system by using control means, namely the proportional time-delay (PTD) control (Wang and Chau, 2008) and the chaotic speed reference (CSR) control (Wang and Chau, 2009), which not only need no mechanical auxiliary but also provide a flexible control of chaotic motion. Simulation and experimental results are both used to illustrate the validity of this chaotic compactor.

10.1.1 Compactor System

Figure 10.3 shows the configuration of the vibratory compactor system where the motor shaft is mounted by a solid eccentric mass (Yoo and Selig, 1979). Compared with the mechanical design that uses three interlinked eccentric masses, this compactor has the advantage of simple structure. The dynamical model

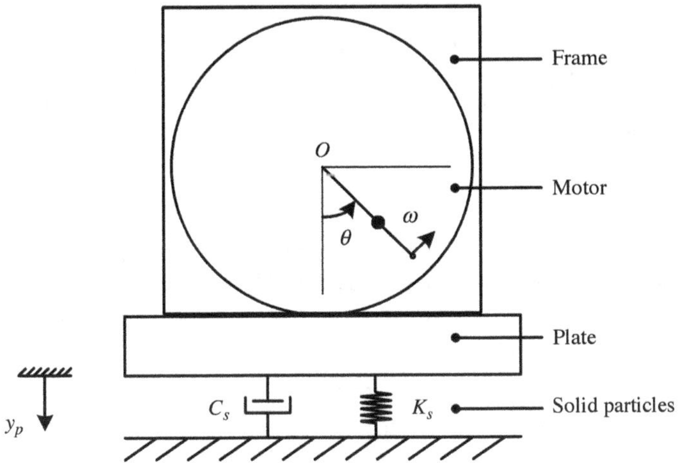

Figure 10.3 Vibratory compactor system

of this compactor can be expressed as:

$$\frac{\mathrm{d}i}{\mathrm{d}t} = -\frac{K_E}{L_a}\omega - \frac{R_a}{L_a}i + \frac{V_{in}}{L_a} \tag{10.1}$$

$$\frac{\mathrm{d}\omega}{\mathrm{d}t} = -\frac{B}{J}\omega - \frac{K_T}{J}i + \frac{M_e gr \sin\theta}{J} \tag{10.2}$$

$$\frac{\mathrm{d}\theta}{\mathrm{d}t} = \omega \tag{10.3}$$

$$\frac{\mathrm{d}\omega}{\mathrm{d}t} = \dot{\omega} \tag{10.4}$$

$$\frac{\mathrm{d}y_p}{\mathrm{d}t} = \dot{y}_p \tag{10.5}$$

$$\begin{aligned}(M_p + M_e)\frac{\mathrm{d}\dot{y}_p}{\mathrm{d}t} &= M_e r\omega^2 \cos\theta + M_e r\dot{\omega}\sin\theta - K_s y_p \\ &\quad - C_s\dot{y}_p + (M_p + M_e)g\end{aligned} \tag{10.6}$$

where θ and ω are respectively the rotational angle and rotational speed; V_{in}, i, K_T, K_E, L_a, and R_a are respectively the feed-in voltage, armature current, torque constant, back EMF constant, armature inductance, and armature resistance of the PMDC motor; B and J are respectively the viscous damping coefficient and rotor inertia; M_e and r are respectively the weight and equivalent radius of the eccentric mass; g is the gravitational constant; y_p is the vertical displacement of the plate; M_p is the total weight of the motor, the frame, and the plate; and K_s and C_s are respectively the stiffness and damping of the solid particles (Gethin *et al.*, 2001).

10.1.2 Chaotic Compaction Control

Figure 10.4 shows the diagram of the PMDC drive system. The current control is realized by PI control, and the commanded feed-in voltage V_{in}^* is realized by the DC–DC converter. In order to generate chaotic motion in the PMDC drive, there are two possible control methods, namely the PTD control and the CSR control.

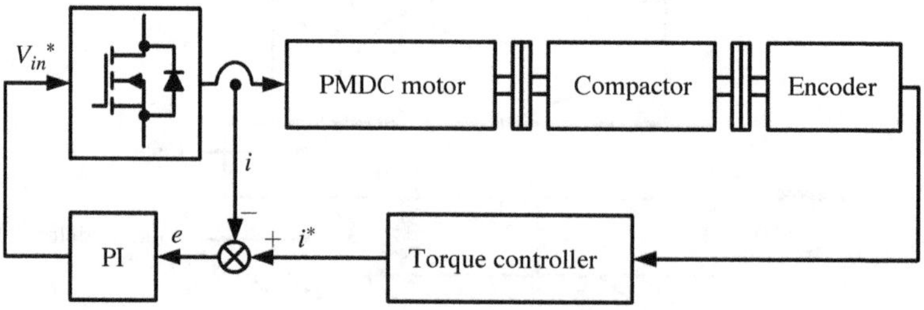

Figure 10.4 PMDC drive system for electric compaction

As detailed in Chapter 7, the control law of the PTD control is designed as:

$$i^* = \frac{\alpha J}{K_T} \sin[\sigma\omega(t-\tau)] - \frac{\beta J}{K_T}(\omega - \bar{\omega}) \tag{10.7}$$

where i^* is the current reference, σ is the time-delay gain, τ is the time-delay constant, α and β are proportional gains, and $\bar{\omega}$ is the nominal rotational speed.

As detailed in Chapter 7, the CSR control is to force ω to follow a chaotic speed reference ω^*. The corresponding control law is represented as:

$$i^* = K_p(\omega - \omega^*) + K_i \int_0^{+\infty} (\omega - \omega^*)\mathrm{d}t \tag{10.8}$$

$$\omega^* = \tilde{\omega} + \Delta\omega\xi_k \tag{10.9}$$

$$\xi_k = A\xi_{k-1}(1 - \xi_{k-1}) \tag{10.10}$$

where $\tilde{\omega}$ is the base speed, $\Delta\omega$ is the speed boundary, and $\{\xi_k\}$ is the chaotic sequence generated by the well-known Logistic map with a control parameter A.

The chaotic compactor is first compared with a conventional compactor. It adopts the constant feed-in voltage (CFV):

$$V_{in}^* = \bar{V} \tag{10.11}$$

where $\bar{V}$ is a constant voltage. Additionally, the chaotic compactor is also compared with an advanced compactor which uses the sinusoidal speed reference (SSR):

$$\omega^* = \tilde{\omega} + \Delta\omega[1 + \sin(2\pi f t)]/2 \tag{10.12}$$

where f is the varying frequency.

10.1.3 Compaction Simulation

In order to compare the performance of chaotic and nonchaotic compactors, they adopt the same PMDC motor with the key parameters listed in Table 10.1. Also, they should provide the same average rotor speed in order to enable a fair comparison. Given $M_e = 0.296$ kg, $M_p = 3.404$ kg, $r = 31.5$ mm, $\bar{V} = 21.7$ V, $f = 0.5$ Hz, $\tilde{\omega} = 130$ rad/s, $\Delta\omega = 30$ rad/s, $\alpha = 400$, $\beta = 10$, $\tau = 1$ s, $\sigma = 1$, $\bar{\omega} = 170$ rad/s, $A = 4$ and the eccentric mass inertia $J_e = 9.12 \times 10^{-5}$ kgm^2, the simulated rotor speed waveforms of the vibratory

Table 10.1 Key parameters of PMDC motor

Torque constant K_T	0.2286 Nm/A
Back EMF constant K_E	0.2286 V/rad/s
Armature resistance R_a	3.42 Ω
Armature inductance L_a	3.4 mH
Viscous damping coefficient B	8×10^{-5} Nm/rad/s
Rotor inertia J	1.588×10^{-4} kgm^2

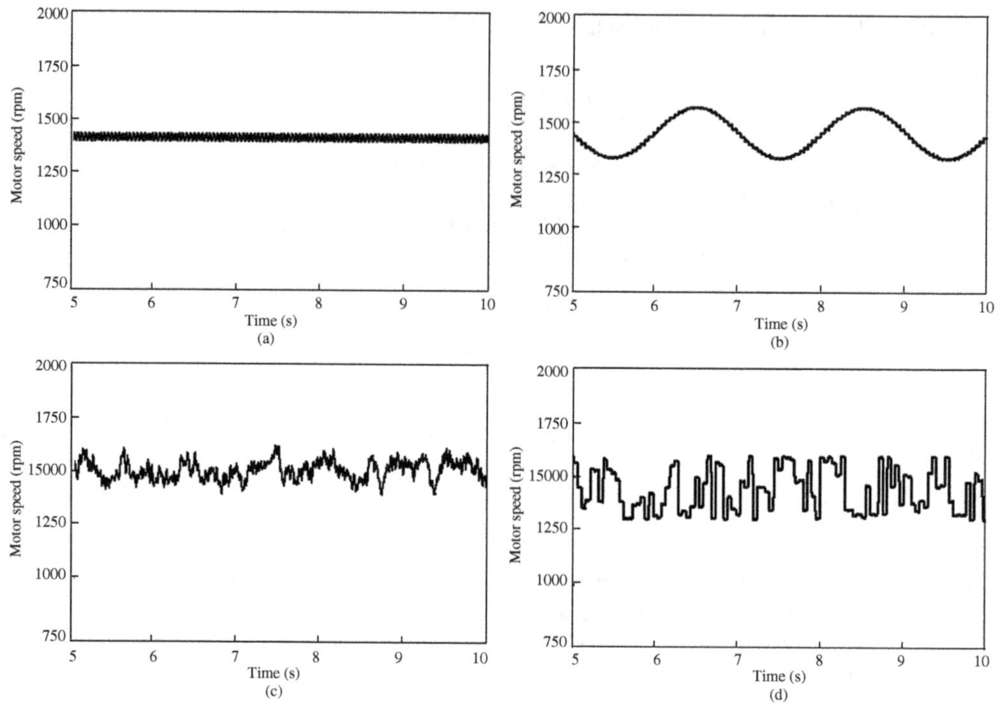

Figure 10.5 Simulated motor speed waveforms. (a) CFV. (b) SSR. (c) PTD. (d) CSR

compactor with different control schemes are shown in Figure 10.5. It can be found that they have the same average value.

The performance of the compactors using CFV, SSR, PTD and CSR control schemes is assessed by simulation. The average compaction energy density (ACED) is a well-accepted indicator to assess the compaction effectiveness of vibratory compactors (Tran and Muro, 2004). This ACED is defined as:

$$E = 2\beta_v\left(M_c + \frac{F_{vm}}{2}\right)\frac{f_c N_c}{V_h W_h t_s} \tag{10.13}$$

where β_v is the vertical amplitude of displacement measured on the terrain surface, M_c is the total gravity of the compactor, F_{vm} is the maximum vertical exciting force, f_c is the frequency of compactor motion, N_c is the number of compaction passes, V_h is the horizontal speed of vehicle movement, W_h is the horizontal compaction width and is orthogonal to the vehicle movement, and t_s is the thickness of soil. As (10.13) only describes the average compaction energy density for periodic motion, it cannot be directly used for chaotic motion. Thus, (10.13) is modified by setting β_v equal to y_p and t_s equal to r_0, defining L as the fixed horizontal movement of compaction vehicle and W as the horizontal compaction width, and then integrating the compaction energy density for a period T. Hence, it yields:

$$E = \frac{1}{T}\int_{t_0}^{t_0+T} f(t)\mathrm{d}t \tag{10.14}$$

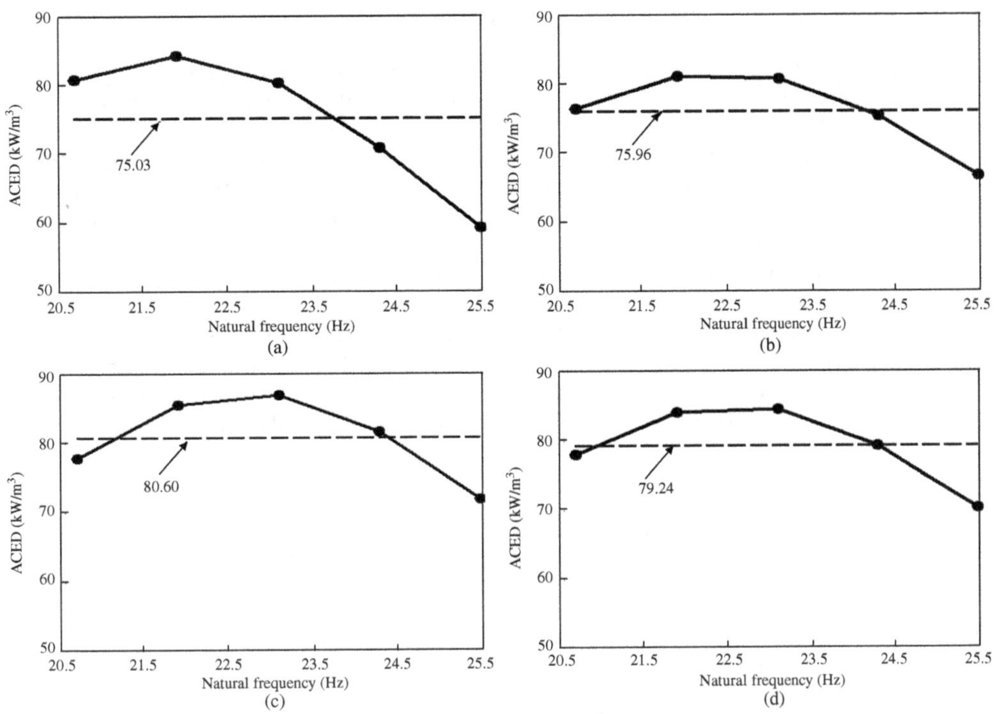

Figure 10.6 Simulated compaction performances. (a) CFV. (b) SSR. (c) PTD. (d) CSR

$$f(t) = \begin{cases} \dfrac{(M_p + M_e)g\dot{y}_p}{LWr_0} & (\dot{y}_p > 0) \\ 0 & (\dot{y}_p \le 0) \end{cases} \tag{10.15}$$

where L, W, and r_0 are respectively the length, width, and thickness of the solid particles under compaction. During the simulation, L is 150 mm, W is 150 mm, and r_0 is 63 mm.

Since practical soil and concrete materials have solid particles with different sizes and hence different natural frequencies f_n, five sets of K_s and C_s are chosen to represent five different particles for simulation. Namely, K_s are chosen as 62.55, 69.97, 77.80, 86.05, and 94.72 kN/m, while the corresponding C_s are chosen as 234.06, 247.55, 261.04, 274.53, and 288.02 N/m/s. Thus, according to $f_n = \sqrt{K_s/(M_p + M_e)}/2\pi$, the natural frequencies are distributed between 20.7 and 25.5 Hz.

The ACED of the compactors using different control schemes under the same power consumption are simulated, as shown in Figure 10.6. The average ACED values of the compaction with five different solid particles are 75.03 kW/m^3 with CFV control, 75.96 kW/m^3 with SSR control, 80.60 kW/m^3 with PTD control, and 79.24 kW/m^3 with CSR control. This indicates that the PTD and CSR compactors have better compaction performance than the conventional CFV compactor and the advanced SSR compactor. Quantitatively, the chaotic CSR compactor offers better ACED than the conventional CFV compactor by 5.6%.

10.1.4 Compaction Experimentation

In order to prove the effectiveness of the electrically chaotic compactor experimentally, the whole system is prototyped, as shown in Figure 10.7. The parameters of the compactor are the same as that for simulation. The mixed solid particles are composed of soybeans, small green beans, small red beans, big red beans, and rice, which will provide a wide range of natural frequencies for compaction. The control schemes are implemented by the dSPACE CP1104 control board. It generates proper PWM signals to drive the MOSFET-based DC–DC converter, which in turn controls the PMDC motor. The voltage and current feedbacks of the motor are respectively measured by a voltage sensor LEM LV25-P and a current sensor LEM LA25-NP, which are also used to compute the power consumption.

Firstly, based on an encoder coupled with the motor shaft, the motor speed waveforms of the compactor using different control schemes are measured as shown in Figure 10.8. The measured nonchaotic results agree well with the simulation results in Figure 10.5. It should be pointed out that the measured chaotic speed waveforms using PTD and CSR controls cannot fully match the simulated results, which is actually

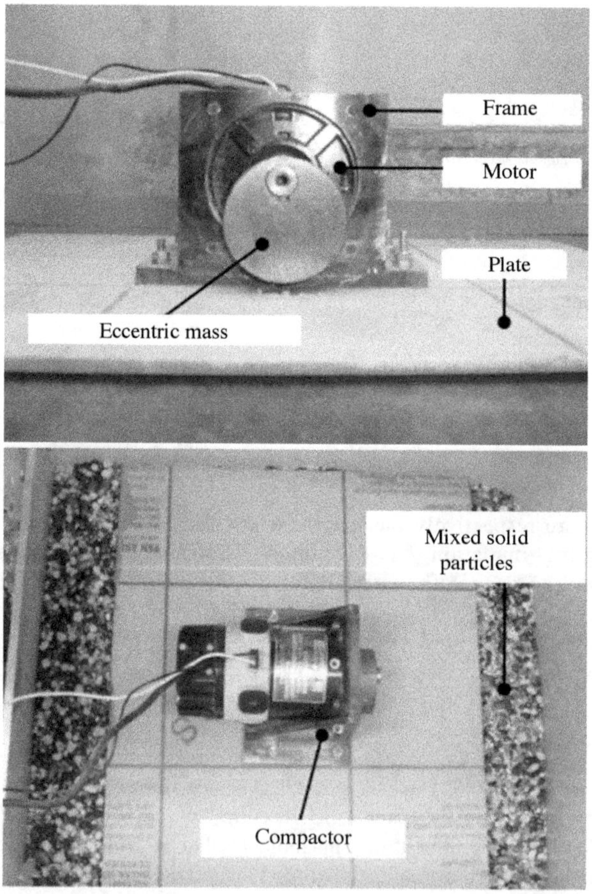

Figure 10.7 Prototype of vibratory compactor system

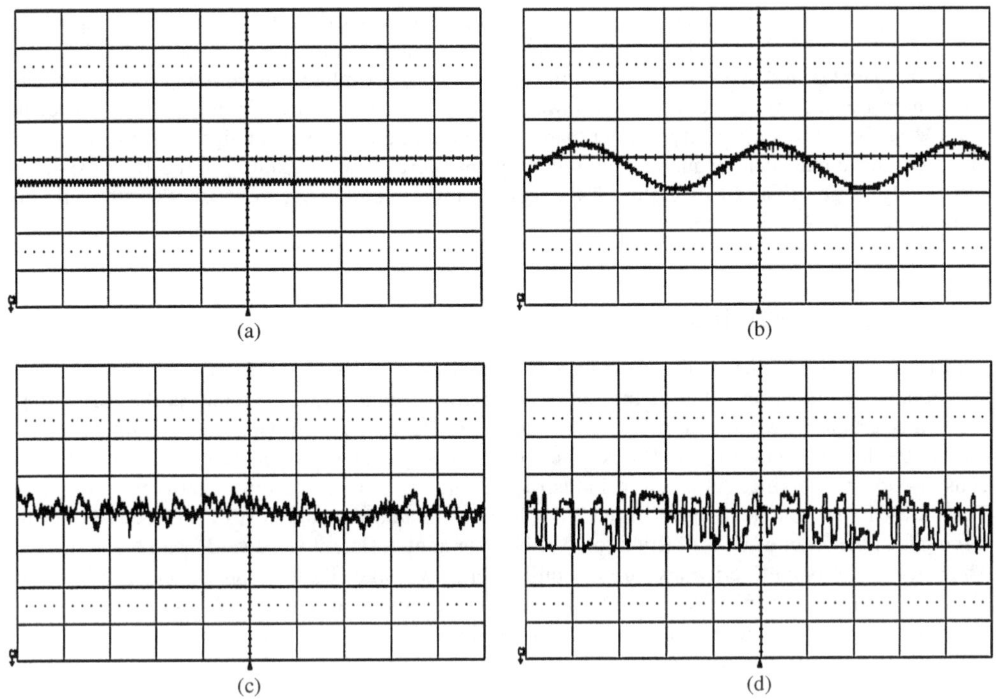

Figure 10.8 Measured motor speed waveforms (500 ms/div; 191 rpm/div). (a) CFV. (b) SSR. (c) PTD. (d) CSR

the random-like nature of chaos. Nevertheless, the measured chaotic motion boundaries closely match the simulated motion boundaries.

Secondly, the compactors using different control schemes are used to press the mixed solid particles. The compaction of each control is carried out five times, and the average fall height of the solid particles under compaction is physically measured as shown in Table 10.2. During the experiments, the energy consumption is calculated by online integrating the measured motor voltage and current based on the dSPACE control board, while the fall height is directly measured by using a vernier caliper. The outcomes are 23.4 mm with CFV control, 24.5 mm with SSR control, 25.1 mm with PTD control, and 25.2 mm with CSR control. For all compaction processes, the energy consumption is kept at 1 kJ, and the initial height of the solid particles is set at 95 mm. Thus, it experimentally verifies that the chaotic PTD and CSR compactors can offer better compaction performance than the conventional CFV compactor and advanced SSR compactor. Quantitatively, the CSR compactor can create larger fall height than the CFV compactor by 7.3 per cent.

Table 10.2 Comparison of compaction performance

	Initial height (mm)	Final height (mm)	Fall height (mm)
CFV	95	71.6	23.4
SSR	95	70.5	24.5
PTD	95	69.9	25.1
CSR	95	69.8	25.2

In order to scale up this chaotic compactor system to the industrial level, the rated torque of the motor should be scaled up while keeping the rated speed near the natural frequencies of solid particles. Then, the weights and sizes of the eccentric mass, frame, and plate can be magnified accordingly. Also, there are two basic criteria to be followed: firstly, the resistive torque generated by the eccentric mass $M_e gr$ should be sufficiently smaller than the rated motor torque so that high-performance control of the eccentric mass can be guaranteed; secondly, the corresponding exciting force $M_e \omega^2 r$ should be sufficiently larger than the weight of the compactor system $(M_p + M_e)g$ so that the vibratory compaction can be implemented.

10.2 Chaotic Mixing

Industrial mixers are one of the most important mixing devices in the food, drug, chemical, and semiconductor industries. One of the major problems of conventional mixers is the formation of segregated regions when mixing fluids with low Reynolds numbers (Dong, Johansen, and Engh, 1994). The occurrence and persistence of these segregated regions require extensive energy consumption for mixing. As a result, industrial mixers are among the most ineffective equipment. The industrialists and academics in the USA have estimated that the cost of ineffective industrial mixing is in the order of US$ 1 billion to 10 billion per annum (Harnby, Edwards, and Nienow, 1992). The effects of this ineffectiveness are not only energy wastage (Raynal and Gence, 1997), but can also be disastrous; for example, a nuclear-chemical waste explosion in Russia has been attributed to improper mixing of volatile compounds (Alvarez-Hernández *et al.*, 2002). Although mixing can be improved by increasing the rotational speed, such high-speed operation generally consumes additional energy and is sometimes impractical. Particularly, some shear-sensitive materials for biotechnological applications, such as proteins and other macromolecules, are readily damaged by high shear rates when adopting high rotational speed. Thus, the improvement of mixing is highly desirable and justifiable.

In order to improve the mixing process, time-varying rotation has been proposed (Lamberto *et al.*, 1996), aiming to destroy the segregated regions formed in the mixing process. Among various time-varying schemes, bidirectional rotations with different frequencies are adopted for industrial mixing. In addition, the bidirectional rotation can be further modulated by a sinusoidal wave. The use of time-varying rotation to improve mixing is due to the principle that the flow is continuously perturbed, hence preventing the formation of coherent segregated regions. Conceptually, it is similar to kneading the bread dough where it is stretched and folded repeatedly to create a well mixed result. In recent years, time-varying rotational mixing has been further extended to chaotic mixing because chaos inherently offers the properties of stretching and folding which match with the aforementioned requirement of a good mix (Ottino, 1989). Although there are many studies on chaotic mixing, only a few approaches are considered to be practical. By properly designing the radius of the chamber wall, the radius of twin rotors and the gap size, and then separately controlling the rotational speeds, a twin-screw mixer was developed for chaotic mixing (Jana and Sau, 2004). By moving one of the vanes in the central impeller upward by half the vane height and one adjacent vane downward by the same distance, a perturbed three-impeller design was proposed to create chaotic motion for mixing (Alvarez-Hernández *et al.*, 2002). By continuously varying the angle of the impeller shaft with respect to the vertical axis, a chaotic mixer was also created (Fountain *et al.*, 2000). However, all these chaotic mixers generate the desired chaotic motion by mechanical means, thus suffering from two fundamental problems – complexity and inflexibility. Firstly, the complex configurations significantly increase the cost of the system, enlarge the hardware size, and reduce the operation reliability. Secondly, the inflexible designs definitely limit the applicability and generality since the mixing materials and the mixing tanks are prone to changing. In order to fundamentally solve these problems, the desired chaotic motion for mixing should be produced by electrical means (Chau *et al.*, 2004).

Therefore, the purpose of this section is to newly chaoize a drive system, hence resulting in a controllable chaotic motion, for application in industrial mixing processes. Compared with the aforementioned mechanical means, the chaotic drive system not only produces the desired chaotic mixing but also offers the advantages of mechanical simplicity, high flexibility, and high controllability. As discussed before, the chaotic drive system can be a DC drive system (Ye and Chau, 2007) or an AC drive system (Ye and Chau, 2005a), while the corresponding chaotic motion can be produced by various control-oriented means (Ye and Chau, 2005b) or design-oriented means (Ye and Chau, 2005a). Therefore, in this section, the discussion is focused on the use of a PMDC drive system and the associated means of control.

10.2.1 Mixer System

Figure 10.9 shows the basic configuration of an industrial mixer system, which is composed of a tank stirred by a 6-bladed Rushton turbine with six equally spaced vertical vanes. Notice that the Rushton turbine is only a typical example, and other impellers such as the propeller, paddle, and helical ribbon can be used.

A PMDC motor is used as the agitator, which can be modeled as:

$$\frac{\mathrm{d}}{\mathrm{d}t}\begin{pmatrix}\omega \\ i\end{pmatrix} = \begin{pmatrix} -\dfrac{B}{J} & \dfrac{K}{J} \\ -\dfrac{K}{L} & -\dfrac{R}{L}\end{pmatrix}\begin{pmatrix}\omega \\ i\end{pmatrix} + \begin{pmatrix} -\dfrac{T_l}{J} \\ \dfrac{U}{L}\end{pmatrix} \tag{10.16}$$

$$T = Ki \tag{10.17}$$

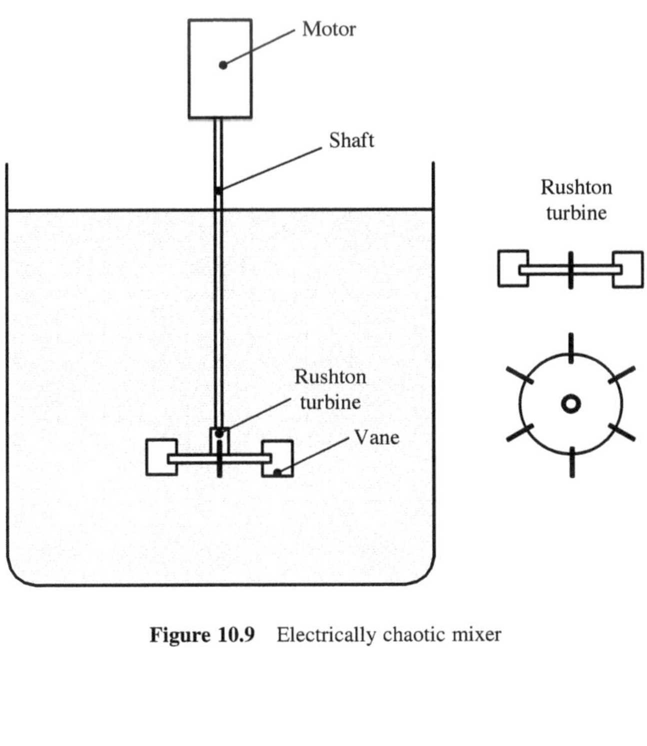

Figure 10.9 Electrically chaotic mixer

where B is the viscous damping coefficient, J is the rotor inertia, K is the torque constant, L is the armature inductance, R is the armature resistance, T is the motor torque, T_l is the load torque, U is the DC supply voltage, ω is the motor speed, and i is the armature current.

10.2.2 Chaotic Mixing Control

As detailed in Chapter 7, the PMDC drive system can be chaoized by a time-delay feedback control, as governed by:

$$T^* = \mu \xi B \sin\left(\frac{\omega(t-\tau)}{\xi}\right) \tag{10.18}$$

$$i^* = \frac{T^*}{K} \tag{10.19}$$

where T^* is the torque command, i^* is the current command, μ is the torque parameter, ξ is the speed parameter, and τ is the time-delay parameter. It should be noted that all three parameters are adjustable to achieve the desired chaotic motion.

Based on the above derivation, both the armature current and rotor speed of the PMDC motor are used as feedback signals, while the torque command T^* calculated by (10.18) is used to generate the current command i^* for pulse width modulation (PWM) control. Figure 10.10 shows the corresponding control system. Firstly, the measured speed feedback is delayed by a preset value of τ. Then, the delayed speed is fed into the torque control block in which proper values of μ and ξ are preset. Hence, it generates T^* and then i^*. Subsequently, the difference between i^* and the measured current feedback is fed into the current control block in which a simple PI control is adopted. Hence, it generates the desired duty ratio for the full-bridge PWM converter that provides bidirectional current control of the PMDC motor.

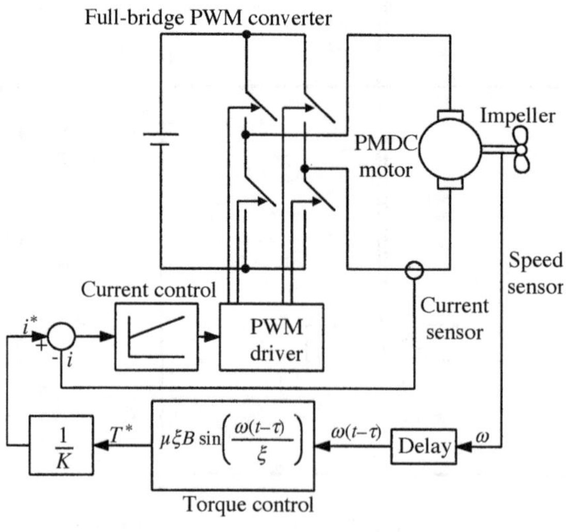

Figure 10.10 Chaotic mixing control

Table 10.3 PMDC motor parameters

Supply voltage U	24 V
Torque constant K	0.05 Nm/A
Armature resistance R	1.1 Ω
Armature inductance L	0.4 mH
Viscous coefficient B	2.2×10^{-3} Nm/rad/s
Rotor inertia J	1.0388×10^{-5} Nm/rad/s^2

10.2.3 Chaotic Mixing Simulation

In order to conduct computer simulation, realistic system parameters are adopted. Table 10.3 summarizes some practical data of the PMDC motor. A detailed analysis of the effects of μ, ξ, and τ has been discussed in Chapter 7. When selecting $\mu = 1.4$, $\xi = 10$, and $\tau = 1.5$ s, the motor speed and armature current waveforms are as shown in Figure 10.11. It can be seen that the motor

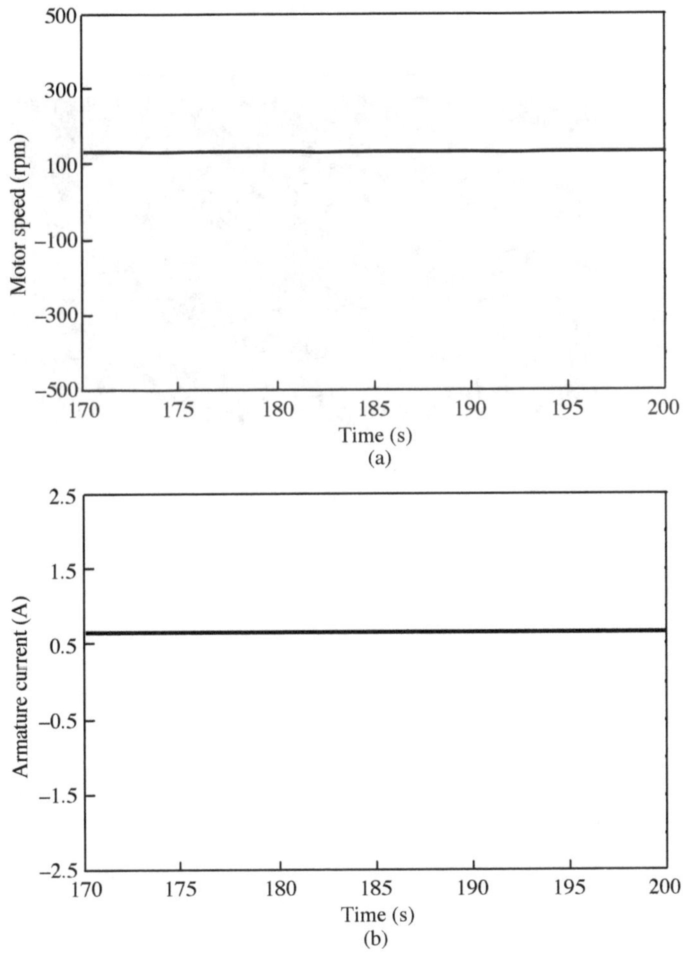

Figure 10.11 Normal waveforms when selecting $\mu = 1.4$, $\xi = 10$ and $\tau = 1.5$ s. (a) Motor speed. (b) Armature current

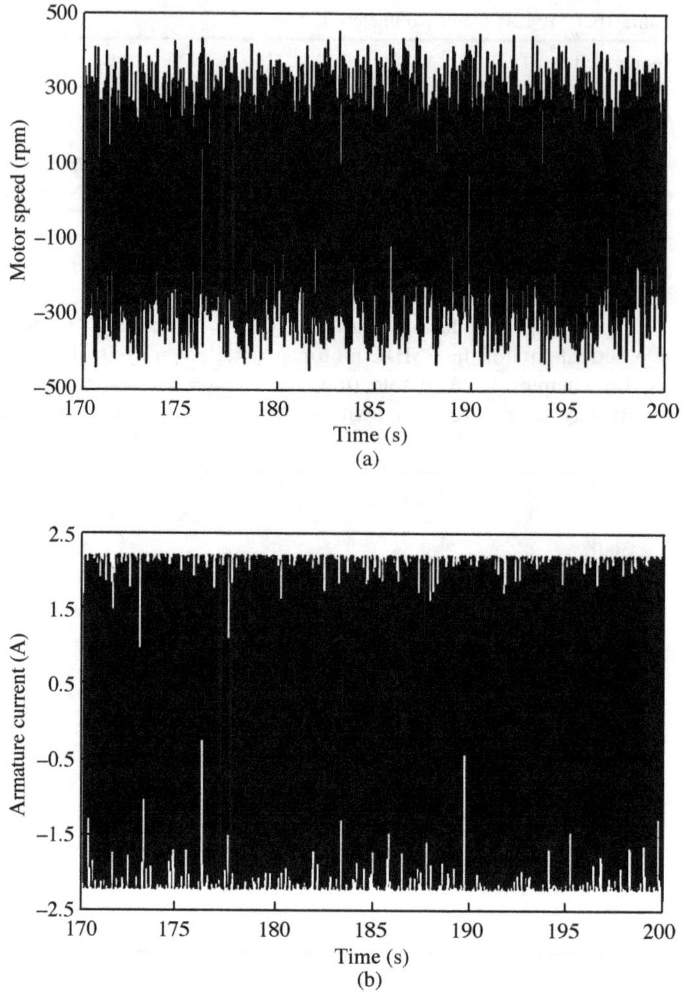

Figure 10.12 Chaotic waveforms when selecting $\mu = 5, \xi = 10$ and $\tau = 1.5$ s. (a) Motor speed. (b) Armature current

operates at a fixed point (which is equivalent to the normal or so-called period-1 operation). When selecting $\mu = 5$, $\xi = 10$, and $\tau = 1.5$ s, the motor exhibits chaotic motion. The corresponding speed and current waveforms are shown in Figure 10.12. It can be found that both the amplitude and direction of the motor speed and armature current change with time, thus presenting ergodicity. It is this character of ergodicity that differentiates chaotic mixing and normal constant speed mixing. When selecting $\mu = 5$, $\xi = 20$, and $\tau = 1.5$ s, the corresponding chaotic speed and current waveforms are as shown in Figure 10.13. It can be seen that ξ can be used to adjust the speed range of the chaotic motion, which is why it is called the speed parameter. Furthermore, when selecting $\mu = 5$, $\xi = 10$, and $\tau = 0.01$ s, the corresponding chaotic speed and current waveforms are shown in Figure 10.14. It can be seen that a change in the time-delay constant has little effect on the

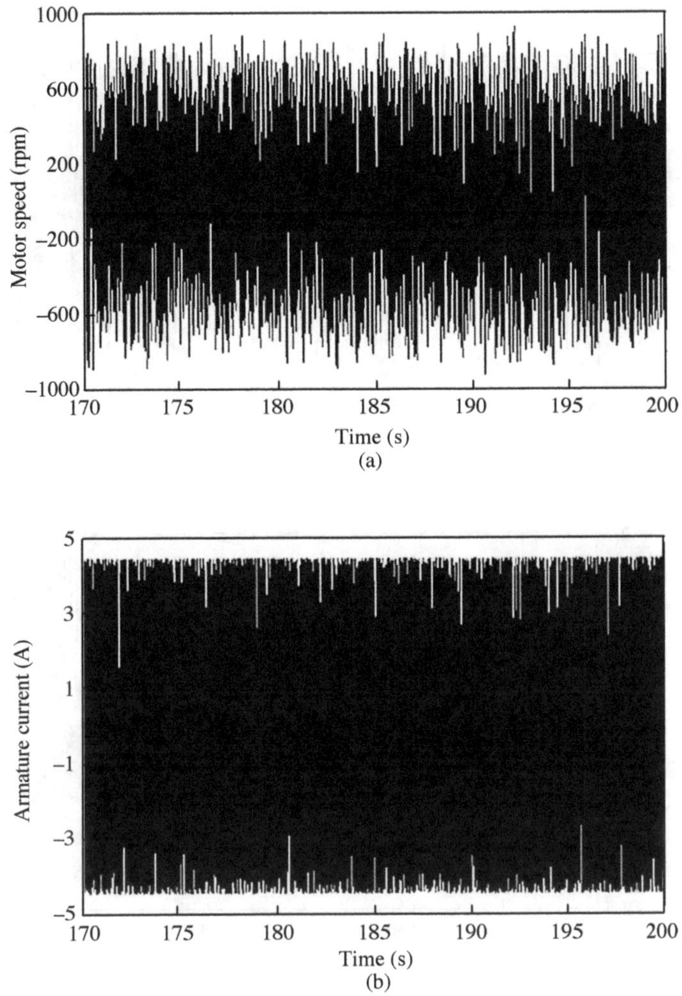

Figure 10.13 Chaotic waveforms when selecting $\mu = 5, \xi = 20$ and $\tau = 1.5$ s. (a) Motor speed. (b) Armature current

speed range. The significance of this parameter lies in the realization of the system. If the parameter is too large, the refresh rate of the control system is too low; and if the parameter is too small, it requires too much computer resource.

It should be noted that the above simulation results are under a no-load condition, because the mathematical modeling of fluid dynamics during mixing is too complicated to be handled. The corresponding fluid dynamics involve a set of partial differential equations – known as the Navier–Stokes equations – which can only be solved by using such sophisticated methods as the 3-D finite element method (Fountain *et al.*, 2000). Therefore, the simulation results are used to illustrate the chaoization of the drive system, whereas the experimental results will be used to illustrate the chaotic mixing process in which the mixing load is taken into account.

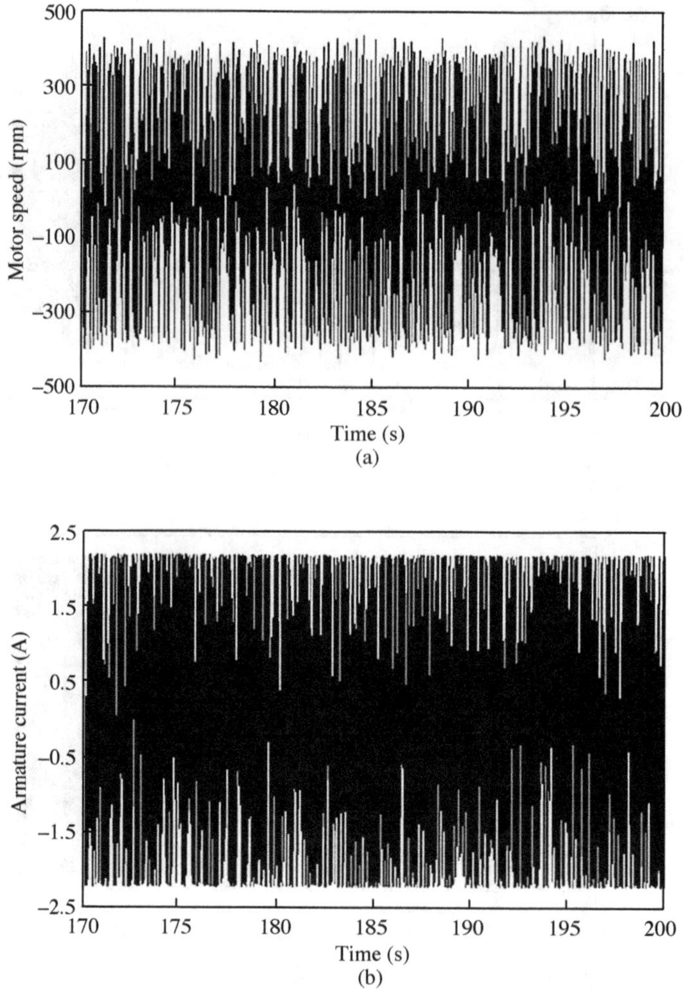

Figure 10.14 Chaotic waveforms when selecting $\mu = 5$, $\xi = 10$ and $\tau = 0.01$ s. (a) Motor speed. (b) Armature current

10.2.4 Chaotic Mixing Experimentation

The mixing system consists of a tank and an impeller spun by a digitally controlled electric drive. The motor is mounted vertically on a stand with its shaft positioned at the center of the tank. The shaft is mounted through a holding plate, which ensures consistent positioning between the experiments and minimizes oscillations of the shaft tip. As shown in Figure 10.15, the adopted motor is a PMDC motor with the parameters listed in Table 10.3; the stand is a drill holder, and the tank is a 1-litre glass beaker.

It should be noted that, when using a mixture with a low or moderate Reynolds number, it is particularly difficult to achieve an effective mix since the corresponding flow is laminar, whereas a mixture with a high Reynolds number can easily achieve an effective or so-called turbulent mix (Chaté, Villermaux, and

Figure 10.15 Experimental mixing system

Chomaz, 1999). Thus, a viscous solvent (light corn syrup) is purposely adopted so that the mixing effectiveness can be evaluated.

There are many methods to evaluate the mixing processes, which can be divided into two categories: intrusive and nonintrusive. The intrusive methods include a probe or tracer that is put in the stirred tank to measure flow velocities, but they disturb the flow patterns that the investigators intend to measure. The nonintrusive methods, such as the laser Doppler anemometer (Mavros, 2001) and the acid-base neutralization reaction (Ascanio *et al.*, 2002), are more attractive since they do not disturb the flow patterns. Increasingly, the acid-base neutralization reaction offers the advantages of simple arrangement and low cost. This method is therefore adopted to evaluate the chaotic mixing.

Firstly, the tank is filled with 200 ml of light corn syrup, 5 ml of pH indicator solution (universal indicator), and a 5-ml solution of 1 N HCl. The solution is mixed until a uniform red color is observed since it is acidic. Next, another well-mixed solution (dark green color) – 100 ml of light corn syrup, 2.5 ml of pH indicator solution, and 2.5 ml of 1 N NaOH – is added into the tank. Although the whole solution is acidic as there is twice as much acid as base, there are dark green regions since diffusion is limited by the viscous solvent (light corn syrup).

The beginning of the mixing process of the acid-base solution is set as time zero. The drive is controlled by a dSPACE digital controller to realize various mixing methods, including constant-speed, periodic, and chaotic motions. The whole experiment is recorded by using a camera focused at the impeller.

To assess whether chaotic mixing is more effective than other mixing methods, it is equivalent to evaluate whether chaotic mixing can offer a more homogeneous mixture under the same amount of energy consumption. The experiment is designed to compare the mixing time needed to achieve homogeneity based on the same input power. Firstly, the chaotic mixing experiment is conducted. The input voltage and

Table 10.4 Comparison of different mixing methods

Mixing Method	Time (s)
Constant-speed	360
Rectangularly bidirectional	33
Sinusoidally bidirectional	33
Chaotic	30

current are recorded online until homogeneity is achieved. Hence, the average input power is calculated. Then, the other mixing methods – namely, the conventional constant-speed mixing, the rectangularly bidirectional mixing, and the sinusoidally bidirectional mixing – are also conducted under the same input power for comparison.

As tabulated in Table 10.4, the chaotic mixing ($\mu = 30, \xi = 20$, and $\tau = 1.5$ s) takes 4.2 W to fully mix up the aforementioned acid-base solution within 30 s; whereas the constant-speed mixing at 600 rpm requires 360 s, the rectangularly bidirectional mixing with a magnitude of 600 rpm and a frequency of 0.5 Hz requires 33 s, and the sinusoidally bidirectional mixing with an amplitude of 850 rpm and a frequency of 0.5 Hz also requires 33 s, all under the same power of 4.2 W. Thus, it quantitatively verifies that chaotic mixing has the definite advantages of shorter mixing time and, hence, lower energy consumption than the others, particularly the conventional constant-speed mixing.

The improvement is expected since the constant-speed mixing process involves the formation of a segregated region which is the major obstacle for effective mixing, whereas the chaotic mixing process essentially prevents the formation of segregated regions. Figure 10.16 shows the grayscale

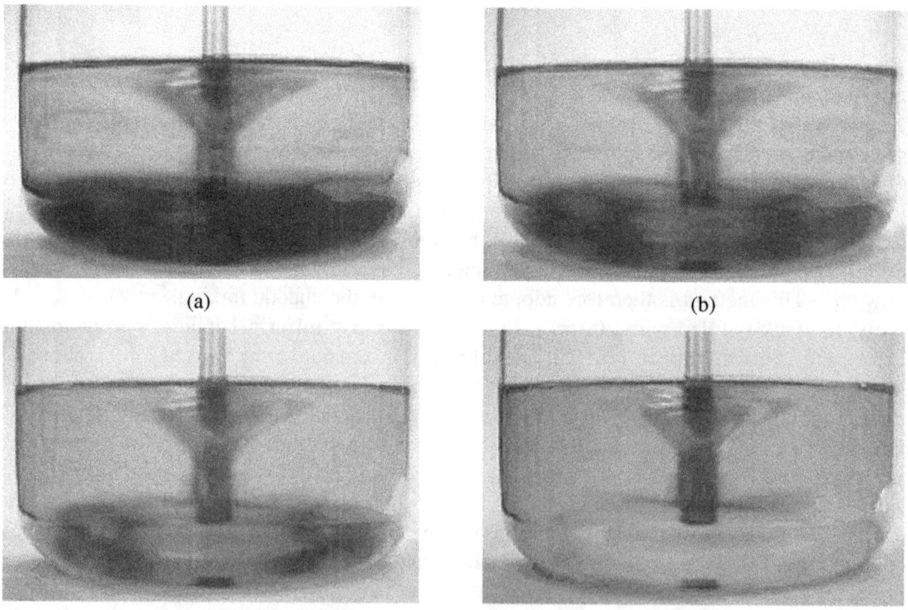

Figure 10.16 Constant-speed mixing. (a) After 5 s. (b) After 10 s. (c) After 20 s. (d) After 360 s

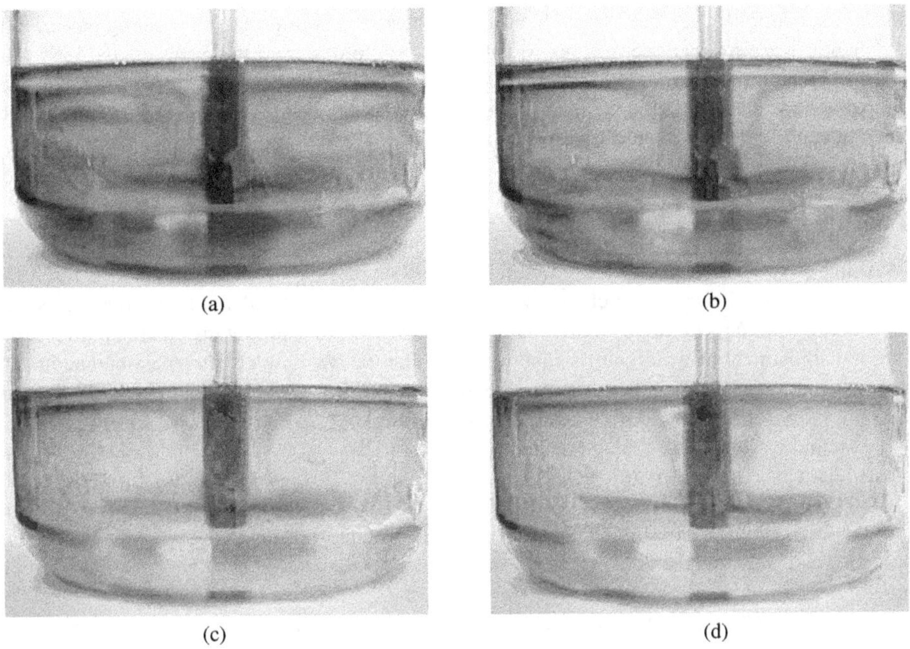

Figure 10.17 Chaotic mixing. (a) After 5 s. (b) After 10 s. (c) After 20 s. (d) After 30 s

level of the constant-speed mixing process, which illustrates that the segregated region (like a donut) persists for a long time. On the other hand, Figure 10.17 shows the grayscale level of the chaotic mixing process, which confirms that no segregated region occurs. For the two bidirectional mixing processes, there are segregated regions formed during the positive or negative sessions, but the change of direction leads to the destruction of larger donuts and, then, the formation of smaller ones. Therefore, these two bidirectional mixing processes still take a longer time than the chaotic mixing, although showing significant improvement over the conventional constant-speed mixing.

As mentioned, the segregated region is the key obstacle for the conventional constant-speed mixing process to achieve effective mixing. It has been identified that the size of this region depends on the Reynolds number of the mixture – in general, the lower the Reynolds number, the larger is the segregated region (Alvarez-Hernández *et al.*, 2002). Thus, chaotic mixing is particularly attractive for mixing those highly viscous fluids that have a low or moderate Reynolds number.

10.3 Chaotic Washing

Commercial and domestic washing machines are among the most energy-consuming devices. Actually, the Welsh Consumer Council has revealed that about one-third of the energy used in the UK is consumed in the home, and the washing machine is one of the major energy-consuming home appliances. Thus, an improvement in washing equipment is highly desirable and justifiable. There are two major kinds of washing machine, namely the clothes-washer and the dishwasher. The former is almost indispensable for all households, while the latter is highly desirable for all restaurants.

Similar to chaotic mixing, the use of chaotic water current can improve the washing power. In recent years, chaotic washing has been proposed for both the clothes-washer (Wang *et al.*, 1996) and the dishwasher (Nomura, Wakami, and Kazuyuki, 1996). However, they generate the desired chaotic motion by mechanical means, thus suffering from complexity and inflexibility. In order to solve these problems, the desired chaotic motion should be produced by electrical means.

10.3.1 Chaotic Clothes-Washer

For washing clothes, Goldstar produced a chaotic washer in 1993, which was claimed to be the first application of chaos theory to the clothes-washer. Its key is to utilize a small pulsator (which stirs the water) that rises and falls randomly as the main pulsator rotates, thus producing chaotic motion. Hence, it enables the adjustment of water whirls chaotically so that the washing effectiveness can be improved – namely, increasing the washing power and preventing the twisting of clothes. Consequently, an improved version was proposed in 1994, as shown in Figure 10.18. The key is to use two fan motors to force the water from the second washing tank through the induce holes into the first washing tank while the first washing tank is rotated by another motor, hence producing chaotic flow water for washing clothes (Wang *et al.*, 1996). Although the above chaotic clothes-washers can provide better performance than a traditional washer, the corresponding mechanisms to produce the desired chaotic water current are complicated and thus costly, leading to offset the merits of the use of chaotic clothes-washing for commercial washers.

As discussed in Chapter 7, the desired chaotic motion can be easily generated by electrical means. Namely, various electric drive systems can be chaoized to provide various chaotic motions by using control-oriented or design-oriented approaches. An electrical chaotic clothes-washer has recently been proposed (Ye, Chau and Niu, 2006). As shown in Figure 10.19, the pulsator of the washing tank is simply spun by a chaotic drive system, hence directly producing the desired chaotic water current for clothes-washing. Rather than using a PMDC drive, the shaded-pole induction drive is adopted by

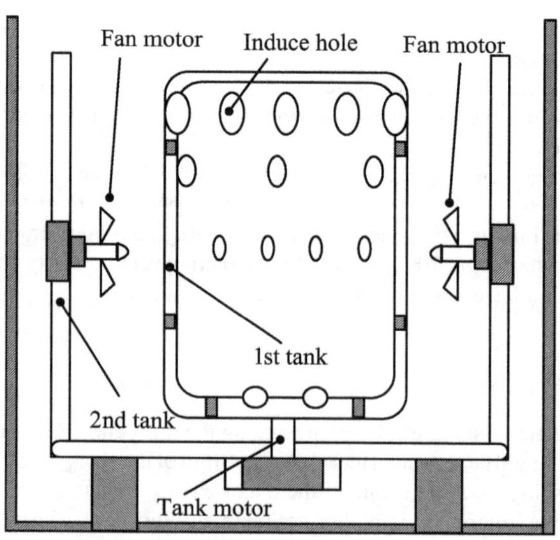

Figure 10.18 Mechanical chaotic clothes-washer

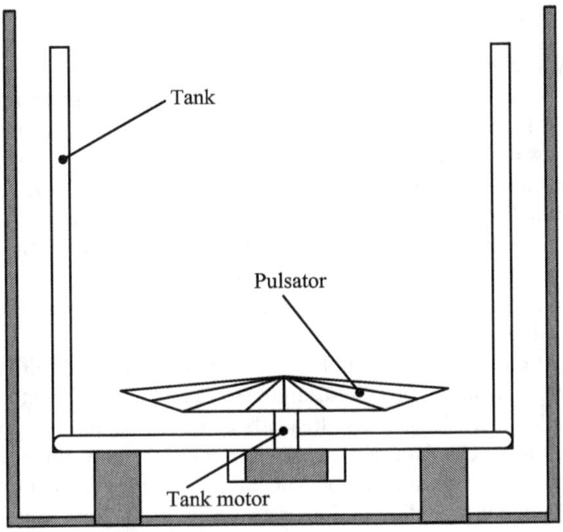

Figure 10.19 Electrical chaotic clothes-washer

this clothes-washer, since it offers higher efficiency, smaller size, more durability, and is maintenance free. Since the motor parameters are essentially fixed after production, the operating parameters – namely, the supply voltage and frequency – are selected as the bifurcation parameters. This shaded-pole induction drive system and its bifurcation diagrams have been thoroughly discussed in Chapter 7.

When selecting the supply voltage as 220 V and the frequency as 10 Hz, the shaded-pole induction drive can generate the desired chaotic speed for clothes-washing as shown in Figure 10.20. It should be noted that the pattern of chaotic motion can be altered by selecting different sets of supply voltage and frequency parameters.

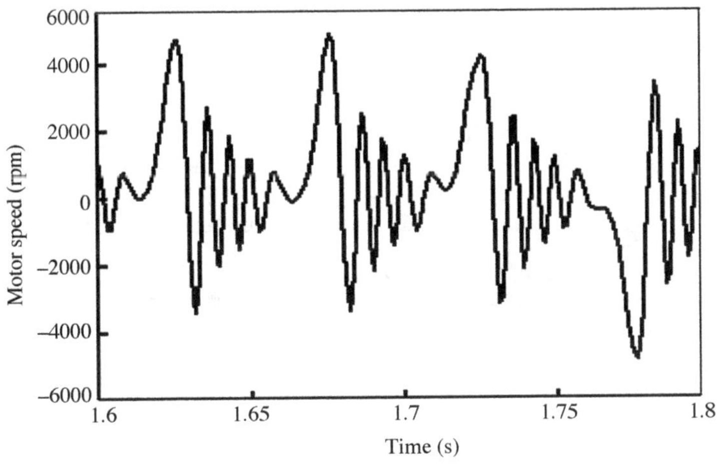

Figure 10.20 Chaotic motor speed for clothes-washing

10.3.2 Chaotic Dishwasher

The first commercially available dishwasher was introduced in 1960. With ever-increasing demands on our living standards, the use of dishwashers is becoming more and more attractive. A conventional dishwasher is shown in Figure 10.21, which consists of a dish basket, a rotatable link with multiple nozzles, and an electric water pump. As depicted in Figure 10.22, the water is pressurized by the pump, then goes through the inside of the link to eject water streams via multiple nozzles. Those nozzles ejecting upward water streams function to wash the dishes, whereas one particular nozzle is purposely designed to eject a horizontal water stream toward the wall. Due to the reaction force of the horizontal water stream, the link rotates on the pivot so that the upward water streams can cover a wide area to clean the dishes. However, the simple rotation of the link confines the coverage of the water streams, hence limiting the effectiveness of dishwashing.

Based on the anticipation that chaotic water streams should offer better dishwashing performance, a chaotic rotatable link mechanism was proposed, aiming to generate chaotic water streams for dishwashing (Nomura, Wakami, and Kondo, 1995). It has already been identified that a double pendulum can exhibit chaotic motion. Thus, the mechanism consists of two links, namely link-1 and link-2 as shown in Figure 10.23. Link-1 rotates on the central pivot, whereas link-2 rotates on a pivot mounted at an end point of link-1. Both the shape and weight distribution of the two links are not symmetrical with respect to the central pivot. A nozzle at an end of link-1 is designed to eject a horizontal water stream, while another nozzle at an end point of link-2 is designed to eject an upward-slant water stream. These two nozzles produce the reaction forces to rotate the links. Since the two links and the water streams affect each other in their motions, the mechanism exhibits a complex motion. When the shape of the links and the direction of the nozzles are properly designed, the complex motion is found to be chaotic (Nomura, Wakami, and

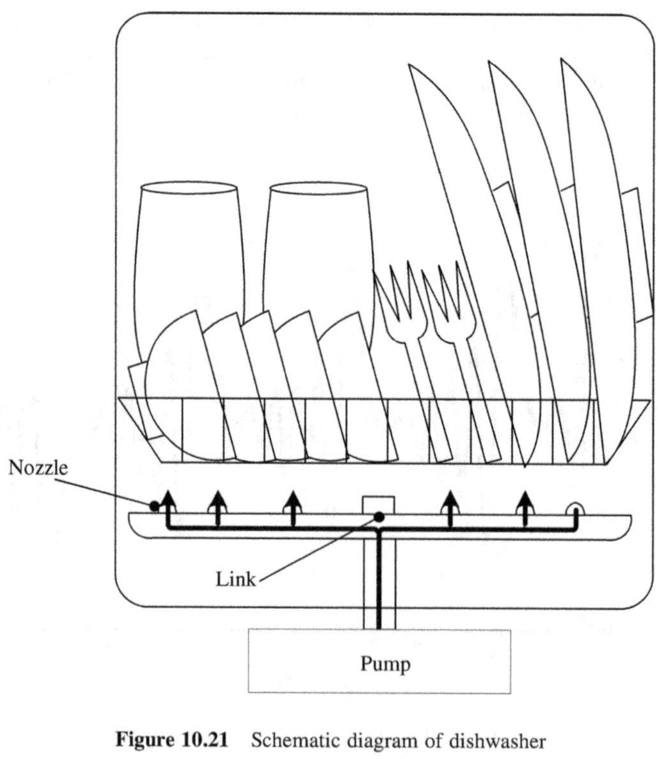

Figure 10.21 Schematic diagram of dishwasher

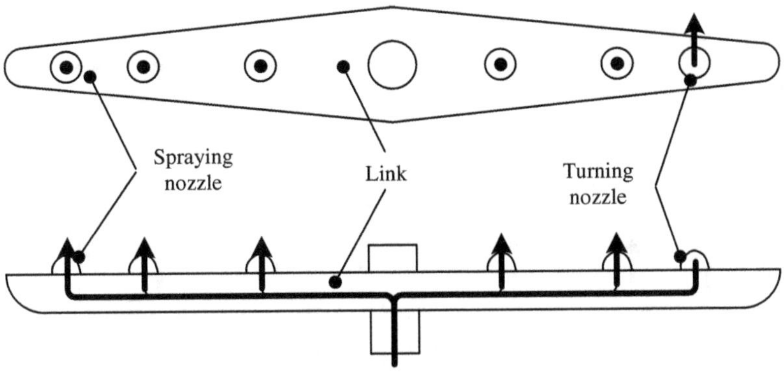

Figure 10.22 Structure of conventional rotatable link

Kazuyuki, 1996). Compared to a conventional dishwasher, this chaotic dishwasher can improve the effectiveness of dishwashing by 11 per cent.

Instead of using a complicated two-link mechanism to generate chaotic motion for dishwashing, an electric drive system can perform the same task more easily. Figure 10.24 shows the structure of an electrically chaotic rotatable link, which consists of a conventional link coupled with an electric drive and an electric water pump. The water is first pressurized by the pump, then goes through the inside of the link to eject water streams via multiple nozzles. All nozzles eject upward water streams to wash the dishes, while the motor chaotically rotates the link so that the upward water streams can effectively clean the dishes. Compared with the aforementioned chaotic motion mechanically generated by a two-link mechanism, the electrically generated chaotic motion has the following advantages:

- The rotatable link is simple in structure and can be independently controlled by the electric drive system. Notice that there is no need to spend horizontal or upward-slant water streams by this system.

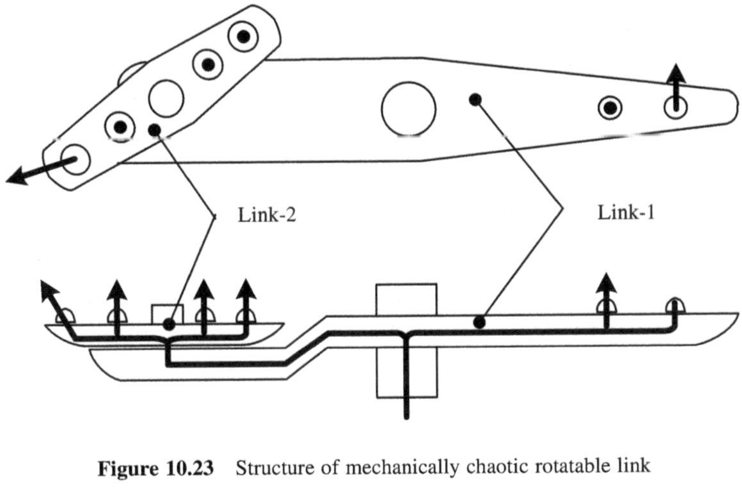

Figure 10.23 Structure of mechanically chaotic rotatable link

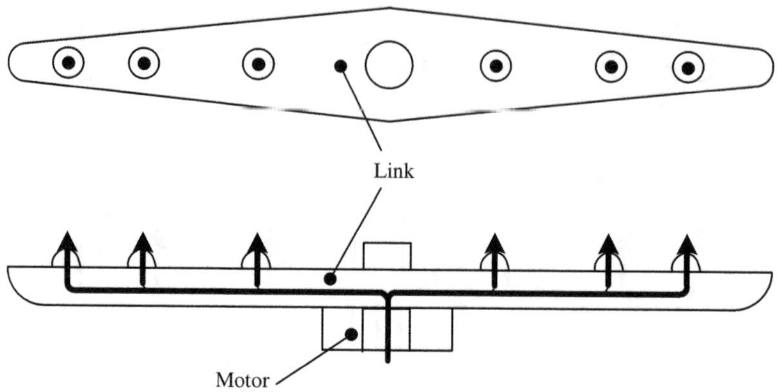

Figure 10.24 Structure of electrically chaotic rotatable link

- By using the electric drive system, the generated chaos motion is well defined and can be analytically formulated. Notice that the aforementioned two-link mechanism involves a complicated interaction between the links and water streams, and the corresponding complex motion is loosely found to be chaotic based on numerical time-series analysis of the measured data.
- The electric drive system can offer different types of chaotic motion, including bidirectional rotation, which are selectable for dishwashing different kinds of tableware. Notice that the two-link mechanism does not offer the necessary flexibility for different chaotic motions, or selectivity for different tableware.

10.4 Chaotic HVAC

HVAC is a well-accepted acronym that stands for heating, ventilating, and air-conditioning, which is particularly important in the design of industrial and office buildings – the so-called building services. In recent years, chaotic motion has been employed by various HVAC systems.

There is a general perception that the room temperature should be kept constant at the optimal value to ensure comfortableness. Thus, the building services engineers have made a great effort, at large expense, attempting to achieve a constant room temperature. However, thermal comfort research has revealed that people prefer imperceptible temperature swings when resting, whereas they desire artificial temperature swings when working (Wyon, 1973). Analysis has also shown that people choose to work harder when there are temperature swings, since they feel that the air is 'fresher'.

In order to provide appropriate temperature swings about the optimal set point for heating, a chaotic kerosene fan heater was developed in 1992, which was claimed to be the first consumer electronic product in the world using chaos theory (Katayama *et al.*, 1993). In this kerosene fan heater, the operating temperature is controlled in such a way that three types of chaotic temperature swing patterns about the optimal set point are selectable. These chaotic patterns are based on an intermittent chaos which can be represented by a simple one-dimensional Poincaré map (Procaccia and Schuster, 1983):

$$\begin{cases} x(t+1) = x(t) + ux(t)^z & (0 \le x(t) < 0.5) \\ x(t+1) = ax(t) + b & (0.5 \le x(t) < 1.0) \end{cases} \tag{10.20}$$

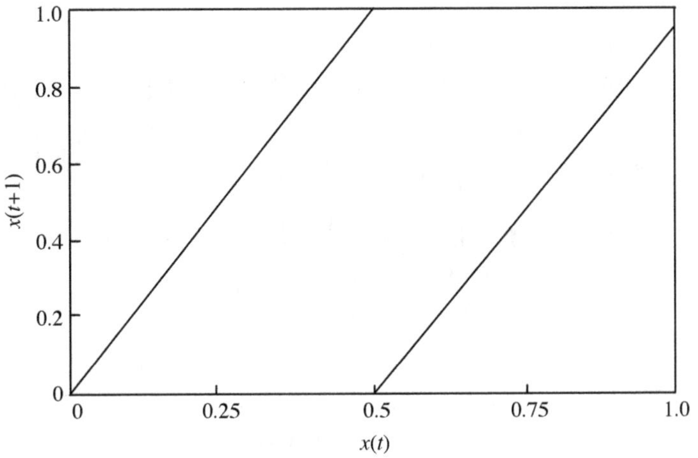

Figure 10.25 Trajectories of intermittent chaos

where u, a, and b are constants, and z is the parameter to control the intermittent period. By selecting $u = 1.4$, $a = 2$, and $b = -1$, the trajectories with $z = 1.5$ is plotted in Figure 10.25. The corresponding time series are shown in Figure 10.26. It can be found that the designer can simply tune z to generate the desired time series.

To assess the effectiveness of this chaotic kerosene fan heater, psychological experiments are conducted with about 30 persons (Kuwata *et al.*, 1996). Each person in the experimental room is requested to estimate the comfortableness of the room temperature according to a thermal sensation every five minutes. Table 10.5 lists the commonly used seven-point psychological scale for the measurement of thermal sensation (Fanger, 1970). As a result, the sensation is much more comfortable with chaotic temperature control than that with fixed temperature control.

Following the spirit of the aforementioned kerosene fan heater, chaotic temperature control can readily be extended to electric fan heaters and air-conditioners. The key is to electrically generate chaotic rotation of the electric drive in such a way that the rates of warming/cooling air flow, and hence temperature swings, are chaotic to enhance comfortableness. Figure 10.27 shows the schematic diagram of a chaotic fan heater. It consists of a direct resistance heating element, a single-phase AC fan motor and a power controller. This fan motor can be chaoized to provide chaotic rotation by control-oriented approaches such as using time-delay feedback control or design-oriented approaches such as selecting specific applied voltage and frequency (Gao, Chau, and Ye, 2005). On the other hand, Figure 10.28 shows the schematic diagram of a chaotic air-conditioner. It consists of a compressor, an internal heat exchanger (evaporator) coupled with an internal fan motor, an expansion valve, an external heat exchanger (condenser) coupled with an external fan motor, and a control unit. The internal fan motor is chaoized to provide the desired chaotic rotation so that a chaotic temperature swing can be obtained.

For chaotic mixing, a chaotic mixer can enhance the mixing effectiveness in terms of mixture homogeneity and energy consumption. Extending this concept from liquid to air, a chaotic fan can readily be used to enhance the ventilation effectiveness. In addition to chaoizing the AC fan motor, the chaotic nature can be further enhanced by chaoizing the direction of air flow – that is, the DC servomotor of swinging vanes can be further chaoized.

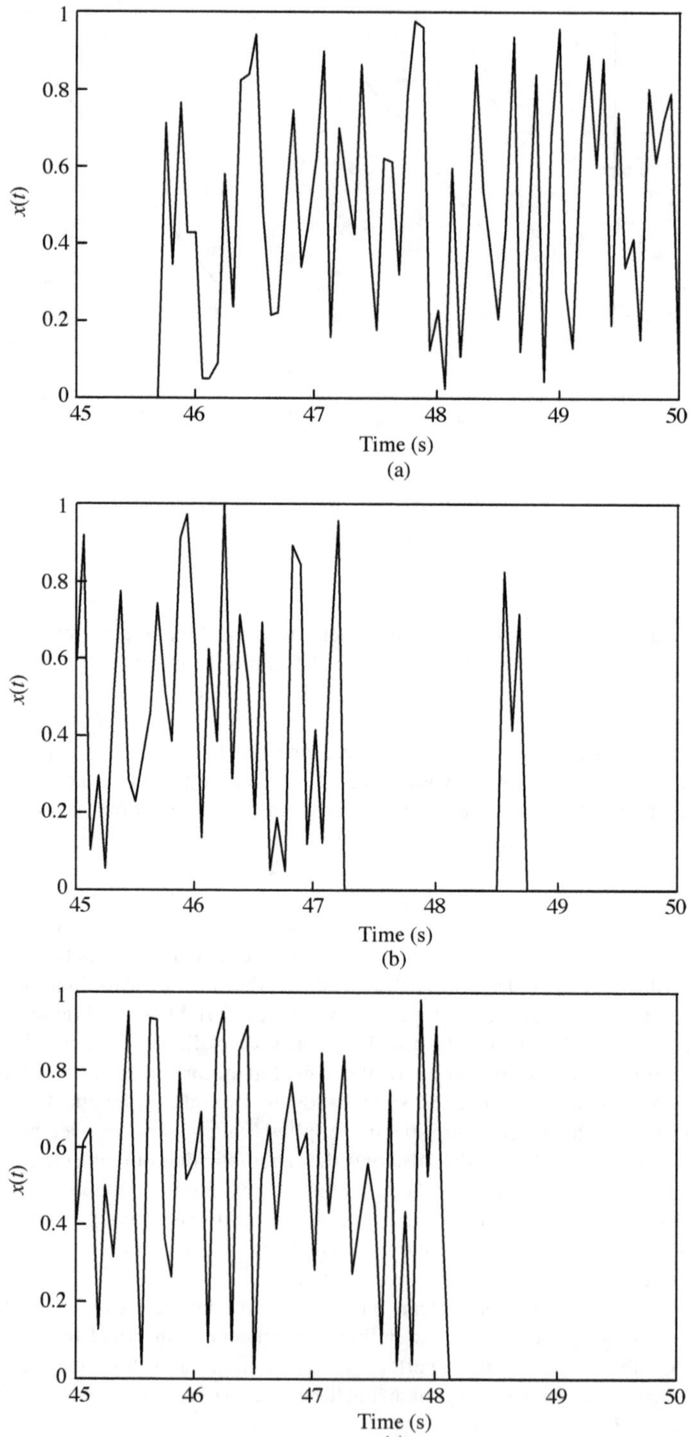

Figure 10.26 Time series of intermittent chaos. (a) $z = 1.5$. (b) $z = 2$. (c) $z = 2.5$

Table 10.5 Psychological scale for thermal sensation

Scale	Sensation
−3	Cold
−2	Cool
−1	Slightly cool
0	Neutral
+1	Slightly warm
+2	Warm
+3	Hot

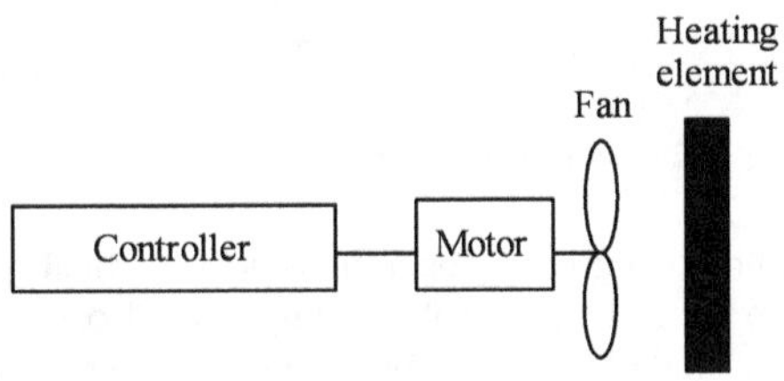

Figure 10.27 Schematic diagram of chaotic fan heater

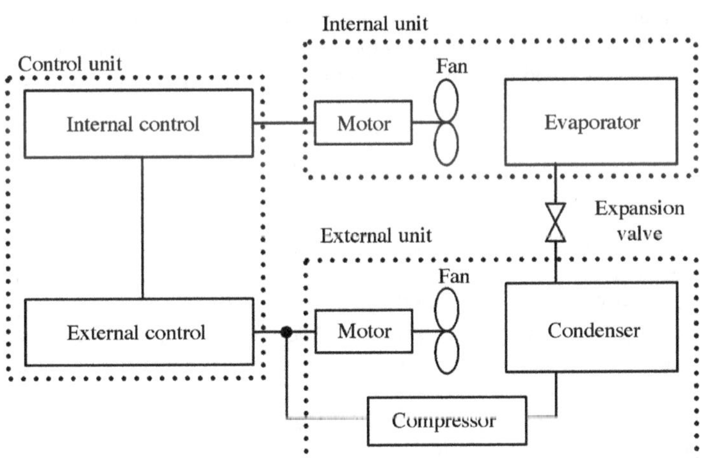

Figure 10.28 Schematic diagram of chaotic air-conditioner

10.5 Chaotic Grinding

Grinding is one of the most important operations in production and manufacturing engineering. It is performed by abrasive particles removing unwanted material to attain the desired geometric and surface properties. The abrasive particles are bonded together to form a revolving body, known as a grinding wheel

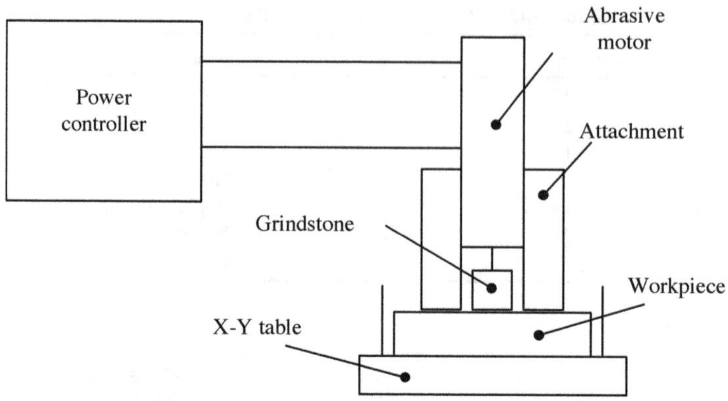

Figure 10.29 Schematic diagram of chaotic abrasive grinder

or grindstone. One of those important grinding operations is the vertical spindle surface grinding. It is an abrasive machining operation for stock removal grinding, and also for surface finish grinding within certain desired tolerances. The goal of stock removal grinding is to remove the unnecessary material as quickly or as cheaply as possible, whereas the goal of surface finish grinding is to provide the quality of surfaces (Shaw, 1996).

Because of the random-like nature of chaotic motion, it is anticipated that the grindstone of a chaotic grinder can provide a more uniform contact surface with the workpiece and hence a better grinding performance. The first application of chaotic motion to grinding was proposed in 1998 (Ito and Narikiyo, 1998). It applies chaos to a vertical spindle surface grinder, as shown in Figure 10.29. The abrasive motor is a DC motor, offering a rated output of 21 W, and a maximum speed of 12 000 rpm under no-load. The motor is controlled in such a way that its rotational speed is governed by:

$$\begin{aligned} \dot{x} &= \sigma(-x+y) \\ \dot{y} &= -xz+rx-y \\ \dot{z} &= xy-bz \end{aligned} \tag{10.21}$$

where x and y are the horizontal displacements on the X–Y table, z is the rotational speed, and the relevant parameters are selected as $\sigma = 10$, $r = 28$, and $b = 8/3$. As shown in Figure 10.30, the resulting speed chaotically fluctuates between 4580 and 12 000 rpm with a period of around 1.06 s. The grindstone is made of aluminum oxide (Al_2O_3). The relevant grit size is designated by the screen number S (number of openings per linear inch), namely $S = 120$. The grinding process is under a wet condition and a constant pressure of 650 g.

There are two important parameters by which to evaluate the grinding performance – namely, the abrasive efficiency and the surface roughness. The abrasive efficiency is the ratio of the volume removed to the electric power consumption. The surface roughness is usually based on the peak-to-valley roughness R_t on the European continent and in Japan, or the centerline average roughness R_a in the UK and the USA (Shaw, 1996).

To assess the advantages of chaotic abrasive machining (chaotic speed variation at 4580–12 000 rpm with a period of around 1.06 s), it is compared with different grinding schemes:

- constant speed at 12 000 rpm (the maximum value);
- constant speed at 4580 rpm (the minimum value);
- sinusoidal speed variation (4580–12 000 rpm with a period of 1 s);
- random speed variation (4580–12 000 rpm with a division of 1 s).

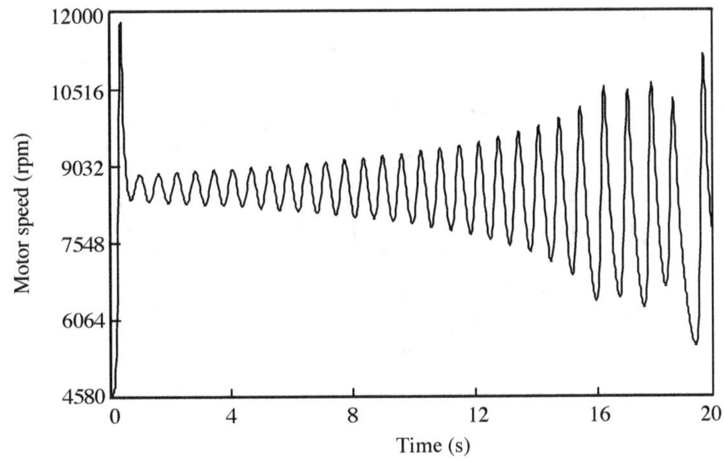

Figure 10.30 Chaotic rotation of abrasive motor

As revealed by experiments (Ito and Narikiyo, 1998), chaotic abrasive machining is better than the others in both the amount of volume removed and the abrasive efficiency. In particular, its abrasive efficiency is twice that of constant-speed grinding at 12 000 rpm. Selecting a plane with $R_a = 2.3$–$2.7\,\mu$m as the workpiece, the chaotic abrasive machining offers the best surface roughness, namely $R_t = 50$–100 nm.

Furthermore, in order to observe the better uniformity benefiting from chaotic grinding, a system arrangement is adopted as shown in Figure 10.31, which functions to compare the grinding trajectories. In the system, a permanent magnet synchronous motor (PMSM) is used to drive the workpiece while the grinding wheel is driven by another constant-speed DC motor. By operating the PMSM at constant speed and chaotic speed, the abrasive trajectories on the workpiece under constant-speed grinding and chaotic grinding are shown in Figure 10.32. It can be observed that the constant-speed grinding trajectory can only cover a limited surface area of the workpiece, whereas chaotic grinding can densely spread over the whole surface of the workpiece. This result illustrates that chaotic grinding not only grinds the workpiece surface more evenly, but also removes the workpiece material more effectively, thus achieving better surface uniformity and higher abrasive efficiency.

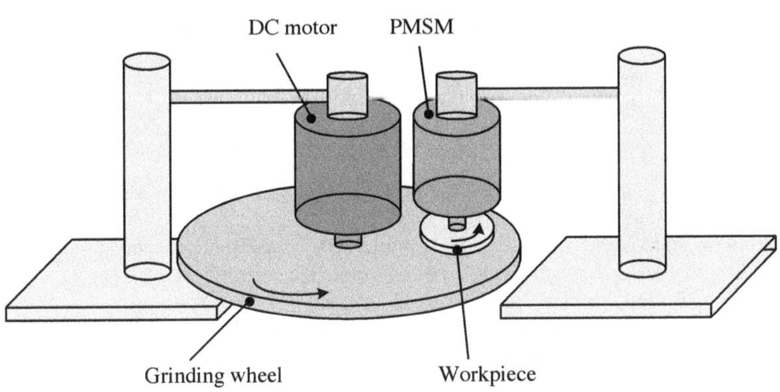

Figure 10.31 Arrangement of surface grinding

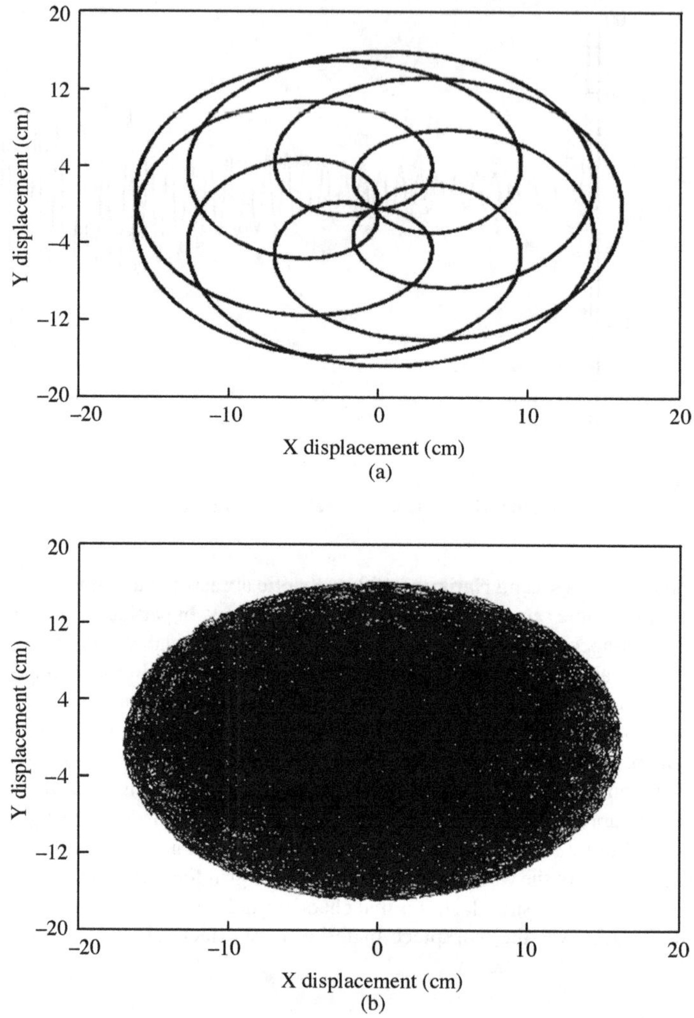

Figure 10.32 Comparison of grinding trajectories. (a) Constant-speed grinding. (b) Chaotic grinding

References

Alvarez-Hernández, M.M., Shinbrot, T., Zalc, J., and Muzzio, F.J. (2002) Practical chaotic mixing. *Journal of Chemical Engineering Science*, **57**, 3749–3753.

Ascanio, G., Brito-Bazán, M., Brito-De La Fuente, E. *et al.* (2002) Unconventional configuration studies to improve mixing times in stirred tanks. *Canadian Journal of Chemical Engineering*, **80**, 558–565.

Chaté, H., Villermaux, E., and Chomaz, J.-M. (1999) *Mixing*, Kluwer, New York.

Chau, K.T. and Wang, Z. (2005) Application of chaotic motion to industrial compactors. Proceedings of International Conference on Electrical Machines and Systems, pp. 1644–1649.

Chau, K.T. and Wang, Z. (2006) Design of permanent magnets to chaoize doubly salient PM motors for electric compaction. *Journal of Applied Physics*, **99**(80R306), 1–3.

Chau, K.T., Ye, S., Gao, Y., and Chen, J.H. (2004) Application of chaotic-motion motors to industrial mixing processes. Proceedings of IEEE Industry Applications Society Annual Meeting, pp. 1874–1880.

Dong, L., Johansen, S.T., and Engh, T.A. (1994) Flow induced by an impeller in an unbaffled tank – I. Experimental. *Chemical Engineering Science*, **49**, 549–560.

Fanger, P.O. (1970) *Thermal Comfort – Analysis and Applications in Environmental Engineering*, McGraw-Hill, New York.

Fountain, G.O., Khakhar, D.V., Mezic, I., and Ottino, J.M. (2000) Chaotic mixing in a bounded three-dimensional flow. *Journal of Fluid Mechanics*, **417**, 265–301.

Gao, Y., Chau, K.T., and Ye, S. (2005) A novel chaotic-speed single-phase induction motor drive for cooling fans. Proceedings of IEEE Industry Applications Society Annual Meeting, pp. 1337–1341.

Gethin, D.T., Ransing, R.S., Lewis, R.W. *et al.* (2001) Numerical comparison of a deformable discrete element model and an equivalent continuum analysis for the compaction of ductile porous material. *Computers and Structures*, **79**, 1287–1294.

Harnby, N., Edwards, M.F., and Nienow, A.W. (1992) *Mixing in the Process Industries*, Butterworth-Heinemann, Oxford.

Ito, S. and Narikiyo, T. (1998) Abrasive machining under wet condition and constant pressure using chaotic rotation (in Japanese). *Journal of the Japan Society for Precision Engineering*, **64**, 748–752.

Jana, S.C. and Sau, M. (2004) Effects of viscosity ratio and composition on development of morphology in chaotic mixing of polymers. *Polymer*, **45**, 1665–1678.

Katayama, R., Kajitani, Y., Kuwata, K., and Nishida, Y. (1993) Developing tools and methods for applications incorporating neuro, fuzzy and chaos technology. *Computers and Industrial Engineering*, **24**, 579–592.

Kuwata, K., Kajitani, Y., Watanabe, M., and Katayama, R. (1996) Chaos simulator as a developing tool and a research environment for applications of chaos engineering. *Journal of Network and Computer Applications*, **19**, 45–66.

Lamberto, D.J., Muzzio, F.J., Swanson, P.D., and Tonkovich, A.L. (1996) Using time-dependent rpm to enhance mixing in stirred vessels. *Chemical Engineering Science*, **51**, 733–741.

Long, Y.J. (2001) Chaotic dynamics and compaction engineering. Proceedings of International Conference on Soft Soil Engineering, pp. 143–147.

Mavros, P. (2001) Flow visualization in stirred vessels, a review of experimental techniques. *Transactions of the Institution of Chemical Engineers, Part A*, **79**, 113–127.

Nomura, H., Wakami, N., and Kazuyuki, A. (1996) Time-series analysis of behavior of a two-link nozzle in a dishwasher. *Electronics and Communications in Japan*, **79**, 88–97.

Nomura, H., Wakami, N., and Kondo, S. (1995) Non-linear technologies in a dishwasher. Proceedings of IEEE International Conference on Fuzzy Systems, pp. 57–58.

Ottino, J.M. (1989) *The Kinematics of Mixing: Stretching, Chaos, and Transport*, Cambridge University Press, Cambridge.

Procaccia, I. and Schuster, H. (1983) Functional renormalization-group theory of universal 1/f noise in dynamical systems. *Physical Review A*, **28**, 1210–1212.

Raynal, F. and Gence, J.N. (1997) Energy saving in chaotic laminar mixing. *Journal of Heat and Mass Transfer*, **40**, 3267–3273.

Shaw, M.C. (1996) *Principles of Abrasive Processing*, Oxford University Press, Oxford.

Tran, D.T. and Muro, T. (2004) Effect of an innovative vertical vibro-tracked vehicle on soil compaction. *Journal of Terramechanics*, **41**, 1–23.

Wang, B.H., Koh, S.B., Ahn, S.K., and Roychowdhury, S. (1996) Chaos Washing Machine and a Method of Washing Thereof. United States Patent 5560230.

Wang, Z. and Chau, K.T. (2008) Anti-control of chaos of a permanent magnet dc motor system for vibratory compactors. *Chaos, Solitons and Fractals*, **36**, 694–708.

Wang, Z. and Chau, K.T. (2009) Design, analysis and experimentation of chaotic permanent magnet DC motor drives for electric compaction. *IEEE Transactions on Circuits and Systems II: Express Briefs*, **56**, 245–249.

Wyon, D.P. (1973) The role of the environment in buildings today: thermal aspects (factors affecting the choice of a suitable room temperature). *Build International*, **6**, 39–54.

Ye, S. and Chau, K.T. (2005a) Design of permanent magnets to chaoize PM synchronous motors for industrial mixer. Proceedings of IEEE International Magnetics Conference, pp. 723–724.

Ye, S. and Chau, K.T. (2005b) Destabilization control of a chaotic motor for industrial mixers. Proceedings of IEEE Industry Applications Society Annual Meeting, pp. 1724–1730.

Ye, S. and Chau, K.T. (2007) Chaoization of DC motors for industrial mixing. *IEEE Transactions on Industrial Electronics*, **54**, 2024–2032.

Ye, S., Chau, K.T., and Niu, S. (2006) Chaoization of a single-phase induction motor for washing machines. Proceedings of IEEE Industry Applications Society Annual Meeting, pp. 855–860.

Yoo, T.S. and Selig, E.T. (1979) Dynamics of vibratory-roller compaction. *ASCE Journal of the Geotechnical Engineering Division*, **105**, 1211–1231.

Index

Chaos in Electric Drive Systems: Analysis, Control and Application, First Edition. K.T. Chau and Zheng Wang.
© 2011 John Wiley & Sons (Asia) Pte Ltd. Published 2011 by John Wiley & Sons (Asia) Pte Ltd.

www.ingramcontent.com/pod-product-compliance
Lightning Source LLC
LaVergne TN
LVHW062115200926
848484LV00007B/1672